Genetic Basis and Methods of
PLANT BREEDING

Genetic Basis and Methods of
PLANT BREEDING

Sultan Singh
Former Professor

I.S. Pawar
Wheat Breeder

CCS Haryana Agricultural University,
Hisar - 125 004

CBS

CBS Publishers & Distributors Pvt. Ltd.

New Delhi • Bengaluru • Chennai • Kochi • Kolkata • Mumbai
Hyderabad • Uttarakhand • Nagpur • Patna • Pune • Jharkhand

ISBN: 81-239-1300-1

First Edition: 2006
Reprint: 2007, 2014, 2019

Published by **Satish Kumar Jain** and produced by **Varun Jain** for
CBS Publishers & Distributors Pvt. Ltd.,
4819/XI Prahlad Street, 24 Ansari Road, Daryaganj, New Delhi - 110002
delhi@cbspd.com, cbspubs@airtelmail.in • www.cbspd.com
Ph.: 23289259, 23266861, 23266867 • Fax: 011-23243014

Corporate Office: 204 FIE, Industrial Area, Patparganj, Delhi - 110 092
Ph: 49344934 • Fax: 011-49344935
E-mail: publishing@cbspd.com • publicity@cbspd.com

Branches:
• *Bengaluru:* 2975, 17th Cross, K.R. Road, Bansankari 2nd Stage,
  Bengaluru - 70 • Ph: +91-80-26771678/79 • Fax: +91-80-26771680
  E-mail: cbsbng@gmail.com, bangalore@cbspd.com
• *Chennai:* No. 7, Subbaraya Street, Shenoy Nagar, Chennai - 600030
  Ph: +91-44-26681266, 26680620 • Fax: +91-44-42032115
  E-mail: chennai@cbspd.com
• *Kochi:* Ashana House, 39/1904, A.M. Thomas Road, Valanjambalam,
  Ernakulum, Kochi • Ph: +91-484-4059061-65
  Fax: +91-484-4059065 • E-mail: cochin@cbspd.com
• *Kolkata:* 6-B, Ground Floor, Rameshwar Shaw Road, Kolkata - 700014
  Ph: +91-33-22891126/7/8 • E-mail: kolkata@cbspd.com
• *Mumbai:* 83-C, Dr. E. Moses Road, Worli, Mumbai - 400018
  Ph: +91-9833017933, 022-24902340/41 • E-mail: mumbai@cbspd.com

Representatives:

• Hyderabad: 0-9885175004   • Nagpur: 0-9021734563
• Patna: 0-9334159340   • Pune: 0-9623451994
• Jharkhand: 0-9811541605   • Uttarakhand: 0-9716462459

Printed at:
India Binding House, Noida, UP (India)

Preface

The progress witnessed in recent decades in agriculture is the result of the evolution of high yielding, disease resistant and photoinsensitive varieties of crop plants by plant breeders in collaboration with scientists of some related disciplines. In fact, in the last 4-5 decades, some branches of science have grown with a very high speed and have generated new opportunities for the agricultural scientists. For example, advancement in the discipline of biotechnology has provided unconventional and rapid means of transferring genes across sexual barriers and thus has tremendously widened the scope of hybridization between unrelated species. Now, it seems that there are endless tasks which can be accomplished by the judicious application of biotechnological techniques. Similarly, plant breeding methodology has now taken a new turn. In addition to the strengthening, modification and refinement of conventional breeding methodology, some non-conventional methods (alien genetic transfers through special techniques, haploid breeding, etc.) are being used to exploit maximum genetic variability present in the germplasms of different species. Use of such techniques signifies the collection, documentation and conservation of genetic resources *in situ* and *ex situ* both by conventional and non-conventional methods. The present book provides detailed and up-to-date information about basic plant breeding concepts and the advancements mentioned above in easily understandable language.

The subject matter of the book has been divided into five sections. The first section deals with general topics. The genetic basis of breeding self- and cross-fertilizing crops has been discussed in section II. Plant breeding methodology (line breeding, hybrid breeding, population breeding and clone breeding) with modifications and refinements has been described in section III. The classification of breeding methods followed in the present book is based on the propagational category of resulting varieties. Section IV includes plant breeding techniques (mutation breeding, polyploidy breeding, haploidy breeding and distant hybridization). Some recent trends in plant breeding (use of biotechnology in crop improvement and patenting of plant materials) have been discussed in section V.

In the preparation of this book, the authors have consulted many books and journals on the subject (Principles of Plant Breeding by R.W. Allard, Plant Breeding : Principles and Methods by

B.D. Singh, Principles and Practice of Plant Breeding by J.R. Sharma, Breeding Field Crops by J.M.Poehlmen and D.A. Sleper, Plant Breeding : Theory and Practice by V.L. Chopra, etc.) and owe sincerely to the writers of these books.

The authors are thankful to the Vice-Chancellor, CCS HAU, Hisar for granting permission for getting the book published by an outside agency.

The authors are also thankful to CBS Publishers & Distributors, New Delhi for their keen interest shown in the publishing of the book.

We are thankful to our family members, friends and colleagues who inspired us to write this book.

Sultan Singh
I.S. Pawar

Contents

SECTION II
Genetic Basis of Breeding Crops
A. Principles of Breeding Self-fertilizing Crops

B. Principles of Breeding Cross-Fertilizing Crops

SECTION III
Plant Breeding Methods
A. Line Breeding
I. Selection Breeding

II. Recombination Breeding

B. Hybrid Breeding

C. Population Breeding

D. Clone Breeding

SECTION I
General Topics

CHAPTER 1

Introduction

In the modern age of science and technology when the horizons of knowledge are expanding in the field of agricultural sciences and some other disciplines in a spectacular manner, the research in these branches of science has at the present juncture reached a stage where inter-disciplinary collaboration is inevitable in a number of areas. The applied science of plant breeding which aims at the development of improved varieties of crops for human and animal use, should no more be pursued as an independent scientific discipline. In fact, to become a competent and successful plant breeder, one must not only know about the botanical, agronomical and economical aspects of his crop but must also have a working knowledge of genetics, statistics, physiology, phytopathology, molecular biology, etc. The past experience shows that crop cultivars which possessed tolerance/resistance to various biotic and abiotic stresses in addition to their high genetic worth in terms of important agronomic traits, have been major contributors towards increased global agricultural production. Geneticists, statisticians, plant physiologists and molecular biologists can help in setting the objectives of plant breeding programmes more clearly and in further improving and identifying potential genotypes.

Although there are a number of critical prerequisites for greater crop productivity like improved farming system, proper guidance to farmers and the availability of inputs and markets, the most critical requirement for maximizing crop yield is the genetic improvement of crop plants. The crop cultivars with high input use efficiency and response are expected to produce superior yield in terms of quantity and/or quality of the produce. Of course, a well balanced combination of improved practices of irrigation, fertilization and pest control has its own role in maximizing crop yield. Therefore, both the genetic potential of the crop cultivar and the management of crop breeding are necessary for harvesting maximum yields.

Definition

From times immemorial, man has been trying, consciously or unconsciously, to conquer nature by evolving plant types according to his own will. This man directed evolution has been named as 'plant breeding' and has been variously defined.

Vavilov (1935, see also Vavilov, 1951) defined plant breeding as 'the evolution directed by the will of man'.

Stebbins (1957) defined plant breeding in terms of natural evolution and opined that 'plant breeding is merely a continuation of the natural evolution of the crop plant species, changing its course in the direction of greater use to mankind.'

According to Frankel (1958), 'plant breeding is the adjustment of plants to the service of man.

Smith (1966) defined plant breeding as 'an art and science of improving genetic pattern of plants in relation to their economic use.'

In Riley's (1978) opinion, plant breeding is 'a technology of the production of varieties of crop plants adapted to man's needs.' He in 1979, defined plant breeding as a technology which has the practical purpose of producing material objects–the variety or cultivar–which offer advantages on pre-existing varieties in terms of absolute yield, stability of yield, agronomic convenience or quality of the market produce.'

According to Poehlman and Sleper (1995), 'plant breeding is the art and the science of improving the heredity of plants for the benefit of human kind.'

Furthermore, since the success of obtaining better genotypes will depend on the nature and extent of genetic variability present in the base material, plant breeding may also be defined as 'the purposeful management of variability.'

With the advent of modern biotechnology, new opportunities (like use of fascinating techniques for plant varietal improvement and the transfer of genes across sexual barriers resulting in the development of transgenic plants) for crop improvement arc quickly developing and thus the definition of plant breeding given by Riley is becoming increasingly relevant. The new biotechnological techniques will make the use of biodiversity (a key ingredient for sustainable agricultural development) more specific and purposeful.

Although it is very difficult to establish a date when man became a plant breeder, plant breeding in the beginning was entirely an art as every thing at that time depended on the skill and the judgement of the selector. But as the knowledge progressed in classical genetics and other related disciplines, plant breeding gradually developed into a science and now in the modern age of science, this discipline has become more of a science and a technology and less of an art. Accordingly, now plant breeding may be defined as an 'art, science and technology of producing varieties of crop plants best adapted to man's needs.'

Objectives

In addition to the technical knowledge about the subject and his crop, a good plant breeder must also be able to manage things properly using modern principles of management. For example, he must be able to take decisions at his own particularly under changed situations during the course of programme, formulate and carry out breeding plans independently by making best use of physical resources and manpower, take full advantage of successes and to face consequences of failures. Efforts must be made to develop the most useful information at the least cost.

To maximize the effectiveness of the programme, the first job of the plant breeder must be to decide the objectives of the programme prioritywise, assigning greatest attention to the most important objective. If the objectives are not clearly defined even the best plant breeding method may not give best results. Scientists of allied disciplines may help in determining major objectives like development of crop cultivars with improved yield (both the quantity and quality) and stability of performance and possessing resistance to diseases and insect pests. Further, a thorough knowledge of the crop (floral structure, per cent outcrossing, time of anthesis, mechanisms enforcing self or cross pollination like self-incompatibility, male sterility, protogyny, protandry, monoecy, dioecy, somatoplastic sterility and cleistogamy, degree of success in emasculation and crossing, input use efficiency and response, etc.) and some other factors like artificial inoculation, growing breeding material in off-season nurseries and laboratory tests for lodging resistance, winter hardiness, etc., help in attempting crosses, carrying out selection more effectively and thus making the programme more successful. The physical structure of the breeding programme, planning and control methods, and decision making must relate to the objectives. Both the long-term (strategies) and short-term (tactics) objectives should be included in the programme. Feedback mechanisms are a necessary requirement to check the performance of progenies, accuracy of records and compare experimental varieties with standard checks. All said and done, decision making is the central activity of the plant breeder and it should be problem solving and not a problem creating activity particularly at critical junctures.

Nevertheless, the ultimate aim of the plant breeder is to evolve better crop cultivars at the least cost and at the least hazards to the environment. This betterment may be with regard to increased quantity or quality of the produce, resistance to diseases and insect pests, more stability over a wide range of environments, photo- and thermo-insensitivity, production of even more curious forms (novelties), high input use efficiency and response, etc. A number of techniques have been developed by agricultural scientists in their search for increased yield and improved quality in crops and at least in some cases, they have achieved a great success. However, behind all kinds of procedures (conventional breeding, variants of conventional methods, breeding techniques like mutation breeding, polyploidy breeding, haploidy breeding, sophisticated biotechnological techniques, advance statistical and biometrical designs and analyses, etc.) which they have used, there have been mainly three basic philosophies : defect elimination, breeding for yield and breeding for model plants or ideotypes (plant types).

Defect Elimination

In defect elimination, cultivar lacking in a particular character, for example, susceptibility to a particular disease, late maturity, weak straw (cereals), weak malting performance (barley), poor flavour, red skin (potato), fragile skin (tomato), etc., is improved by incorporating that character into the cultivar. There are numerous examples where a very promising crop cultivar has gone out of cultivation only because of a single defect. Susceptibility of wheat variety Kalyansona to rust forced this cultivar to go out of cultivation. This variety was at one time most promising and was occupying a very large area all over the world. Co 1148, a wonder variety of sugarcane having best genetic potential not only for the first crop but also for a ratoon crop, probably also going to have the same fate because of its increasing susceptibility to red-rot disease. This approach is, therefore, all the more important to enhance the life of a crop cultivar.

The transfer of character may be done using backcross method, pedigree method or through some biotechnological technique or a new gene for that specific purpose may be created using physical or chemical mutagen. If the two varieties/species (donor and receipient) are easily crossable and the character under transfer is governed by genes with individually distinguishable effects, the backcross method can be used effectively. A dominant allele can be transferred more easily than a recessive allele. The transfer of a recessive allele requires selfing after every backcrossing or after every two backcrosses. A number of genes have been transferred using backcross method in many crop plants like wheat (incorporation of rust resistant genes, Norin 10 dwarfing genes, etc.). Pedigree method has also been used to transfer disease resistance particularly when the inheritance of resistance is bit complex and improvement is sought for other characters also.

Some biotechnological techniques, however, provide opportunity for the transfer of genes across sexual barriers. Using these techniques, some genes like gene for herbicidal resistance have been transferred and transgenic palnts have been developed and many more are in process of development.

A number of attempts have been made to utilize artificial mutagenesis in plant breeding for creating new variability in important characters, transferring genes of interest like genes determining disease resistance, high lysine content and self-compatibility (that is, mutations yielding loss of function), and for producing ever more curious forms in ornamentals (novelties) and quite a good success has been achieved in many cases.

Breeding for Yield

If the character to be improved is governed by many genes, particularly a complex trait like yield which is a prime quantitative character usually with low heritability, backcross method, mutation breeding, simple selection procedures or fascinating biotechnological techniques are not of much use. Two kinds of approaches have been suggested to bring improvement in yield.

1. Improvement through direct selection for yield : Here, the plant breeder tries to improve yield without considering the basis of improvement. This approach may be successful in those breeding materials where heritability is moderate to high. But many plant breeders and geneticists have reservation about the degree of success to be achieved through this approach. Nevertheless, the approach may be made more effective by increasing the frequency of desirable alleles in populations through successive intermating and selection.

2. Improvement through components of yield : In this approach, improvement in yield is done by improving its component characters rather than making direct selection for yield. Plant density is perhaps the most important component trait of yield since this trait is common to all crops and the measurement of yield is based on the production per unit area. Except this character, the components of yield will generally vary according to the crop. For example, in case of cereals like wheat and barley, yield will be the product of $p \times t \times g \times w$ (p is the number of plants per unit area, t the number of ear bearing tillers, g the number of grains per spike and w the grain weight). In case of cotton, on the other hand, seed cotton yield will be the product of the number of branches per plant, number of bolls per plant and boll size. However, some of the component traits may be negatively correlated, e.g., plant density with tiller number and the number of grains per spike with grain weight. In such situation, it is advisable to improve one character, keeping the other character constant.

In addition to components of yield, there are some factors limiting yield. Most of these factors are environmental and interact with genotype. Now it is amply clear that the interaction of environmental factors with genotype (G×E interaction) is not an exception but a rule. The crop cultivars showing high response to controllable environments (fertilizer dose, etc.) and low response to uncontrollable environments (atmospheric temperature, etc.), should be preferred.

Breeding for model plants

An ideotype is a plant model (required plant structure and developmental sequence of a plant) which can best suit a particular environment for producing maximum economic return. The approach of breeding of model plants is not so new as it is often supposed. Long back, Bailey (1895) had opined that the plant breeder must have a specific plant design in his mind before starting a breeding programme to evolve a variety. However, Donald (1968) introduced this approach in a more systematic manner and emphasized that plant breeding programmes should not be confined to defect elimination and breeding for yield only but a plant type for a specific purpose must be kept in mind while evolving a variety. This approach (breeding for model plants) enables the plant breeder to answer the question, "Why a variety yields better or worse than the other ? Resistance to lodging, presence of awns in wheat and barley and erect foliage are some of the model characters. Some other physiological model characters are size of photosynthetic apparatus, efficiency of photosynthetic apparatus, efficiency of transmitting tissue between the source and sink, capacity of sink and life span or persistency of leaves and awns in wheat and barley. Since photosynthesis is the ultimate physiological limitation of crop production, maximum production in crops can be achieved only if the utilization of solar radiation is as high as possible. A better plant type with erect and persistent foliage will certainly have a higher conversion of solar energy into chemical energy as compared to that of a traditional plant type.

In spite of very high success achieved in some crop plants, there is still a big gap between the theoretical maximum of food production and its practical maximum already achieved. Theoretically, 218 tonnes of dry matter can be produced per hectare per year provided all factors affecting crop production like carbondioxide, water, nutrients and sunlight are available to plants in optimum quantity. By now, despite our best efforts, we have been able to produce only 20 tonnes dry matter per hectare per year. This is supported by another fact that the theoretical maximum of the conversion of light energy into chemical energy by photosynthetic system is 18 per cent, while most of the important crops represent less than 3 per cent conversion of the available visible radiation (beet sugar 1.5 per cent; cereal grain-2.0 per cent; potato tubers - 2.4 per cent). Buringh *et al.* (1975) put the theoretical expectations of absolute maximum world production potential still higher. According to them, this potential (about 50,000 million tonnes per year in grain equivalent of a standard crop) is about 38 times more than the present average world production (about 1,300 million tonnes per year).

Importance

Man, directly or indirectly, depends on plants for his almost all kinds of needs, particularly for food. The speed and direction of improvement in crop plants, therefore, greatly depend on the demand of food. When there is an increase in human population, automatically there is more demand of food followed by more enthusiastic involvement of scientists in increasing food production. At such times, maximum importance is attached to the crop plant which is expected to give maximum food return.

The cereals like rice, wheat, barley, maize, millet and sorghum constitute a major part of human diet. These cereals constitute about half of the protein and energy consumed by human beings. In developing countries like India, this contribution by cereals may be upto 70 per cent of the total energy consumption. Other crops like potato, sweet potato, sugarcane, banana, common bean, soybean and groundnut also have a good contribution (about 25 per cent) towards human diet. No doubt, numerous crop varieties have been developed by plant breeders and these varieties have made significant contribution in crop production. In some cases, the production has been increased manifold.

Increase in the yield in any crop may be attributed to two major factors (1) introduction/ development of high yielding varieties (that is, better genetic potential), and (2) improvement in cultivation techniques (that is, management). The importance of plant breeding may be realised more clearly by comparing the relative contribution of these two factors. The calculations made by Silvey (1978) in England clearly show that the contributions of high yielding varieties towards increased yields in case of wheat and barley were 67 per cent and 50 per cent, respectively. Similar results have been obtained in maize by Russell (1974) and Duvick (1977) for about 40 years. In case of maize, the contribution of high yieding varities was found to be 57-63 per cent. It is, therefore, clear from these studies that though the improvement in cultivation techniques has its own importance in increasing yields of crops, the major contribution in this direction has been made by the high yielding varieties.

The service of plant breeder is not only confined to the evolution of high yielding varieties but he is also concerned with the development of varieties which are better suited to mechanical harvesting, processing, storage, etc. The new developments in making agricultural machinery, use of fertilizers, etc. are related to the evolution of new high yielding cultivars.

Accomplishments

Undoubtedly, the accomplishments of plant breeders in the twentieth century have been too many and benefits from their work have been sensational. The development of semi-dwarf, photoinsensitive and input efficient and responsive varieties of grain crops (which spearheaded green revolution) and their adoption all over the world in a revolutionary manner resulted in severalfold increase in the production of these crops and ultimately in huge foodgrain reserves. However, development of noble canes through distant hybridization, evolution of intra- and interspecific hybrids in cotton, commercial hybrids in pearlmillet, sorghum, castor, sunflower, etc., development of rabi maize and sorghum, spring season (Boro) rice, summer grain legumes (mungbean, urdbean, etc.) and a large number of tobacco varieties as well as better cultivation practices are in no way less important than the green revolution in wheat and rice. The development of hybrid rice as a commercially viable proposition has led to a major change in the breeding strategy of self-fertilized crops and opens a new door for the plant breeders dealing with such crops. Similarly, introduction of crops to new areas provides opportunity for consolidating yield gains. The evolution of a number of pulse crop varieties with proven inbuilt resistance to various diseases constitutes a landmark in the history of pulse improvement and would go a long way in enhancing and stabilizing production of these crops in different agroclimatic regions and seasons.

Challenges Ahead

Notwithstanding dramatic contributions of plant breeders towards human well-being in the past, there is no scope for being complacent as the vast agricultural potential still remains to be harnessed and in accomplishing this they have to face endless difficulties in the time to come. In fact, the job of modern plant breeder has now become more exciting, more rewarding and increasingly challenging. His job has become more exciting as the transfer of genes across sexual barriers has now become feasible through the use of some novel biotechnological techniques (protoplast fusion, recombinant DNA technology, etc.) resulting in the production of transgenic plants. As a consequence, genetic engineering which represents attempts to add, subtract, replace or modify genes, is on the brink of revolutionizing the traditional concept of man about God and creation. A new era in the production of crop varieties with intellectual property rights has started and thus the job of the plant breeder has become more rewarding. Nevertheless, the job of plant breeder is becoming increasingly difficult since :

(1) human population is increasing with alarming dimensions and unless checked the danger of population overtaking food supplies predicted by Malthus in 1798 may be realized at any time;

(2) expansionist agriculture is no longer possible and the only option left with the scientists is to have vertical intensification through productivity upgradation;

(3) a large part of the yield potential of many important crops has already been exploited through the use of modern agricultural techniques;

(4) the gains from the dwarfing genes approach are now levelling off;

(5) the protection of crops from the unholy triple alliance of pathogens, pests and weeds is becoming complex day by day;

(6) we are likely to experience a serious eco-imbalance in some parts of the world in next 4-5 decades due to the depletion of protective ozone layer in the upper layer of the atmosphere (stratosphere) and growing accumulation of green house gases in the lowest layer of the atmosphere (troposphere);

(7) if the scientists of developing countires with richest biodiversity on earth fail to keep pace with the new advancements in biotechnology and remain undemonstrative, monopolistic control of a few companies over transgenic plants can create problems in future; and

(8) conscious or unconscious, open or surreptitious, legal or illegal loss of precious and irreplaceable genetic wealth with an unprecedented magnitude all over the world due to the reckless devastation of natural habitats of various plant species would not only entail the destruction of earth's eco-system, biodiversity and integrity of the climate system but also would decimate the economics and marginalise the culture of tribals. If this severe biological extinction, an irreversible process, is allowed to go unhindered it may prove to be the single most strategic threat to the global food system and can cause immeasurable loss and bring unimaginable calamities to the living beings leading to horrifying consequences. We must remember that sometimes seemingly unimportant and non-consequential things later on prove to be highly consequential. Therefore, every effort must be made to save the earth, humanity and future. Of course, sustainable agriculture would provide basis for sustaining future.

Future Breeding Strategy

To meet the food requirements of relentlessly increasing human population, it is imperative to raise genetic yield ceiling of crops, stabilize yield potential already attained and to improve input use efficiency and response of crop species.

Raising Genetic Yield Ceiling

The yield potential can be enhanced through the use of genetically diverse parents in hybridization programme, better breeding methods, untapped variability, sophisticated biometrical techniques, exploiting physiological traits and producing commercial hybrids of self-fertilized and often cross-fertilized crops (Singh and Pawar, 1998). The first step in this direction is to collect and conserve crop genetic resources and utilize this precious genetic wealth in the best possible way. The exchange of knowledge, ideas and materials worldwide would certainly help in developing and spreading potential genotypes of crop plants. However, involvement of private sector is imperative for continued investment and competition in agricultural research. The advent of Intellectual Property Rights (IPR) and Plant Variety Rights (PVR) has provided new opportunities for private investment in plant breeding research. Such involvement of private sector has shown good results in Europe. Nevertheless, incorrect use of IPR and PVR and plant quarantine regulations may drastically hamper future progress. The monopolistic control of a few companies over transgenic plants can cause severe blow to the economics of developing countries like India where farmer to farmer seed exchange is more important than seed sales by private seed companies.

The choice of an appropriate breeding procedure can greatly enhance the breeding efficiecny and effectiveness. Since, different methods of breeding have their own merits and demerits depending upon the objective of the breeder and the genetic make-up of the material to be handled, the use of a method which combines merits of two or more methods would be more desirable than using a single conventional method. The shuttle breeding system which involves growing of F_2, F_5 and F_6 generations under optimal conditions of moisture and fertility with selection of single plants in F_2 and F_5 and of superior lines in F_6 and growing of F_3 and F_4 generations under reduced moisture and low fertility conditions with modified bulking in both generations, is being successfully used by the CIMMYT scientists. The method combines the merits of pedigree method (in optimal environments) and bulk method (in stress environments) as well as combines input effciency and input responsiveness using alternate sites of contrasting conditions. As excellent computer facilities are now available, the pedigree trial method and modified pedigree method can be successfully used by the plant breeders. Some serious drawbacks of single seed descent method can be overcome by using multiple seed descent method. One or more cycles of selective intermating in between selfing series are expected to check fixation of linkage blocks in early generations, foster gene recombinations, accumulate desirable genes in the experimetnal material and elevate population mean, degree of heritability and the extent of genetic advance.

As it is well known, landraces are the basic building blocks of modern varieties of crop plants. The identification and use of landraces which have not yet been fully exploited would broaden the genetic base of the future varieties. There is a vast amount of variability present in the wild relatives of crop species which has not yet been utilized or has been under-utilized. For example, the D genome of wheat (contributed by *Triticum tauschii*) controls a number of important traits, but only a part of the variability present in different accessions of this genome has so far been utilized in bread wheat breeding. However, well planned efforts are being made at CIMMYT to include under-utilized germplasm in wheat breeding programme. Similarly, utilization of untapped variability present in the wild relatives of other crop species can bring significant improvement in their yield levels.

Several innovations, extensions, modifications and refinements have been made in biometrical genetic procedures for the analysis of quantitative characters. These procedures can be used to identify potential combiners, decide most appropriate selection method, measure genetic divergence and phenotypic stability, determine magnitude of additive, dominance and epistatic components of genetic variation and their relative sensitivity to environmental change and to delineate potential regions (environments) for each crop to maximize productivity and production. Principal component analysis has been used to assess regional and temporal variations in wheat yield in Western Australia and identify less variable regions for growing bulk of crop to maximize yield (Goodchild and Boyd, 1975) (For details see Singh and Pawar, 2005).

Some physiological mechanisms like stomatal conductance, leaf photosynthetic rate, crop canopy temperature, membrane leakage and ability to use incident solar energy in the early part of growing season, can affect the yield potential of crops. Therefore, a better understanding of these basic physiological mechanisms can play a key role in exploiting yield potential. The results of research done at CIMMYT show that increase in stomatal conductance is positively related to lower canopy temperature, higher rate of leaf photosynthesis and increased yield. Fortunately, there is good variability for stomatal conductance in many crop species. Canopy temperature can be easily measured and can be used as a tool in plant breeding programmes aimed at to increase yield potential. Membrane leakage and canopy cooling have also been found as useful parameters for selecting genotypes to grow in warmer environments. *Triticum boeoticum* has very high leaf photosynthetic rate and rye and triticale have more efficient use of incident solar energy in the early part of growing season. Some wheat genotypes have 2-3 degree celcious lower canopy temperature than others. Further, it is well known that plant species with C_4 pathway (maize, pearlmillet, sorghum, sugarcane etc.) possess a much higher potential rate of biomass production than the species with C_3 pathway (wheat, barley, chickpea etc.). Genetic engineering techniques can be used to improve the efficiency of C_3 species either by reducing/eliminating oxygenase activity of rubisco or by transferring elements of the C_4 pathway to C_3 plants.

Although the type of variety (homozygous line or a hybrid) to be developed by the plant breeder in a particular crop species mainly depends on the genetic basis of heterosis (that is, the relative role of dominance and overdominance towards heterosis), the hybrids not only allow the use of overdominance and unfixable epistasis (homozygote × heterozygote and heterozygote × heterozygote interactions) but also represent the quickest way of accumulating the maximum number of favourable dominant genes in the same genotype. Also, the past experiene with hybrid crops suggests that hybrids, in general, are less sensitive to environmental change than the homozygous varieties and thus hybrids perform better in stress environments. In cross-fertilized crops, hybrid varieties can be developed without facing any serious difficulty. On the contrary, plant breeders have encountered several technical difficulties in producing hybrids in self-fertilized crops. However, the development of commercially viable rice hybrid technology in China has opened a new door for the plant breeders. In China, hybrid rice has occupied more than 55 per cent of the area under rice cultivation with an yield advantage of 1.0 to 1.5 tonnes per hectare. The Chinese experience led the Indian Council of Agricultural Research to launch massive programme in 1989 to develop hybrid technology in rice and strengthen hybrid research programme in crops like pigeonpea, cotton, sunflower and pearl millet. Four rice hybrids have so far been developed for general cultivation in Andhra Pradesh (APRH1 and APRH2), Karnataka (KRH1) and Tamil Nadu (TNRH1) states. Some other promising rice hybrids may be soon released both by public and private sectors. Efforts are going on in Australia, China, Mexico (CIMMYT) and Japan to develop commercially viable wheat hybrid technology. Two wheat hybrids named XN901 through CMS system and Jinhua 1 through CHA system have been released for commercial cultivation in China during 1996 and 1997, respectively (Zhang and Huang, 1998). The use of Veery varieties of SIMMYT in the development of wheat hybrids may throw some promising hybrids with multiple resistance to the diseases of wheat. For producing commercial hybrids in self-fertilized and often cross-fertilized crops, we should identify parents capable of producing F_1 hybrids possessing at least 10-15 per cent economic heterosis, search usable cytoplasmic male sterility with high female fertility, identify alternative systems of producing hybrids and restructure the floral characteristics of the parental stock by exercising selection for greater anther extrusion, large anthers with more pollen grains, greater pollen longevity, greater stigma receptivity, etc.

Stabilizing Yield Potential

Yield stability over time and space is determined mainly by tolerance or resistance to various biotic and abiotic factors. Multilocation testing, expansion of the genetic base for disease resistance and pyramiding genes for durable resistance are other key factors in achieving this goal. Depending upon the situation, genes for disease resistance can be incorporated using a conventional breeding procedure or using bitechnological tools like restriction fragment length polymorphisms (RFLPs) and random amplified polymorphic DNA (RAPDs). The transfer of alien genetic variation to a number of crop species has played a significant role in their origin and improvement.

The factors that destabilize the yield should be identified and all efforts must be made to minimize the effect of such factors. Appropriate choice of varieties with desired level of resistance, use of high quality seeds, balanced use of nutrients and adoption of integrated pest management techniques can substantially bring down the yield losses. The genetic vulnerability can be considerably reduced by releasing and promoting a mosaic of varieties in order to minimize the risk of disease epidemics, deploying varieties with different genetic background, understanding and building a durable resistance base and by conducting better disease surveillance and having better understanding of disease epidemiology. Such steps are now being taken by a number of countries and not only the breeders and seed companies but also many progressive farmers are co-operating in maintaining stable productivity of varieties by rapidly replacing the ones that become susceptible. For achieving this, a close watch on the evolution and spread of diseases is necessary. A number of sources of durable resistance have now been identified and are being exploited extensively to develop cultivars endowed with inherent mechanism of genetic homeostasis. Such approach can assist in maintaining higher yield levels under optimum as well as under sub-optimum environments. Several varieties of crops have a number of attractive features like good level of resistance for diseases, consistently good performance under wide range of growing conditions and excellent quality traits.

Improving Input Use Efficiency

To make crop farming profitable and sustainable, the new varieties should have higher input use efficiency and response and possess tolerance to minor element deficiencies and toxicities. Quite a good piece of work has been done at CIMMYT on this aspect and the results show that semi-dwarf varieties of wheat possess about 20 per cent greater nitrogen use efficiency as compared to tall varieties at all levels of nitrogen availability. Similarly, the wheat varieties with 1B/1R translocation possess higher efficiency to extract phosphorus as compared to the varieties having only 1B.

Furthermore, the new varieties should be tolerant to minor element deficiencies and toxicities. Zinc deficiency and boron deficiency as well as toxicity are becoming common problems in some areas.

Fortunately a good amount of variability exists among semidwarf accessions for nitrogen and phosphorus use efficiency.

CHAPTER 2

Evolutionary Patterns in Plants

Evolution may be broadly defined as a change in the genetic make-up of populations at the level of race (formation of a new race, subspecies, incipient species or ecotype), species (speciation), genus, family, class and so on. The changes brought by environment in the form of adaptive modifications (development of usually adaptive phenotypes in response to regularly occurring environmental influences in the normal habitats of a species) and morphoses (development of rarely adaptive or unadaptive phenotypes in response to environmental stimuli which occur rarely or never in the normal habitats of a species), without a change in the genotype, do not constitute evolution.

A number of theories (the theory of inheritance of acquired characters by Lamarck, germplasm theory of Weismann, the theory of natural selection by Darwin, the mutation theory of de Vries, etc.) have been put forward to explain organic evolution. According to Lamarck, who tried to explain evolution in terms of adaptive modifications, there is a continuous increase in the size of living organisms and their component parts, new need and movement determine the production of new organs, the use and disuse would decide the development and degeneration of organs, respectively, and the changes brought about by these three principles will be inherited by the progeny. However, his theory could not stand the test of time as it was criticized by a large number of scientists before and after his death.

The first criticism against Lamarck's theory came through the work of Weismann who by cutting the tail of mice continuously for more than 20 generations proved that mutilation did not reduce the tail length. However, mutilation should never be confused with the disuse of the organ. Weismann postulated that there are two parts of individual's body, the germplasm and the somatoplasm and that the characters present in germplasm are heritable and those present in somatoplasm are non-heritable.

Darwin believed that individuals of a species multiply in a geometric fashion, variation exists among individuals and natural selection acts on this variability to select best fitted ones. He classified variations found in populations into two classes—continuous variations (showing continuous grading in a character) and discontinuous variations or sports (appearing suddenly and showing discrete classes). Darwin attached little significance to sports as a possible cause of evolution since he thought that such variations will mostly be harmful. According to him, continuous variations played a major role in evolution. Natural selection acts on these variations and they attain perfection ultimately but gradually. He, therefore, attached maximum importance to natural selection as a force behind the evolution of species. Nevertheless, a bit different slogan 'survival of the fittest' was given by Spencer.

After the realization that the knowledge of the mechanism of heredity can greatly help in understanding the process of evolution, Darwin put forward a hypothetical model in an attempt to explain the principles of heredity. He called it as the theory of pangenes. Pangenes, according to him, are minute particles produced by all organs of an individual, carry information about the organs, travel through the blood stream, ultimately reach the gametes and form the same organs in the next generation from which they were produced. This theory was later on discarded since it was similar to the theory of Lamarck.

Although Darwin was able to recognize natural selection as one of the major patterns of evolution, Darwinism suffered from several drawbacks :

(1) Darwinism was confined to the already existing variations and did not take into account the creation of new hereditary variations due to mutations. Darwin advocated for the survival of the fittest but was unable to see the possibility of the arrival of the fittest.

(2) In fact, natural selection operates mainly through the relative fitness or differential reproduction of genotypes rather than through the death of the unfit and the survival of the fittest as was conceived in Darwin's natural selection or struggle for existence.

(3) Darwin emphasized the importance of continuous variations but did not pay attention towards non-heritable (environmental) variations which are an important part of the continuous variations but play little role in evolution.

(4) By proposing the theory of pangenesis, Darwin indirectly accepted the inheritance of acquired characters which in the present day and time can hardly be accepted by any one.

(5) Certain cases of overspecialization like evolution of huge dinosaurs and giant deer of Ireland (both now extinct) are unexplainable on the basis of Darwinism (continuous variation and natural selection).

Considering that sudden heritable changes (gene mutations) are the building blocks of evolution, Hugo de Vries proposed the mutation theory of evolution. According to this theory, these sudden changes in gene are entirely responsible for evolution. However, mutation should never be considered as the whole truth of evolution since other factors (natural selection, distant hybridization, etc.) are equally important for the evolutionary process. This theory took care of some of the drawbacks associated with Darwinism. The theory, for the first time, indicated the possibility of the arrival of the fittest and presented explaination about the cases of overspecialization.

Based on the information from numerous species hybrids, Lotsy (1916) put foward the theory of evolution by hybridization and said that the importance of mutations in evolution has been overemphasized. He advocated that recombination between preexisting genes plays most important role in evolution. This theory got support from Du Rietz (1930), Camp (1944, 1945), Anderson (1949) and Stebbins (1950).

But none of these theories gives complete information about the patterns of evolution. Nevertheless, after the basic principles of genetics were well known to the scientists of this area, there was significant progress in the understanding of the factors involved in the process of organic evolution. As a result, the modern synthetic theory of evolution was proposed based on the work of and got support from Fisher (1930, 1932, 1939), Wright (1930, 1931a, b, 1932, 1940a, b, 1949), Haldane (1933, 1937, 1939), Dobzhansky (1940, 1947, 1951), Mayr (1940, 1942, 1947, 1949, 1963) and Stebbins (1940, 1945, 1949, 1950, 1970, 1971).

Since evolution is a process of change, its description may be divided into two following parts:

(1) The factors affecting the process of evolution, that is, the factors affecting the genetic composition of populations (or statics of evolution).

(2) The mechanisms involved in the evolutionary process, that is, the interaction of (1) factors in speciation (or dynamics of evolution).

According to the modern synthetic theory, the basic factors involved in the evolutionary process are gene mutations, changes in chromosome number and structure, distant hybridization, selection and reproductive isolation. The first three of these factors supply raw material for evolution by providing genetic variability and the last two factors act upon this variability and give direction to the evolutionary process. In addition to these five factors, migration of genetic material from one population to another and random fluctuations in gene frequencies in effectively small population, that is, a restriction of the population size (genetic drift, random drift or Sewall Wright's effect) also sometimes play role in the process of evolution.

The mechanisms of evolution generally appear at three levels :

(1) Gene mutations, chromosomal changes and distant hybridizations constantly and unremittingly generate genetic variability and thus provide raw material for evolution.

(2) Selection acts upon this raw material and gives new shapes to the existing populations.

(3) The genetic diversity attained at (1) and (2) levels is fixed by various prezygotic and post-zygotic isolating mechanisms.

A brief description of the factors involved in evolutionary process and their interaction in speciation would make the things more clear.

Gene Mutations

Some investigators put all kinds of changes in chromosomes under a single head, mutations. According to them, mutations may involve change in a single gene (gene mutation or mutation proper), small or large segment of chromosome (structural changes in chromosomes–duplications, deficiencies, inversions and translocations), full chromosome (aneuploidy–monosomics, nullisomics, trisomics and tetrasomics) or whole set of chromosomes (euploidy–haploidy and polyploidy). Here, gene mutations and chromosomal changes will be dealt separately.

Genetic diversity is a pre-requisite for evolution to occur and that gene mutation, a sudden heritable change, is the only factor which constantly and unabatedly goes on generating new hereditary variants and is thus the basic cause of Mendelian variation. Of course, genetic diversity is not fortuitous, rather it is an outcome of a long historical process of development. Factors like genetic recombination, distant hybridization and chromosomal changes though increase genetic variability but are incapable of creating new genes. Gene mutations are mostly recessive and are not expressed unless they come in homozygous condition in the next generation after self fertilization or both alleles of the locus mutate simultaneously and similarly (an extremely rare occurrence). However, there are some mutations of dominant type which express themselves in the same generation.

Gene mutations act in three general ways :

(1) They by themselves and through Mendelian recombinations (through crossing over and segregation and independent assortment of genes) increase genetic variability and thus provide opportunity to both the nature and man to select better genes or gene combinations through natural selection and artificial selection, respectively.

(2) They bring changes in gene and genotypic frequencies of populations. Nevertheless, mutations with normal rates (5×10^{-5} or less) do not play a major role in changing population structure unless some specific mutagen is used which can drastically change the mutation rate.

(3) Some macromutations (with large and multiple effects) may bring drastic changes in the phenotype of a species. A few of such mutations have so large effects that even the taxonomic status of the species is changed.

The effect of a mutation on population structure, however, depends on its size (micro or macro), occurrence (recurrent or non-recurrent) and its kind (reversible or non-reversible). Both the micro and macro mutations have played important role in the evolution of crop species. According to Darwin (1859), mutations with small effects have played relatively greater role in evolution than the mutations with large effects. However, the former type of mutations have manytimes indistinguishable effects.

However, if there is a single mutation (a case of non-recurrent mutations), the chance of survival of the mutant allele is very small since the probability of elimination of such allele increases with each generation from the time of the occurrence of mutation. According to Fisher (1930), if the number of generations (n) is large, the probability of the survival of a single mutant allele is $2/n$. Therefore, such an allele can rarely establish itself in a natural population.

If, on the other hand, the allele A_1 goes on mutating to a_1 (recurrent non-reversible mutations), the frequency of a_1 will increase with each generation until A_1 is completely replaced by a_1. If the initial frequency of A_1 is p_o and the rate of mutation of A_1 to a_1 is m, the frequency of A_1 allele after n generations will reduce to $p_o (1-m)^n$ and even if the value of m is small, the allele A_1 will ultimately be eliminated and a_1 will be fixed.

Further, if there are recurrent reversible mutations, that is, mutations occur in both the directions, the Δq (the net amount of change in the frequency of a_1) will be $m_1 p_o - m_2 q_o$ (where m_1 is the rate of mutation of A_1 to a_1 and m_2 is the rate of mutation of a_1 to A_1). This relationship indicates that the value of Δq will be larger if the difference between the frequencies of the two alleles is more and the frequency of the abundant allele will decrease rapidly. However, as the difference between the two allelic frequencies becomes narrower, the value of Δq also goes down. A point may come when $m_1 p = m_2 q$ or $\Delta q = 0$ (that is, a mutational equilibrium between the frequencies of A_1 and a_1). In such situation, $q = m_1 /(m_1+m_2)$ and $p = m_2 /(m_1+m_2)$. Nevertheless, mutational equilibrium should never be confused with Hardy-Weinberg equilibrium as the former is an equilibrium of the gene frequencies themselves under opposite mutation pressures and is basically more stable than the latter which refers to a relative stability with regard to the genotypic and phenotypic frequencies in populations on the basis of assigned gene frequencies. Further, since the rate of mutation is usually very low, the gene frequencies require a sufficiently large number of generations to reach their mutational equilibrium frequencies.

Macro mutations in some cases have been found with multiple effects. In case of maize, for example, a single spontaneous mutation at Tu–tu locus on chromosome IV is mainly responsible for changing the old pod corn to modern domesticated corn. Similarly, a gene mutation has resulted in the evolution of nutritionally superior (high lysine and tryptophan content) opaque-2 maize (Mertz *et al.*, 1964). But opaque-2 maize varieties have been found less acceptable than normal varieties in both the developed and developing countries. The reasons for this less acceptability are mainly three: (1) Opaque-2 varieties are less yielder (about 10 per cent less) than the normal maize varieties, (2) they are more susceptible to ear and kernel rots and (3) they have different appearance, texture and taste. However, if sincere efforts are made for developing high yielding and resistant opaque-2 varieties in developing countries like India where maize is consumed by human beings directly, this type of maize may find a better place. In India, three nutritionally superior opaque-2 composites, namely, Shakti, Rattan and Protina were developed in 1971 and recommended for cultivation in different regions of the country. With a reasonable level of yield potential, these composites have substantially increased content of lysine (about double) and tryptophan (more than two and a half times) but considerably reduced content of zein (an alcohol soluble harmful chemical constituent and having very high positive correlation with total protein but extremely low in lysine and tryptophan, two of the essential amino acids) in endosperm. Also, these composites contain considerably reduced content of leucine (one of the non-essential amino acids with poor nutritional value). Another mutant gene associated with the favourable improvement of amino acid pattern in maize is floury-2 (Nelson, 1965) which has not only brought a substantial improvement in lysine and tryptophan fractions but has also caused a considerable increase in the methionine content. In maize, recessive mutants for higher sugar content (sugary, su su) and non-purple aleurone colour (aa) have also been found. However, the sugary gene and dominant allele for purple aleurone (A) have close linkage with shrunken shape of kernels (sh.sh.).

The suppression of pairing between homeologous chromosomes (diploidization) is considered to be an important event in the evolution of polyploid wheats. A single gene mutation in the long arm of chromosome V of B genome of wheat is responsible for such suppression of pairing.

Another example where macro mutations have played an important role in the evolution of cultivated species is the evolution of cauliflower, heading cabbage and broccoli from wild cabbage.

There are many other examples where spontaneous as well as induced mutations have played significant role in the evolution and further improvement of crop species. For example, the use of spontaneously occurring and induced dwarfing genes in several crops (wheat, rice, sorghum, etc.) has revolutionized the world agriculture. The extensive use of dwarfing genes, namely Rht_1, Rht_2 (from Japanese cultivars and situated on the chromosome 4A and 4D, respectively) and Rht_8 (from Yugoslav cultivars) in wheat breeding programmes to develop short statured varieties with resistant and sturdy straw and a vigorous and healthy root system, has helped a lot in controlling lodging (bending or breaking-over of culm) and saving grain losses in this important cereal. These height reducing genes also increase grain yield in wheat by increasing the number of tillers and number of grains per plant (Poehlman and Sleper, 1995). By producing intervarietal hybrids in wheat and comparing them with their isogenic parents, Gale, Salter and Law (1986) demonstrated overdominance for grain yield due to the pleiotropic effect of Rht_3 gene on plant height, number of grains per tiller and grain weight. This gene has large and opposing effects on grain number and grain weight and thus the overdominance for grain yield could be explained on the basis of multiplicative interaction between these two component traits.

The large scale exploitation of useful spontaneous as well as induced dwarfing mutations in rice breeding programmes led to the development of modern semi-dwarf varieties which made a significant contribution towards green revolution in rice-growing regions (both tropical and subtropical) of the world. The foundation of the use of mutant dwarfing genes in rice was laid in 1956 with the development of the variety Taichung Native 1 by incorporating a semi-dwarf recessive gene sd_1 from a spontaneous semi-dwarf mutant. Originally, this gene (sd_1) was found in a farmer's field in Taiwan and was named as Dee-geo-woo-gen. In 1976, a semi-dwarf variety of rice named Calrose 76 was released in USA by artificially inducing a semi-dwarfing gene through X-rays treatment. Fortunately, this induced gene is allelic to the spontaneously occurring semi-dwarfing gene sd_1. The induced gene not only reduced plant height in rice but also increased grain yield. Later on, some semi-dwarfing (named sd_2 and sd_4) non-allelic to sd_1 were discovered and incorporated in rice varieties. In addition to semi-dwarfing genes, an induced mutant for waxy endosperm and a partially dominant gene for earliness have been discovered and used in rice breeding programmes. As a result, a waxy semi-dwarf variety (named Calmochi 202) has been developed by crossing Calrose 76 with a waxy variety (named Calmochi 201). Efforts are being made to discover mutant genes for male sterility and apomixis and exploit them for developing male sterile lines (particularly photosensitive male sterile lines) and true breeding F_1 hybrids, respectively (Poehlman and Sleper, 1995).

The development of viable and productive dwarf varieties in sorghum is the result of the extensive utilization of spontaneous mutants carrying dwarf genes. One more advantage associated with these dwarf varieties is that they can be machine harvested. The use of dwarfing genes in sorghum breeding programmes began with the development of Dwarf Milo cultivar in USA from the seed of a mutant plant found in a farmer's field of standard Milo variety. In sequence, a double Dwarf Milo cultivar was developed from a second dwarf mutant plant found in the field of Dwarf Milo. In addition to these two mutants, several mutant plants found later in other varieties of Milo and Kafir became an important part of the US basic breeding stock of sorghum and provided basis for the development of dwarf sorghums for sorghum-growing regions of the world (Central America, Argentina, Africa, India, etc.).

Chromosomal Changes

No doubt, structural changes in chromosomes, aneuploidy and haploidy have their own importance in the evolutionary process. For example, translocations have played a significant role in the evolution of some species of *Gossypium*. Some inversions have helped in establishing polymorphic types in some species. Completely homozygous lines and varieties have been developed using haploidy. However, our main emphasis here will be to discuss the role of polyploidy in the evolution of cultivated plant species.

Polyploidy has not only been working as a very powerful agent in the spontaneous creation of new plant species from times immemorial, but some chemical agents (colchicine and other chemicals) have been used for overcoming hybrid sterility and the artificial production of polyploids. Polyploidy has also been used as a tool to know the putative ancestors of already existing polyploids by producing experimental polyploids and comparing them with the naturally occurring ones. As soon as a polyploid is formed, it becomes reproductively isolated from its parents.

A set of chromosomes (or a genome) is the basic number of chromosomes (or monoploid set of chromosomes or x number of chromosomes) of a species or a group of related species and thus in a genome there is only one of each kind of chromosome. The number of chromosomes present in the gamete is termed as haploid chromosome number or n number of chromosomes and, similarly, the number of chromosomes present in the diploid individual is called as diploid number or 2n chromosome number. In a truly diploid species, n = x but in an amphidiploid or allopolyploid, depending upon the level of polyploidy, n is equal to 2x, 3x and so on.

Polyploidy can be defined as the occurrence of more than two full sets of chromosomes in a species. Broadly, polyploidy is of two types : (a) autopolyploidy and (b) allopolyploidy. The former is the reduplication of genome of the same species, while in the latter type, genomes from different species are involved. It is the allpolyploidy which has played greater role in the evolution of crop plants. Allopolyploids behave just like normal diploid species. Wheat, triticale, mustard, cotton, tobacco, oats, groundnut and sugarcane are some examples of such polyploids.

Two probable explanations have been given about the spontaneous formation of allopolyploids:

(1) Spontaneous occurrence of wide crosses followed by spontaneous somatic chromosome doubling.

(2) Spontaneous formation of unreduced gametes during meiosis in male, female or both followed by union of such gametes at the time of fertilization, that is, through misadventures in meiosis.

The first explanation, a classical one, does not seem to be based on a very sound reasoning.No doubt, wide crosses do occur in nature but the chances of simultaneous occurrence of distant hybridization and spontaneous chromosome doubling are extremely rare and thus formation of allopolyploids through this route is less likely. The hybrids of wide crosses generally show abnormal meiosis followed by very low seed set. On the contrary, the second explanation seems to be more likely for the spontaneous formation of both the allopolyploids and autopolyploids as there are more opportunities for higher level of fertility in this route. Once the gametes are formed, reduced or unreduced, their subsequent union is an almost normal process. However, for the origin of allopolyploids (or amphidiploids), the occurrence of both the polyploidy and distant hybridization (before or after the doubling of chromosome number) is essential. The detailed information about different allopolyploid species is given under wide crosses in this chapter.

Autopolyploidy has been found useful in producing seedless fruits and vigorous plants with higher yield. It is generally believed that autotetraploidy is relatively more important to agriculture than other ploidy levels, cultivated potato (*Solanum tuberosum*), alfalfa, coffee, several forage grasses and a number of ornamental plant species are autotetraploids and are better than their related diploid species in many respects. However, in case of sweet potato (*Ipomaea batatas*), autohexaploidy appears to be the optimum level of ploidy. The degree of tolerance to repetitions of whole chromosome sets in some cases is still higher where maximum vigour is associated with octaploids or even higher polyploids. For example, in strawberry, octaploids possess maximum vigour.

On the lower side, autotriploidy appears to be the optimum ploidy level in case of banana, sugarbeet, watermelon, apples, pears, grapes, etc. Triploid bananas possess greater vigour as compared to their corresponding diploids and tetraploids. Similarly, roots of triploid sugar-beet plants are larger in size and more sugary than those of diploid and tetraploid plants, in order. Autotriploid watermelons which are being grown at a commercial scale in Japan, have uniform sweetness right from the centre to the rind. Autotriploid apples, pears and grapes also present a similar situation regarding sweetness and some other favourable characteristics. These examples indicate that each group of plants has its own optimum level of ploidy.

Autotriploidy in bananas, watermelons, pears, apples and grapes offers an additional advantage of seedlessness. Since autopolyploids behave abnormally at meiosis, multivatent and univalent formation followed by production of non-functional gametes is a regular feature of such polyploids and thus there is no regeneration of species unless they are maintained through vegetative propagation or fresh seed is produced by crossing tetraploids with diploids as in case of watermelon. Therefore, the advantage of this type of polyploidy can more easily be taken in case of vegetatively propagated plant species (potato, sugarcane, banana, pear, apple, grape, etc.).

However, it is not always possible to make a sharp distinction between auto- and allopolyploidy. The results of careful studies indicate that some plant forms which were regarded as autopolyploids earlier have now been put under allopolyploids. In addition to the advantages discussed above, polyploidy provides opportunity for more number of mutations by increasing the total number of genes in the species, increases adaptability as an amphidiploid possesses the environmental tolerance of both parent species and offers a genetic system which allows self-fertilizing species to make use of homozygous genomic heterosis. Hexaploid bread wheat, for example, possesses three different genomes (A, B and D) in homozygous condition. But because of the homeologous nature of the chromosomes, there are several common loci in the three genomes. In such situation, different alleles are in homogygous condition in each genome separately, but intergenomically the condition is heterozygous. Heterosis exhibited in this way has been named as homozygous genomic heterosis.

Wide Crosses and Introgression of Genes

In wide crosses (interspecific and intergeneric hybridizations) the characters of two or more species are combined together. But since the species involved in such hybridizations usually have morphologically and genetically dissimilar chromosomes, the F_1s produced from them are mostly sterile and thus need doubling of chromosomes before behaving as a normal species except where the species can be maintained through vegetative propagation. However, if both the species involved in hybridization have unreduced gametes, the union of such gametes may produce a new normally behaving species.

A number of crop plants (tetraploid and hexaploid wheats, tetraploid mustards, tetraploid cottons, tetraploid tobaccos, tetraploid groundnut, etc.) have evolved through distant hybridization and are natural amphidiploids. In case of hexaploid wheats, three different diploid species, namely, *Triticum monococcum* or *T. urartu* or *T. boeoticum* or *T. aegilopoides* or *T. thaoudar* (contributor of A genome), *T. searsii* or *T. speltoides* or *Aegilops speltoides* (contributor of B genome) and *T. tauschii* or *T. squarrosa* or *Ae. squarrosa* (contributor of D genome) have contributed their genomes. The hybridization between the species carrying A and B genomes resulted in the evolution of tetraploid wheats. Tetraploid wheat later on crossed to diploid wheat and gave rise to hexaploid or bread wheat.

On the basis of geographical, cytological, genetical and biochemical proofs, it is now fairly accepted that tetraploid cottons originated from hybrids between two diploid species, one from old world (*Gossypium herbaceum*) and the other from new world (*G. raimondii*). The two tetraploid cultivated species of cotton, *G. hirsutum* and *G. barbadense* , have a monophyletic origin. *G. tomentosum* is a tetraploid wild cotton found in Hawaii.

The ancestors of cultivated tetraploid tobacco, *Nicotiana tabacum* have also been identified fairly accurately. The putative parents of *N. tabacum* are two diploid wild species *N. sylvestris* and

N. tomentosa. The evidence in favour of this evolutionary path comes from the pairing behaviour shown by these species. Twelve chromosomes of *N. sylvestris* pair with 12 chromosomes of *N. tabacum* and similarly 12 chromosomes of *N. tomentosa* pair with remaining 12 chromosomes of *N. tabacum*, that is, hybrid of *N. tabacum* and *N. sylvestis* and that of *N. tabacum* and *N. tomentosa* each produce 12 bivalents and 12 univalents at meiotic metaphase I. Further, though considerable chromosome differentiation may have taken place in all the three species, the amphidiploid of *N. sylvestris* and *N. tomentosa* resembles to *N. tabaccum* in a number of characteristics.

Various species of mustard (*Brassica*) have been extensively studied for their chromosome homology and the fertility of interspecific hybrids produced by crossing diploid and tetraploid species. Their genomic relationships clearly suggest that the three naturally occurring amphidiploids *Brassica juncea*, *B. napus* and *B. carinata* have evolved through distant hybridizations between the three diploid species, *B. campestris* (A genome with n=10), *B. nigra* (B genome with n=8) and *B. oleracea* (C genome with n=9) and show a triangle as proposed by U (1935). The artificial synthesis of amphidiploids, DNA estimations and protein profiles have confirmed that the amphidiploid *B. juncea* derived from the cross between *B. campestris* and *B. nigra*, amphidiploid *B. napus* from the cross between *B. compestris* and *B. oleracea* and the amphidiploid *B. carinata* derived from the cross between *B. nigra* and *B. oleracea* and these three amphidiploids contain AABB (2n=36), AACC (2n=38) and BBCC (2n=34) genomes, respectively.

The ancestors of tetraploid groundnut are not so accurately known as in case of tobacco and mustard. The section Arachis of the genus *Arachis* consists of two tetraploid species (*Arachis hypogea* and *A. monticola*) with AB genomes, seven diploid species (*A.villosa*, *A. duranensis*, *A. cardenasii*, *A. chacoense*, *A. correntina*, *A. stenosperma* and *A. spegazzinii*) with A genome, one diploid species (*A. batizocoi*) with B genome and one diploid species (*A. spinaclava*) with D genome. Out of the eleven species, only *A. hypogea* is cultivated. Interspecific hybrids between *A. hypogea* and diploid species with A or B genome generally produce 10 bivalents and 10 univalents at metaphase I. Different A genome species show complete homology of the 10 chromosomes and thus interspecific hybrids between them form 10 bivalents during meiosis and are fully fertile. However, hybrids between *A. batizocoi* and A genome species produce several univalents at metaphase I indicating a good degree of non-homology between chromosomes of different species. In view of this behaviour of the chromosomes, Stebbins (1957) called *A. hypogea* a segmental allopolyploid. Since 10 chromosomes of each of the A genome species show complete homology with 10 of the 20 chromosomes of *A. hypogea*, it is difficult to know the real progenitor of A genome of groundnut. However, some scientists consider *A. cardenasii*, *A. duranensis* or *A. villosa* as a putative parent of groundnut. Since *A. batizoioi* is the only species with B genome, this species should be considered as the contributor of B genome. Gregory and Gregory (1976), however, suggest that the two A genome species, *A. cardenasii* and *A. duranensis* could be the contributors of two genomes of *A. hypogea*.

In addition to numerous spontaneous distant hybridizations which resulted in the evolution of many valuable allopolyploid crop species, efforts have been and are being made by plant scientists to develop new man-made crops by attempting crosses between different plant species/genera. Perhaps the most important example of such a man-made crop is *Triticale* (*Triticosecale*) which has been produced by making intergeneric crosses between *Triticum* and *Secale* to combine the quality of wheat with the hardiness of rye. Hexaploid and octaploid triticales have been developed by crossing rye (*Secale cereale*) with tetraploid (*T. durum*) and hexaploid (*T. aestivum*) wheats, respectively.

But because of some inherent drawbacks (cytological instability, shrivelled grains, etc.) associated with triticale, this man-made cereal could not find a place similar to other cereals like wheat, rice and maize in most of the countries of the world. Of course, in some countries, triticale is being grown at a commercial scale.

Crosses between elephant grass (*Pennisetum purpureum*) and pearl millet (*Pennisetum glaucum*) have been successfully made to combine high forage production of elephant grass with drought tolerance and seed production of pearl millet. Similarly, hybrids between sorghum and Sudan grass have been produced to combine the desirable characteristics of these two species.

Karpechenko (1928) produced the well known intergeneric amphidiploid *Raphanobrassica* by crossing radish (*Raphanus sativus*, RR genome with 2n=18) and common cabbage (*Brassica oleracea*, CC genome with 2n=18) to combine root characteristics of radish with leaf characteristics of common cabbage. Later on, several workers artificially produced this amphidiploid and suggested a common name 'radicale' to compare it with 'triticale'. However, till date, the scientists could not combine the desirable characteristics of the two genera.

Efforts have also been made to develop *Aegilotricum* (*Aegilops venticosa*, DDNN × *Triticum turgidum*, AABB) and *Agrotricum* (*Agropyron intermedium*, $E_1E_1E_2E_2NN$ × *Triticum aestivum*, AABBDD) for transferring resistance against the disease eyespot and the rusts, respectively, to wheat.

Anderson and Hubricht (1938) coined the term 'introgressive hybridization' and called it as an infiltration of genetic material of one species into the gene pool of another. Later on, Anderson (1949) attached great evolutionary significance to such hybridization. However, it is difficult to give full account of all those species where gene introgression has occurred since in many cases the size of the genetic material transferred is so small that it becomes very difficult to detect the contamination caused by this kind of hybridization. In any case, the taxonomic status of a species is not impaired by introgression of genes. Nevertheless, there are some specific cases where introgressive hybridization has occurred in nature or some genes have been purposefully introgressed from one species to another by man. There is a firm belief that some genes got introgressed into maize (*Zea mays*) germplasm from its wild relative gammagrass (*Tripsacum*). Further, present day varieties of sugarcane are complex interspecific heterozygous polyploids. Introgression of germplasm from three or more species of *Saccharum* has played a major role in the development of modern varieties of sugarcane. For example, some genes governing resistance to diseases and insect pests have been and are being introgressed into sugarcane germplasm from its wild relative kans (*Saccharum spontaneum*).

Riley (1938) described an striking case of introgression in the populations of two species of *Iris*, *I.fulva* and *I. hexagona* variety *gigantocaerulea*, earlier growing in two ecologically different habitats of Mississippi Delta region (*I. fulva* preferably on clay soils and partial shade but *I. hexagona* in tidal swans and full sun). However, due to the destruction of the forests and the drainage of swams for pastures by man, this ecological isolation got disturbed resulting in numerous hybridizations between these two species. The backcrosses of partially sterile F_1 hybrids to parental species which occur frequently in these man-made habitats, resulted in the transfer of some genes from *I. fulva* to *I. hexagona*. An equally important example of introgression has been described by Heiser (1947) in sunflower. Numerous sunflower populations growing on disturbed soils (man-made habitats) in the form of weeds clearly show introgressions of genetic material between *Helianthus annuus*, *H. petiolaris* and *H. bolanderi*. Stebbins (1950) has reviewed a number of introgressions between different plant species. Quite a large number of modified environments are continuously being created by man by disturbing the natural habitats and thus breaking down the ecological isolation between many species. These man-modified environments now serve as the main pockets of introgressive hybridization as the new ecological conditions help the sympatric populations to undergo adaptive genetic changes more rapidly. However, it is very difficult to know the exact frequency of introgressive hybridization as the data available on this aspect are scanty and manytimes uncritically collected. Sympatric populations may, however, resemble for some character either due to introgression or due to parallel selection of similar genes by similar environments. We must be able to make distinction between these two kinds of resemblances. Further, it is also important to know whether the gene exchange has occurred at population/race level (that is, primary intergradation) or at species level (that is, secondary intergradation or introgression). Primary intergradation should never be confused with introgression. However, one conclusion can be safely drawn that because of some inherent differences between plants and animals, introgressive hybridization occurs more frequently in the former than in the latter (For details see Dobzhansky, 1964).

Although a tussle between two naturally occurring but contrasting processes, sexual reproduction (whose main function is to increase genetic diversity by producing an immense variety of genotypes) and reproductive isolation (which acts as a barrier to gene exchange between populations), goes on continuously, a compromise exists in nature to resolve this conflict. Natural selection preserves adaptive plasticity of populations by regulating gene exchange between populations in such a way that populations do not become completely reproductively isolated. This is the reason why different well established plant species sometimes show easily detectable gene exchange between them. However, during the preservation of adaptive plasticity, a sacrifice in the form of destruction and elimination of some ill-adapted individuals is made.

The amount of genetic material introgressed into one species from another may vary from a single major gene to a group of genes governing one or more characters. There are several examples of introgression of characters into cultivated plant species from their wild relatives (in *Zea mays* from *Tripsacum*, in *Saccharum officinarum* from *S. spontaneum,* etc.). For single gene transfer, the classical example of a radiation induced gene substitution procedure used by Sears (1956) can be cited. In this procedure, a gene governing leaf rust resistance was introgressed into *Triticum aestivum* (a hexaploid bread wheat) from *T. umbellulatum* (a diploid wild grass). In the beginning, a cross was made between *T. turgidum* (tetraploid emmer wheat) and *T. umbellulatum* to produce a triploid F_1, the grass-emmer allohexaploid was crossed with Chinese Spring variety of *T. aestivum* followed by repeated backcrosses with Chinese Spring to produce and then identify leaf rust resistant plants with 43 chromosomes (42 chromosomes of wheat and one chromosome of grass carrying leaf rust resistance gene). These plants (with 43 chromosomes) were treated with x-rays before flowering and the Chinese Spring variety was pollinated by the pollen formed on the X-rayed plants. After screening the progeny thus produced one plant with the absence of undesirable grass plant characteristics but retaining rust resistance was identified. The general features of this plant clearly indicated that a small segment of grass chromosome carrying leaf rust resistance gene was introgressed into a wheat chromosome. Fortunately no undesirable genes were associated with the leaf rust resistance gene. The bread wheat variety developed by this method was called Transfer. This variety became a source of resistance to leaf rust in subsequent wheat breeding programmes. There are several other examples of single gene transfer in crop plants where single desirable genes have been transferred to cultivated species from their wild relatives.

Some biotechnological marker tools like RFLPs (Restriction Fragment Length Polymorphisms) and RAPDs (Random Amplified Polymorphic DNAs) are now being used in many laboratories for incorporating desirable genes in crop plants. Development of transgenic plants is the result of the use of such techniques. For incorporating multiple resistance in crop cultivars, group of genes are being transferred. Wheat-rye translocation 1B/1R carrying linked genes Lr 26/Sr 31/Yr 9 is serving the purpose of incorporating resistance to all the three kinds of rusts (leaf, stem and stripe) in wheat cultivars.

Introgressive hybridization has a good future scope. There are a number of crop species which need transfer of some desirable genes from other species particularly from their wild relatives which possess disease and insect pest resistance genes. Introgression of genes is not restricted to the disease and pest resistance only, but any kind of gene or group of genes may be transferred from one species to another. In triticale, hexaploidy seems to be the optimum level of polyploidy since hexaploid triticales are superior to octaploid triticales. If genes from *Triticum tauschii* responsible for chapati making quality are transferred to hexaploid triticale without changing its ploidy level it can serve a good purpose for improving the quality of hexaploid triticale.

The examples of crop species falling under different types of evolutionary patterns clearly indicate that the evolutionary patterns have not always been independent of each other during the course of evolution of any species. In many cases, either two or more than two patterns were simultaneously or successively involved in the evolution of the species concerned. For example, both mutation and introgression of genes have played role in the evolution of modern domesticated maize. Similarly, both the interspecific hybridization and polyploidy have been mainly responsible for the evolution of tetraploid cottons and tobacco. In the evolution of polyploid wheats, all the three patterns of evolution discussed have been found involved. In addition to wide hybridization and polyploidy, a single gene mutation in 5B chromosome caused tetraploid and hexaploid wheats to behave like normal diploids.

Selection

Selection is the primary force in nature-directed evolution as well as in the man-directed evolution (that is, plant breeding). Whereas the main force behind nature-directed evolution is natural selection, the main force behind plant breeding is artificial selection. As has been mentioned earlier, Darwin attached maximum attention to natural selection as a cause of evolution and raised it to the status of a scientific theory. In artificial selection, we select those characters which are advantageous to us, while the characters favoured during natural selection (primarily progressive adaptation of organisms to their environment which may not be always associated with high yield potential, better quality, etc.) may or may not be useful for man. Further, the speed of improvement during artificial selection is usually higher than that of the improvement through natural selection. Nevertheless, the improvement achieved through natural selection would be longer lasting than the improvement achieved through artificial selection since in the former case the genotype/cultivar evolved has already faced the consequences of natural selection. This is why the wild types are more resistant to diseases and insect pests than the cultivated types. In fact, the plant breeder should not entirely depend on artificial selection as the varieties evolved by him have ultimately to go to farmers' fields where natural selection has to play an important role in the survival and the performance of plants.

The two types of selection, natural and artificial, though differ in the respects mentioned above, have two characteristics in common :

(1) Selection, natural or artificial, acts an already existing genetic variability and is unable to create any new variability.

(2) The effectiveness of selection depends on the extent of variability present in the base material and the degree of heritability of the character under consideration.

Natural selection not only acts on existing variability and gives direction to the evolutionary process but also helps in strengthening and eventually completing the isolating mechanisms.

Selection is not only important from an evolutionary point of view but it can also change gene frequencies in populations. The frequency of any allele in a natural population largely depends on the fitness (adaptive value, selective value, relative reproductive success or relative ability of any genotype to produce surviving offspring) of its carriers. Fitness, in turn, is dependent upon selection which includes a number of simple and complex mechanisms. Different genotypes, generally, show varying degrees of fitness at the same location and the fitness of the same genotype is different at different locations. Since some individuals may not survive under changed environmental conditions while a population may face such a selection pressure by producing better types through recombination, one must consider a population response 'rather than' individual response' to selection. The selection coefficient (that is, the force acting against genotype to reduce its fitness) is inversely proportional to fitness, that is, fitness=1-selection coefficient (s). The value of fitness usually varies between zero and unity. However, in case of overdominance, sometimes, it is more than unity. Further, fitness is different from dominance and a dominant allele may show even zero fitness. The effect of selection on population structure is, therefore, of primary interest to plant breeders. Selection can affect the individuals of the population at any stage of their life cycle.

Selection can affect a genotype at gametic stage or at zygotic stage. If it affects at gametic stage, the recessive allele cannot be shielded by the dominant allele and thus if the fitness of the allele is zero (that is, s=1), it will be eliminated from the population in a single generation. But if s<1, the amount of change in the frequency of the affected allele (Δq) will be $-sq(1-q)/1-sq$ and the number of generations needed for a specified change in the value of the affected allele (sn) will be $\log e\ q_o$ $(1-q_n)/q_n\ (1-q_o)$, where n is the number of generations required to change the frequency of the allele from q_o to q_n.

If selection affects the genotype at zygotic stage, the frequency of the affected allele will change according to different conditions of dominance (no diminance, partial dominance, full dominance or overdominance), selection (in favour or against) and lethality (lethal or non-lethal in homozygous condition). For example, if there is no diminance, selection acts against the allele and the affected allele is not lethal in homozygous condition, the frequency of that allele after one generation will be $-sq\ (1-q)/1-2\ sq$. In case of no dominance, the deleterious allele is not completely shielded from selection in the heterozygous condition. However, the selection coefficient of the deleterious homozygote will be just double to that of the heterozygote if the affected allele is not completely lethal in homozygous condition. Since absence of dominance makes all the alleles available for selection, the frequency of deleterious allele will reduce drastically if the selection coefficient is large.

If there is partial dominance and deleterious allele is not completely lethal in homozygous condition, the selection coefficient of the heterozygote will not be exact half to that of the deleterious homozygote.

If dominance is full and recessive allele is lethal in homozygous condition, the $\Delta q = -q^2/1+q$. That is, if q is large, its value will reduce rapidly but as q becomes smaller the change per generation also decreases because relatively a very small number of recessive homozygotes will be produced in such a condition. It indicates that a recessive allele with lethal effect can remain in population for indefinite period. In case the recessive allele is not completely lethal, the $\Delta q = -sq^2 (1-q)/1-sq^2$. As regards the change in gene frequency per generation, it will be most rapid when q=0.67 and selection will be quite inefficient when q is small. However, when the dominant allele is selected against, the number of generations required to eliminate the dominant allele will mainly depend upon the value of selection coefficient and thus if s=1, the dominant allele will be eliminated from the population in a single generation of selection.

In case the heterozygote is superior to both the homozygotes, $\Delta q = pq (s_1 p - s_2 q)/1 - s_1 p^2 - s_2 p^2$. The value of Δq may become zero if $s_1 p = s_2 q$, that is, the population achieves equilibrium. At equilibrium, $p = s_2/(s_1 + s_2)$ and $q = s_1/(s_1 + s_2)$, where s_1 and s_2 are the selection coefficients associated with the two alleles (say A_1 and A_2). One conclusion can safely be drawn from the foregoing discussion that p and q will reach a stable equilibrium if the values of s_1 and s_2 are constant. If the frequency of any allele departs from its equilirbium value, it will be forced back by selection pressure and this is how the selection preserves genetic variability in natural populations. Should we consider $s_1 = .01$ and $s_2 = .04$, the equilibrium value of q will be .2. If the value of q is below .2, the Δq is positive and vice-versa. Such a stability in the values of s_1, s_2 and consequently in the frequencies of the two alleles has a great evolutionary significance since it results in a balanced polymorphism.

The heterozygote may be inferior than both the homozygotes (that is, a case of negative over-dominance). In such situation, the population does not usually attain equilibrium but the frequency of the abundant allele will go on increasing every generation at the expense of the rarer allele. For example, if the initial frequency of A_1 is .7 and that of A_2 is .3 and the heterozygote does not at all survive, the relative frequencies of the two alleles after one generation of selection would, respectively, be 0.49/0.49+0.09 = 0.84 and 0.09/0.49+0.09=0.16 (approximately). After second generation, these frequencies will be 0.96 and 0.04, respectively. However, if and only if the value of q is 0.5, the population will be in a stable equilibrium since in such a condition the Δq equals zero.

In nature, however, mutation and selection act simultaneously and thus the conclusion drawn on the basis of one factor (either mutation or selection) only may be misleading. The fixation or elimination of an allele will be faster if both these factors exert their pressures in the same direction, but if they oppose each other and if the allelic gain through incoming mutations equals the allelic loss on account of selection, the population will reach a stable equilibrium. The relative effectiveness of the two factors depends on the gene frequency but in different ways. Whereas mutation is most effective when the mutant allele has low frequency, selection is most effective when the allele is abundant in the population. On the basis of the gain through incoming mutations and the loss on account of selection the equilibrium values between mutation and selection under different conditions of dominance and viability can be worked out.

Isolating Mechanisms

Isolating mechanisms is a general term proposed for the curtailment or stoppage of the gene exchange between populations, that is, the barriers which prevent interbreeding of populations. It is needless to emphasize the importance of isolation in organic evolution as there is no race formation and speciation without isolation. The interbreeding of different species or genera (distant hybridization) is though quite efficient in breaking down the harmoneous gene combinations, is also an important cause of evolution. However, in different cases, generally different mechanisms are involved in preventing the gene exchange between populations. Also, more than one mechanism may be responsible for checking the interbreeding of populations. Distance (or geographic isolation) is one important factor for preventing the gene exchange between allopatric populations (races or species inhabiting different geographical areas). But in case of sympatric populations (races or species inhabiting the same geographical area), interbreeding is usually checked by some genetic factors.

A Mendelian population is a heterogeneous group of genotypes sharing in a common gene pool. When heterogeneity of a population becomes easily detectable by open eye or by some method, such population is called polymorphic or polytypic. Similarly, a macro environment is consisted of many micro environments. Different individuals of a population, race or species inhabiting a macro environment, interact with these micro environments. These individual × environment interactions gradually make the population polytypic and ultimately different small groups become adapted to specific spatial or temporal environments, that is, intrapopulational polymorphism in a sympatric population. Further, a plant species is generally found/grown over a wide range of environments (including seveal macro environments) and may become polymorphic due to both the sympatric and allopatric diversity. In some cases, such a polymorphism takes the form of a balanced polymorphism and heterozygotes show adaptive superiority over homozygotes. Heterozygosis for inversions and translocations suppresses recombination between coadapted gene complexes. A number of plant species show presence of inversion heterozygotes (*Secale cereale, Avena* species, *Melandrium* species, *Tradescantia* species, etc.) and translocation heterozygotes (*Pennisetum glaucum, Datura stramonium,* spcies of *Gossypium, Oenothera,* etc.). In addition to chromosomal morphism, genic polymorphism also plays role in creating adaptive polymorphism (a widespread phenomenon in the living world).

If a species (an interbreeding biological unit which satisfies the criteria of reproductive isolation and sympatry and is morphologically distinguishable from other similar units) becomes polytypic, its morphologically distinguishable subpopulations are called as races (populations of a species which differ in some morphological and genetical respects but are potentially capable of intermating with one another), subspecies, incipient species, ecotypes (adaptive races originating due to genotypic response of a plant species to different habitats), geographic races (if they are genetically distinct allopatric populations), etc. Once the subspecies become completely reproductively isolated, they attain the status of independent species. This process of speciation in originally homogeneous populations (species) goes on and new species are formed.

According to Stebbins (1970), isolating mechanisms can be broadly classified into two types: (1) prezygotic mechanisms and (2) post-zygotic mechanisms. Whereas prezygotic mechanisms prevent pollination or fertilization between populations, the postzygotic mechanisms act after the fertilization has occurred resulting in inviable or sterile hybrids or hybrid breakdown. Geographic, ecological, seasonal, mechanical and ethological isolations come under prezygotic isolating mechanisms and hybrid inviability, hybrid sterility and hybrid breakdown come under postzygotic isolating mechanisms. However, Dabzhansky (1964) had classified isolating mechanisms into two broad categories ; (1) geographic isolation and (2) reproductive isolation. The latter includes ecological, seasonal, mechanical, ethological, gametic or gametophytic, hybrid inviability, hybrid sterility and hybrid breakdown. Although some authors have questioned the validity of the basis of this classification (that is, the distinction between geographic and reproductive isolation), Dobzhansky advocated that geographic isolation is caused by spatial factors, while reproductive isolation is caused by biological factors.

Geographic Isolation

As has been discussed above, this type of isolation is independent of all the reproductive isolations, that is, geographically isolated (allopatric) populations may or may not be reproductively isolated. However, geographic isolation is sometimes associated with some reproductive isolation. Further, geographihcally isolated populations after facing consequences of natural selection or genetic drift for 'long periods of time usually become genetically different and ultimately reproductively isolated.

Ecological Isolation

Both the allopatric and sympatric populations may show ecological isolation as the same geographic region usually has several subgeographic areas with different types of soil, climate and living organisms. The populations inhabiting such subgeographic areas are called ecotypes. In the beginning, such ecotypes usually have the ability to exchange genes freely if grown together but later on ecotypic differentiation may become so large that they become incapable of interbreeding under natural conditions, that is, the ecotypes become reproductively isolated and attain the status of independent species. There are numerous examples where different species of the same genus (different species of *Triticum, Hordeum, Gossypium, Quercus, Pinus, Acer,* etc.)are inhabiting the same area or are made to live sympatrically but do not exchange genes under natural conditions, that is, the process of speciation is complete and the populations behave as valid species. On the other hand, there are also examples where populations of the same species (*Antirrhinum glutinosum,* pine trees, etc. growing at different altitudes) adapted to different ecological conditions or inhabiting different geographic areas are unable to maintain their reproductive isolation and produce fertile F_1 hybrids under natural conditions if grown together.

Seasonal isolation – In a number of cases, two populations, potentially capable of producing fertile hybrids, fail to hybridize due to the difference in their flowering time. But such isolation becomes ineffective if there is even slight overlapping in the flowering periods of such populations. Therefore, as an independent isolating mechanism, seasonal isolation is not very important. However, this type of isolation if associated with some other form or forms of isolating mechanisms may ultimately lead to reproductive isolation of the populations. Since time of flowering plays major role in its occurrence, this isolation is also called as temporal isolation.

Mechanical isolation – In plant populations, the mechanical isolation which occurs due to the physical non-correspondence of the floral parts, is perhaps the most effective and most important isolating mechanism. Differenes in colour and structure (shape and size) of flowers and the resemblance of flowers with the female of the fertilizing insect are responsible for this type of isolation. In some entomophilous plant species, the formation of hybrids between populations is prevented because the fertilizing insect species becomes temporarily habitual of visiting flowers of the same colour and structure and thus different species of the same genus attract different fertilizing insect species. Further, in some orchids, there is resemblance between the elaborate structure of the flower and the female insect. The male insect in an attempt to pseudocopulate the flower, effects pollination by shedding pollen on the stigma.

Ethological isolation – This isolation is also called as psychological or sexual isolation and is confined to animal kingdom only. The isolation occurs due to some simple or complex courtship pattern of males and females before mating, causing lack of mutual attraction between the two sexes and thus engendering sexual isolation.

Gametic or gametophytic isolation – This type of isolation occurs because of the lack of attraction between pollen tubes and ovules of different populations/species or the poor viability of the pollen tubes in the sexual tissues of another type. A number of examples of such incompatibility between pollen tubes and sexual tissues of different types (populations) are available. For instance, the pollen tubes of normal (non-sugary) maize travel more rapidly than those of sugary maize if mixture of two kinds of pollen is applied to the silks of normal plant, whereas two kinds of pollen tubes show similar rate of growth if pollen mixture is applied to sugary silks (Mangelsdorf and Jones, 1926). A more clearcut example of differential rate of pollen tube growth is presented by popcorn and other varieties of maize (Demerec, 1929). The pollen tubes of nonpopcorn maize fail to grow in the sexual tissues of popcorn and virtually no seeds are formed in such crosses. On the contrary, popcorn pollen tubes grow easily in the sexual tissues of nonpopcorn varieties. The application of mixture of pollen of popcorn and nonpopcorn to popcorn silks results in the formation of all selfed seeds except a very few hybrid seeds.

A similar differential behaviour of pollen tubes has been found in intra and interspecific crosses of *Datura*. Whereas pollen tubes show normal growth in intraspecific crosses, they show varied behaviour in interspecific crosses. Interspecific crosses between short styled females and long styled males are several times more successful than the reciprocal crosses. The bursting of pollen tubes in the stylar region of another species is another cause of incompatibility between different species of this genus.

Hybrid inviability – The prevention of gene exchange between different species/types is not only restricted to prezygotic variables (the rate of pollen tube growth in foreign styles, style length, the frequency of bursting of pollen tubes before entering embryo-sac and various types of isolating mechanisms) but some postzyotic factors (hybrid inviability, sterility and hybrid breakdown) also play an important role in such prevention. After zygote formation, the development of hybrid may be stopped at any stage due to the lack of genic harmony between the embryo and endosperm or because of the non-adjustment of the chromosomes of the two species. Although disturbances in nuclear metabolism (including DNA replication and RNA synthesis) are said to be the cause of such disharmony, the real cause of hybrid inviability is not yet fully known. If hybrid inviability is due to physiological disturbances, the suppression of the seed development can be overcome by removing the seed coat (*Linum perenne* × *L. austriacum*) or by rescuing embryos and growing them in nutrient solution (*Linum austriacum*×*L. perenne*). Similar techniques have been used to overcome the suppression of seed development in interspecific hybrids of *Datura*.

Hybrid sterility – Even if hybrids are viable and vigorous, they may be eliminated because of sterility. Hybrid sterility which may be due to the failure of bivalent formation at meiosis leading to the breakdown of cell division mechanism and production of few or no functional gametes, small structural differences between parental chromosomes or some other factor, has been variously classified : (1) **genic sterility** (due to the genetic constitution of the individual) and **Chromosomal sterility** (due to the abnormal behaviour of chromosomes at meiosis like failure to form bivalents); (2) **gametic sterility**

(due to the production of degenerate gametes) and **zygotic sterility** (due to the production of inviable zygotes); (3) **haplontic sterility** (due to the lethality at haploid stage, that is, gametic sterility) and **diplontic sterility** (due to the disturbances at the diploid stage of the organism, that is, zygotic sterility); and (4) **segregational sterility** (due to small structural differences between chromosomes of the two parents) and **developmental sterility** (due to developmental differences). In some cases, hybrid sterility is due to the degeneration of male gametophyte after normal meiosis. Genic as well as chromosomal sterility may be interspecific or intra-specific. Hybrid sterlity due to translocations and inversions frequently occurs in the interspecific crosses in *Triticum* and *Pisum*. Asynaptic plants have been found in the interspecific crosses of *Triticum*. Chromosomal sterility has been observed in the intergeneric hybrids between radish (*Raphamus sativus*) and cabbage (*Brassica oleracea*), that is, *Raphanobrassica* (Karpechenko, 1928) and hybrids between polyploid species (*Triticum, Gossypium, Nicotiana*, etc.).

Hybrid breakdown – Viable, robust and fertile F_1 hybrids do not always produce normal progeny in F_2 generation. For instance, the three tetraploid species of cotton, *Gossypium hirsutum* (having high yield potential), *G. barbadense* (a quality cotton) and *G. tomentosum* (wild species), are easily crossable and produce vigorous and fertile F_1 hybrids, but the F_2 progenies of these hybrids do not behave normally. This abnormal behaviour at F_2 stage is due to hybrid breakdown which acts as a bottleneck in developing homozygous varieties with desirable characteristics of the two parents. Although cryptic structural variability is said to be the cause of hybrid breakdown in cotton, the problem remains to be still unresolved.

Migration

Migration is another factor that can introduce new genes into a population and thus can change the gene frequency in the recepient or hybrid population. This change in the gene frequency will mainly depend upon two things :
(1) the difference in the gene frequencies between the donor (migrant) and the receipient populations, and,
(2) the proportion of alleles that is being received by the hybrid population each generation.

If the initial gene frequency in the recepient population is qr, the frequency of the same gene in the donor population is qd, the proportion of newly introduced genes each generation is m, the gene frequency in the recepient population after n generations of migration (provided there is no change in the population size, that is, gain is equal to the loss in gene frequency) will be (qrn–1) (1–m) + mqd and the difference in gene frequency between recepient and donor populations will be qrn–qd. However, the theoretical assumptions of constant population size and continuous flow of genetic material from donor to the recepient population are not manytimes fulfilled by natural and experimental populations. Like mutation and selection, migration is a systematic process where the action in changing the frequencies of genes is dependent on the existing gene frequencies and thus both the magnitude and the direction of change are predictable.

Genetic Drift

Genetic drift is a dispersive process and is thus independent of the existing gene frequencies in the population. As a result, the change in gene frequencies is predictable in magnitude but random in direction. The consequences of the dispersive process are mainly three :

(1) It tends to divide the original population into smaller groups (populations) because there is a more tendency of mating between individiduals which are more close to one another.

(2) Since individuals belonging to the same group become more and more alike, within group genetic variability is reduced.

(3) Homozygotes are increased at the expense of heterozygotes.

To have a better understanding of genetic drift and its consequences on population structure, we must study this process from two viewpoints :

(i) To consider it as a sampling process, that is, the frequencies of different alleles in infinitely large random mating populations remain constant from generation to generation (ignoring effects of directional forces), while the same (gene frequencies) drift drastically in finite populations since such populations will mostly be associated with large standard deviations (chance or random drift); and

(ii) to consider it as an inbreeding process, that is, even though there is random mating among individuals of an effectively small population, more consanguineous matings will take place than in a sufficiently large random mating population.

Although the idea of genetic drift was firstly given by Brooks (1899), precise information about the process as random fluctuations in the frequencies of alleles in finite populations was provided especially by the work of Wright (1921, 1931a,b, 1932). Later on, Wright (1948, 1949), Li (1948) and Lerner (1950) discussed the process in more detail.

Since genetic drift can alter the gene frequencies drastically in effectively small populations and can thus disturb the Hardy-Weinberg law, the study of the process becomes more important for plant breeders as they have to handle such populations more frequently.

The importance of genetic drift as a sampling process can easily be proved by taking into accout two hypothetical populations, one consisted of 4 parents and the other of 4000 parents. If the locus in question is represented by two alleles A_1 and A_2 with their respective frequencies p and q, the standard deviation of proportion (σ) for diploid parents will be pq/2N, N being the number of actual parents used to start a generation. Further, if p=q=0.5, $\sigma = \sqrt{.5 \times .5/8} = \sqrt{.03125} = .1767$, for the first population. For the second population, $\sigma = \sqrt{.5 \times .5/8000} = \sqrt{.00003125} = .0055$. That is, the gene frquency in the second population will fluctuate mostly between .4945 and .5055 (that is, .5±.0055), while the same will fluctuate between .3233 and .6767 (that is, .5±.1767) in the first population indicating that if the same small size is continued each generation, the population will ultimately reach a point of no return (since gene frequency can not go beyond 0 or 1). It means that the next sampling will lead to a further dispersion as each line in the second generation will start from a different gene frequency. However, if the fixation has reached, it does not mean that all the lines will be fixed for the same gene but the fixation will be according to the initial gene frequencies if we consider several such small populations together (Also see Allard, 1960).

As such drift is usually beyond the control of the breeder. However, the breeder may enhance or reduce the rate of chance fixation by exerting selection pressure and/or by increasing or decreasing the populations size.

The inbreeding effect of a small population – Random mating in an effectively small population is nothing but matings between relatives. Thus, inbreeding becomes an important factor affecting the genetic composition of such (finite) random mating populations. If all gametes of the parents unite completely at random (including selfs), the rate of fixation (loss of heterozygosity or decay of heterozygosity) per generation because of inbreeding will be 1/2n where n is the number of actual parents. However, if there is presence of self incompatibility in monoecious plants or the population is consisted of dioecious plants with equal number of male and female parents, the formula for the rate of fixation takes the form 1/2 n + 1 (Allard, 1960). In large populations, 2n and 2n+1 will be approximately equal. Nevertheless, the effective number of parents is not always equal to the total number of individuals in the population. For example, the number of males and females of a population may not be necessarity equal. Such a situation is manytimes found in plant breeding experiments where each of a large number of lines is crossed to a few male testers or vice versa (a few male sterile lines may be crossed with a large number of pollen parents). Effective population size under such a situation can be estimated by the following formula :

$$n_e = \frac{4n_m n_f}{n_m + n_f} \text{ (Allard, 1960)}$$

Where, n_e is the effective population size, n_m the number of males and n_f the number of female parents. We can estimate the effective size of the population by taking hypothetical numbers of the two sexes as 60 males and 740 females -

$$n_e = \frac{4 (60 \times 740)}{60 + 740} = 222$$

The rate of fixation in this population will be $1/2 \times 222 = 1/444$ and not $1/2 \times 800$. The calculation indicates that the effective population size is determined mainly by the rarer sex and is approximately equal to 4 n_m or 4 n_f depending upon whether males or females are rarer. Therefore, the rate of fixation can be approximated by 1/8 n_m or 1/8 n_f. One more point must be made clear that the total number of individuals in a population may be greater than the actual breeding size of the population which further may be greater than the effective size of the population (that is, apparent population size > breeding population size > effective population size).

One conclusion can be easily drawn from the above discussion that whereas fixation proceeds very slowly in large populations, its rate in small populations is significantly high. This indicates that population size must be an important consideration in breeding experiments.

CHAPTER 3

Plant Biodiversity and its Conservation

Biological diversity of plants is the earth's most vital resource and from times immemorial, has unabatedly been providing basic material for human needs by offering scope for natural and artificial selections to tailor genotypes better suited to diverse agro-ecological and socio-cultural needs and is thus a key ingredient for sustainable agricultural development. However, there is vast ignorance about the biodiversity existing on earth and the most disturbing fact is that even before the biodiversity of our planet is fully understood and properly utilized, a considerable portion of the rich biological wealth is lost. The loss of this precious genetic wealth is occurring with an alarming rate all over the world due to the destruction of natural habitats of various plant species. The speed at which the genetic diversity of our commercially grown crops is shrinking is so high that about 25 per cent of the existing diversity could be lost over the coming 30 years. It is not only the loss of species which is causing concern among the naturalists but the loss of genetic diversity within species is becoming more serious as the rate of loss of genetic diversity is relatively higher than that of species extinction. Furthermore, the alarming loss of genetic diversity of our crop plants would increase their vulnerability to pest rampage and widespread disease epidemics. Genetic diversity is, therefore, a prerequisite for the survival of the species. The already alarming situation has been further aggravated by the replacement of old genetically homogeneous varieties and hybrids and by grossly altered land use under growing human pressure. Unfortunately, the severe biological extinction is occurring at a crucial time when the transfer of genes from a species to another unrelated species has become feasible and transgenic plants are being produced using some novel biotechnological techniques and when some gross changes (particularly global warming) are likely to occur in the biosphere in near future because of the growing accumulation of green house gases in the troposphere and due to a rapid deepening of the hole in the protective ozone layer in the stratosphere over Antarctica, the pulsating continent. Therefore, all efforts must be made to increase awareness about the importance and potential value of plant genetic resources and to safeguard the precious and irreplaceable crop biodiversity both *in situ* and *ex situ* to assure our planet's health and well-being. In fact, sustaining the natural resource base would play a key role in sustaining agriculture which, in turn, would act as a major force in sustaining future.

Introduction of Plant Material

Introduction of economic plants into new places has played a significant role in the development of world agriculture. This is how many crop species like wheat, rice, maize and potato have spread over large areas of the world and have become plant wealth of these regions. At its beginning, the plant introduction was mostly haphazard and non-specific and was made, knowingly or unknowingly, by a number of agencies like naturalists, settlers, travellers, traders, explorers, pilgrims and invaders. For example, there is no record available that when crops like mustard (*Brassica juncea*), mung (*Vigna radiata*), apple (*Malus sylvestris*), pear (*Pyrus communis*) and walnut (*Juglans regia*) from Central Asia and jowar (*Sorghum bicolour*), pigeonpea (*Cajanus cajan*), finger millet (*Eleusine coracana*), sesame (*Sesamum indicum*) and Asian cotton (*Gossypium herbaceum*) from Africa were introduced into India. Grapes (*Vitis vinifera*) and cherries were introduced into India from Afghanistan by Muslim invaders before 14th century. But, when the importance of crop introduction was realized, particularly after the discovery of Western Hemisphere, it gradually took form of organized programmes (both adhoc and co-operative) and, as a result, the exchange of plant material was started between the two hemispheres.

Although there must have been innumerable unrecorded cases of shifting crop plant materials from their place of origin to new places, the first recorded attempt in this direction relates to the year 1493 when Christopher Columbus brought from Spain seeds of almost all common crops grown in that country to West Indies. Quite a large number of crop species collected by him found their place in West Indies. The spread of alfalfa or lucerne (*Medicago sativa*) took place rapidly in Chile and Peru after its introduction into South America in early sixteenth century by Spainish explorers. It then gradually spread on the eastern parts of United States via California. A number of crops like potato, sweet potato (*Ipomoea batatas*), ground nut (*Arachis hypogea*), maize, tobacco, chillies (*Capsicum annum*), papaya (*Carica papaya*), guava (*Psidium guayava*), pineapple (*Ananas comosa*) and cashewnut (*Anacardium occidentale*) were introduced into India in sixteenth century by Portuguese. In the later part of the eithteenth century, some plant species viz. litchi (*Litchi chinensis*), loquat (*Eriobotrya japonica*) and tea (*Camellia sinensis*) from China and vegetables like cabbage (*Brassica oleracea*) and cauliflower (*Brassica oleracea*) from the Mediterranean region, were brought to India by East India company. The Turkey wheat (previously grown in Russia) which spread over large areas was introduced into Kansas in 1870s by the settlers of that time.

The plant introduction took an organized shape with the beginning of U.S. Department of Agriculture in 1839. The importance of this programme was further emphasized in United States in the year 1862 by establishing Land Grant Colleges in different states and by treating U.S. Department of Agriculture as a separate organization. Plant materials were exchanged both at national and international levels. A huge number of plant materials were collected by Hansen (1896) from Russia and a separate section was established in 1898 to handle the material collected by him. Similarly, the wheat material collected by Mark Carleton from Russia formed basis of the durum wheat industry in the United States.

The major credit for setting the phytogeographic basis for plant exploration (arranged adhoc or cooperative trips for making collection of plant materials) goes to Russian scientist N.I. Vavilov and his group who performed a number of explorations and collected, introduced and evaluated huge quantity of plant material from all over the world in 1920s and 1930s. According to Vavilov, the natural variability in all cultivated plant species is geographically confined to some specific pockets of the world which he called centres of origin or centres of genetic diversity.

Vavilov also gave the concept of parallel variation in related species. This was named as the 'law of homologous series in variation'. According to this law, related species are expected to possess similar characters. Describing the genetic basis of the law, Vavilov suggested that it gives a clue to characters which are not yet known.

Although Botanical Survey of India collected, evaluated and maintained plant materials of botanical and medicinal values right from its establishment in 1890, the first Indian who could foresaw the importance of germplasm variability in crop improvement was Dr. B.P. Pal (1937). Because of his interest, systematic effort towards the introduction of crop plant species in India was made in the year 1946 by initiating a scheme on plant introduction in the Division of Botany, Indian Agricultural Research Institute (IARI), New Delhi. In view of the importance attached to plant introduction in 1950s, this scheme was expanded in 1956 as the Plant Introduction and Exploration Organization. The importance of plant introduction was realised further and a separate Division of Plant Introduction was established in I.A.R.I. in 1961. In 1976, the reorganization of this division was done and it took the form of an independent unit as the National Bureau of Plant Genetic Resources (NBPGR).

As has been mentioned in the earlier part of this chapter, a number of crop plant species were introduced into India from time to time and many of them found their own niche in this country. However, the introduction of dwarf wheats from Mexico into India in mid nineteen sixties has spectacularly increased wheat production in this country. This phenomenal increase in the production of this cereal may be attributed to some specific characteristics of these varieties like high response to intensive agricultural practices (high doses of fertilizers, more irrigation, etc.) because of their short and stiff straw, photoinsensitiveness, high harvest index and more efficient plant type. The first indication of this increase in the wheat production in India was given in 1962 by Borlaug's material raised at I.A.R.I. farm. The old Indian material, on the contrary, was tall and did not respond favourably to intensive agricultural practices.

The material received from CIMMYT (Centro Internacional de Mejoramiento de Maiz Y Trigo, that is, International Maize and Wheat Improvement Centre) was screened and two varieties, namely, Sonora 64 and Lerma Rojo 64-A, were selected for further testing. After extensive testing of these two varieties under Indian conditions in 1964 (at 155 locations), about 250 tonnes of seed of these varieties were imported from Mexico in 1965. The two varieties were planted at farmer's fields in the different parts of the country in the same year and based on their performance a huge quantity of seed of these two varieties was imported in 1966 from Mexico. The varieties were sown over large areas during 1966-67. In spite of high yield levels shown by these Mexican varieties, they were not liked by the Indian people because of their red colour of kernels.

In addition, the varieties Kalyansona, Sonalika, Safed Lerma and Chhoti Lerma were selected from the advanced generation material introduced from Mexico. Further, use of mutation breeding and hybridization between the Mexican and Indian wheat varieties formed the basis of present day wheat varieties in India.

Procedure of Introduction

Usually no difficulty is encountered in introducing indigenous plant material from one place to another. However, the procedure of introduction of exotic material is lengthy and difficult. In India, there is a set procedure of introducing materials from other countries and involves a number of steps like procurement of plant material (the agency through which introduction is to be made, type of material, packing and despatch, etc.), quarantine, cataloguing and preservation, evaluation, multiplication and utilization. The National Bureau of Plant Genetic Resources (NBPGR), New Delhi is mainly responsible for all kinds of plant introductions in the country.

Procurement of Plant Material

At first, the requirements of the plant materials to be introduced are sent to NBPGR. The Bureau either meets the requirements from its own stock or tries to obtain the germplasm from its counterparts in other countries or other agencies generally through correspondence, in the form of gift, exchange of germplasm, purchased material, etc. The NBPGR also takes part in the activities of IBPGR (International Board for Plant Genetic Resources, established in 1974 at Rome) which is responsible for identifying needs in exploration and collection of germplasm, improving conditions for the storage of collected materials, imparting training to genetic resources workers, exchange of information at international level, improving storage of data and retrieval systems relating to genetic conservation and advising on financial needs in new or deficient areas of activity (IBPGR, 1975). In addition to the collection of germplasm through correspondence, if felt necessary, explorations are organized and expeditions are sent to the places from where germplasm is to be collected. These places are mostly the native habitats of plant species or where maximum genetic variability of any particular species is available. IBPGR took the form of IPGRI (International Plant Genetic Resources Institute) in 1991.

As regards the plant parts to be introduced, it depends on the species. In case of species which are sexually reproduced, introduction of plant material in the form of seed is the easiest and safest way since the seeds create least difficulty in packing and transport and usually have longer viability than other propagules (the part of the plant used for propagation of a particular species). However, in case of vegetatively propagative species, vegetative parts of the plant (root or stem cuttings, buds, bulbs, tubers, suckers, runners, etc.) are introduced which may require some special packing methods. Also, the transport of vegetative plant parts involves more cost because generally they are transported by air mail, whereas the packed seeds are usually transported by surface mail. A phytosanitary (plant health) certificate that the material is clear, pure and free from diseases, insect pests and weed seeds, is also sent along with the material.

Quarantine

Quarantine (keeping material in compulsory isolation for a time to prevent spread of diseases, insects, weeds, etc.) is an important step of plant introduction. After the material is received (either by surface mail at prescribed seaports like Bombay, Calcutta and Madras or by air mail at Delhi International airport), it is put to detailed examination for the presence of pathogens, insect pests and weed seeds. This entry inspection is done by the NBPGR and other recognized laboratories (Division of Mycology and Plant Pathology and Division of Entomology, I.A.R.I. which assist the NBPGR for evaluation of material for diseases and pests). In case the material does not meet the standards of plant protection and quarantine, it is either sent back to the source country or destroyed. If the material is found suitable according to the plant protection and quarantine regulations, it is catalogued and then handed over to the user along with the plant health certificate. If felt necessary, the material may be treated with fungicide and insecticide. Thus, all efforts are made to check the entry of harmful organisms and weeds in the country.

All kinds of plant materials cannot be introduced in a country or exported. Entries of some plant species to India are restricted. For example, the entry of rubber plant and its seed from America and West Indies, sunflower from Argentina, Mexican jumping beans, cocoa from Ceylon, Africa and West Indies and of sugarcane from Philippines, Australia, Fiji, New Guinea and West Indies, is prohibited. The seeds of linseed, cotton, sugarcane and berseem cannot be introduced into India through letter post. Similarly, there are restrictions on the export of certain plant species. For example, propagating mateiral of tea, black pepper and jute cannot be exported from India. Other countries also have put similar restrictions on the introduction and export of certain plant materials.

Cataloguing and Preservation

After getting clearance from quarantine, the introduced material is listed and an accession (entry) number is given to each collection. The name of the species and variety, source country, adaptation and other characteristic features of the material are also recorded. Depending upon the type of collection (whether from the same country or abroad), these are broadly classified as IC (indigenous collection of cultivated material), IW (indigenous wild collection) and EC (exotic collection). After giving a particular number to the material (with prefix IC, IW or EC), it is handed over to the user or included in the stock of the bureau.

For checking alterations in the genetic composition of the conserved stocks during rejuvenation, the seed sample to be used for rejuvenation or multiplication should be large enough so that it is a true representative of the stock from which it has been drawn. Further, subjective selection against variants in the population should be avoided. As far as possible, the plant material should be grown under environmental conditions similar to the original habitats of the species. The material should be properly reidentified before and after harvesting.

Evaluation

Systematic evaluation of new introductions is a pre-requiste for their effective utilization. For this, the performance of newly introduced material is evaluated at different substations of NBPGR or at Central Research Institutes (crops like sugarcane, tea, rice, potato and tobacco for which such institutes exist in the country). In addition to the testing of new introductions for resistance to diseases (which should be done at 'hot spots' of diseases) and insect pests, their genetic potential in respect of important agronomic traits is also assessed. The evaluation should be objective-oriented. Multidisciplinary efforts (teams of scientists from different disciplines like plant breeding, soil science, agronomy, plant pathology, entomology and plant physiology) can help a great deal in such evaluations. Teams of this type are very actively working at several national (ICAR co-ordinated projects, agricultural universities and institutes) and international institutes like IRRI (International Rice Research Institute, Manila, Philippines) and CIMMYT, Mexico.

This evaluation should not be confined to a single year/season or a single location. It has been observed in many cases that introduced materials show poor performance in the beginning but when they are grown again and again in the same environment, their performance is considerably improved, that is, the materials get acclimatized to the new environment (acclimatization is the process which leads to the adaptation of a material to a new environment). Acclimalization is nothing but a consequence of natural selection and thus its extent is mainly determined by the range of genetic variation present in the material. More the genetic variability present in the introduced material, more are the chances of its adaptation to a new environment since certain genes or gene groups may be favoured by natural selection during the due course of time. Thus, segregating materials have better chances of adaptation under new environmental conditions than the homozygous varieties or strains. The other two factors which determine the degree of acclimatization of a new material are the mode of pollination and the duration of the life cycle of the crop. A cross-fertilized crop has better chances of acclimatization than a self-fertilized crop because cross pollination produces new gene combinations some of which may be favoured by natural selection. Similarly, an annual species has advantage over a perennial species with regard to acclimatization since the former would produce several generations by the time the latter produces only one generation. Mutations may be helpful in acclimatization in some cases when the material has been grown for quite a long period.

Multiplication and Utilization

The new introductions which are found promising are multiplied and sent to users for their utilization in any of the following ways :

1. Direct commercial utilization : If the introduced material shows high adaptability in the new environment and is superior to indigenous germplasm for important agronomic characteristics, it can be directly used as a new commercial crop (a number of new crops like sunflower, sugarbeet, maize, soybean, potato and triticale have been introduced in the past into India and have become important crops of the country) or as a new commercial variety (varieties IR8 and Taichung Native 1 of rice, Sonora 64 and Lerma Rojo 64-A of wheat and Bonnevile and Arkel of pea are some examples of this type).

2. Selection of desirable types from introduced material : If the introduced material is segregating or genetically heterogeneous, superior types may be selected from this material in the next generation or any other subsequent generation. The wheat varieties Kalyansona, Sonalika, Chhoti Lerma and Safed Lerma were selected from the advanced generation material introduced from Mexico.

3. Using as a donor : Manytimes, the introduced material does not prove superior to the indigenous material for most of the agronomic characters, but it possesses one or a few desirable characteristics (like resistance to a particular disease or insect pest, efficient plant type, quality characters and male sterility) for which the indigenous material is deficient. Under such a situation, the introduced material may be used in the hybridization programmes to combine all desirable characters in a single variety. Introduction of DGWG dwarfing genes in rice from Taiwan, Norin 10 dwarfing genes in wheat from Japan and the male sterile line Tift 23A of bajra from U.S.A. are some examples of this type.

Disadvantages of Plant Introduction

There are three main disadvantages of plant introduction :

1. Introduction of diseases : Sometimes, some pathogen is also introduced along with the material. For example, late blight of potato from Europe in 1883, coffee rust from Ceylon in 1876 and bunchy top of banana from Ceylon in 1940, were introduced in India along with plant materials.

2. Introduction of insect pests : Some insect pests (potato tuber moth from Italy in 1900, fluted scale of *Citrus* after 1928, wooly aphis of apple, etc.) were also introduced in our country along with plant materials.

3. Introduction of weeds : Like diseases and insect pests, some problematic weeds (*Phalaris minor, Argemone maxicana, Lantana camera,* etc.) were introduced in India along with plant materials and are creating lot of problems to the farmers of the country. For example, *Phalaris minor* which was introduced from U.S.A. and is found in wheat fields, is morphologically similar to wheat in the beginning and thus weeding is not possible. It can be identified only at flowering stage when roguing becomes very difficult.

Many of the introduced diseases and insect pests have now spread almost over the whole country and creating serious problems. However, most of the diseases, insect pests and weeds were introduced in India when rigid quarantine rules were not in force and hence these disadvantages were common. But, now the situation has altogether changed and the materials are thoroughly examined before their entry into country. Under the present situation, therefore, these unwanted pathogens, pests and weeds have little chance to be introduced in the country along with plant materials. However, the carefulness and sincerity of the quarantine officials is of utmost importance in this matter.

Centres of Origin

N.I. Vavilov and his associates, for the first time, recognized centres of origin (the place where a particular species was first domesticated and its wild relatives occur) of crop plants on the basis of tremendous genetic variability of plant species found in some specific areas of the world. Vavilov was of the opinion that the place of greatest diversity in crop plants are also the places of their origin. This idea (of Vavilov), however, has been questioned by some scientists as the genetic diversity in a species at a particular places can also increase because of hybridization between diverse types, that is, gene recombination or because of migration with civilization and thus deserves revision. Now it is agreed by many workers that whereas the term centre of origin is only an interpretation, the term centre of diversity is a biological fact. There are cases where the centre of origin of a crop plant and its centre of diversity are found far away from each other. For example, the centre of origin of wheat is Middle East but it (particularly tetraploid wheat) shows widest variation in Ethiopia where no wild relatives of wheat occur. This indicates that the genetic diversity of wheat in Ethiopia increased after its first domestication in Middle East during the Neolithic agricultural revolution.

In view of the above criticism, Vavilov reconsidered and revised his concept about the centres of origin of crop plants. The real centre of origin of a crop plant was called as a 'primary centre' and where genetic diversity with regard to that species increased after its domestication at its place of origin, was named as 'secondary centre'. It was realized that the best approach to know the centre of origin of a plant species would be to find out the place of occurrence and distribution of its wild relatives on the basis of genetic affinities. In case of a polyploid species, the area of contact between the two involved species (the putative parents) would indicate its place of origin.

In addition to the primary and secondary centres of crop plants, there are some small areas within the large centres of genetic diversity where diverse genes have a higher degree of concentration than the centre of diversity itself. These small areas of relatively greater genetic diversity are called as 'micro-centres' and are more important for studying the evolution of species and for the collection of material.

One more point which is worth discussing here is that the centres of diversity of different crop plants, manytimes, show overlapping, that is, different plant species show a strong correlation in their distribution patterns. For example, Middle East is not a centre of origin of wheat alone, but many other crops like barley, rye, flax, pea and lentil also show great genetic diversity in that area.

Vavilov suggested 6 centres of origin in 1926 (Old World Centres) and two more in 1935 (New World Centres). According to him, the places of origin of all cultivated species are confined to these areas (centres).

Old World Centres of Origin

1. The Chinese Centre : It is the first and the largest independent centre of origin of crop plants. It includes mountainous regions of Central and Western China and adjacent low lands. 136 plant species are said to be originated from this centre. Some are : hull-less barley (*Hordeum hexastichum*) soybean (*Glycine max*), sugarcane (*Saccharum sinense*), hemp (*Cannabis sativa*), radish (*Raphanus sativus*), opium poppy (*Papaver somniferum*), pears (*Pyrus communis*), peaches (*Prunus persica*) and chinese tea (*Camellia sinensis*).

In addition to the above mentioned species, this centre is a secondary centre of origin of plant species like sesame (*Sesamum indicum*), maize (*Zea mays*), rajma (*Phaseolus vulgaris*), turnip (*Brassica rapa*) and cowpea (*Vigna anguiculata*).

2. The Indian Centre

2(a) Main centre : It includes major part of India including Assam and Burma but excludes North West India, Punjab and North West Frontier Provinces. 117 plant species have been listed under this centre. Some of them are : rice (*Oryza sativa*), gram or chickpea (*Cicer arietinum*), pigeonpea (*Cajanus cajan*), urdbean or mash (*Vigna mungo*), mungbean (*V. radiata*), ricebean (*Phaseolus calcaratus*), mothbean (*Vigna aconitifolius*), clusterbean (*Cyamopsis tetragonoloba*), cowpea (*V. anguiculata*), sesame (*Sesamum indicum*), safflower (*Carthamus tinctorius*), sugarcane (*S. officinarum*), brinjal (*Solanum melongena*), tree cotton (*Gossypium arboreum*), black pepper (*Piper nigrum*), *Sorghum* and mango (*Mangifera indica*).

2 (b) The Indo-Malayan Centre : Indo-China and Malay Archipelago regions come under this centre. Fiftyfive plant species are said to be originated from this centre. The important cultivated species are : Banana (*Musa* spp.), pomelo (*Citrus maxima*), coconut (*Cocos nucifera*), sugarcane (*S. officinarum*).

3. The Central Asiatic Centre : Sometimes, it is also called as Afghanistan centre of origin. North-West India (Punjab, The North-West Frontier Provinces and Kashmir), Afghanistan, the Soviet Republics of Tadjikistan and Uzbekistan and Western Tian-Shan regions come under this centre. Fortythree plant species were originated from this centre. Some are : common wheat (*Triticum aestivum*), club wheat (*T. compactum*), shot wheat (*T. spherococcum*), flax (*Linum usitatissimum)* Lentil (*Lens esculenta*), pea (*Pisum sativum*), gram (*Cicer arietinum*), cotton (*Gossypium herbaceum*), broadbean (*Vicia faba*), sesame (*Sesamum indicum*), carrot (*Daucus carota*), walnut (*Juglans regia*) and grape (*Vitis vinifera*).

In addition to the above mentioned species, it is also a secondary centre of rye (*Secale cereale*).

4. The Near Eastern Centre : It is also known as The Asia Minor Centre or The Persian Centre of Origin and includes the interior of Asia Minor, the whole of Transcausasia, and the high lands of Turkmenistan. Eightythree plant species are said to be originated from this centre. Some of the important cultivated species are : Enkorn wheat (*T. monococcum*), durum wheat (*T. durum*), pollard wheat (*T. turgidum*), common wheat (*T. aestivum*), timopheevi wheat (*T. timopheevi*), rye (*S. cereale*), alfalfa (*Medicago sativa*), Persian clover (*Trifolium resupinatum*), cabbage (*Brassica oleracea*), common oats (*Avena sativa*), melon (*Cucumis melo*), fig (*Ficus carica*) and grape (*Vitis vinifera*).

It is also a secondary centre of gram (*Cicer arietinum*), pea (*Pisum sativum*), rape (*B. campestris*), black mustard (*B. nigra*), leaf mustard (*B. japonica*) and turnip (*B. rapa*).

5. The Mediterranean Centre : The borders of Mediterranean sea come under this centre. Eightyfour plant species were originated in this area. Some of them are : durum wheat (*T. durum*), emmer wheat (*T. dicoccum*), spelta wheat (*T. spelta*), hulled oats (*Avena strigosa*), broad bean (*Vicia faba*), lettuce (*Lactuca sativa*), cabbage (*Brassica oleracea*) and clovers (*Trifolium* spp.).

6. The Abyssinian Centre : This centre includes Ethiopia, Eritrea and part of Somali land. Thirty eight plant species are listed under this centre. Some are : durum wheat (*T. durum*), pollard wheat (*T. turgidum*), emmer wheat (*T. dicoccum*), barley (*H. vulgare*), jowar (*Sorghum bicolour*), pearlmillet (*Pennisetum glaucum*), pea (*Pisum sativum*), flax (*Linum usitatissimum*), sesame (*Sesamum indicum*), lentil (*Lens esculenta*), castorbean (*Ricinus communis*), okra (*Hibiscus esculenta*), chickpea (*Cicer arietinum*) and coffee (*Coffea arabica*).

New World Centres of Origin

7. The South Mexican and Central American Centre : Regions of South Mexico, Guatemala, Honduras and Costa Rica come under this centre. Some of the important species which originated from this centre are : maize (*Zea mays*), upland cotton (*G. hirsutum*), common bean or rajma (*Phaseolus vulgaris*), limabean (*Phaseolus lunatus*), pepper (*Capsicum annum*), pumpkin (*Cucurbita melanosperma*), melons (*Cucumis melo*) and guava (*Psidium guayava*).

8. The South American Centre : Vavilov (1935) divided this centre into three centres :

(a) The Peruvian - Equadorean - Bolivian Centre : It includes mainly the high mountainous areas (formerly the centre of the Megalithic or Pre-Inca civilization). Some of the species which originated from this centre are : potato (*Solanum tuberosum*), sweet potato (*Ipomoea batatas*), tomato (*Lycopersicon esculentum*), tobacco (*Nicotiana tabacum*), sea island cotton (*G. barbadense*), papaya (*Carica papaya*) and limabean (*Phaseolus lutanus*).

(b) The Chiloe Centre : It includes the island near the coast of Southern Chile. Potato (*S. tuberosum*) originated from this centre.

(c) Brazilian-Paraguayan Centre : Some plant species which originated from this centre are : peanut or groundnut (*Arachis hypogaea*), pine apple (*Ananas comosa*), rubber tree (*Hevea brasiliensis*) and cassava (*Manihot utilissima*).

All these centres are shown in Figure 3.1.

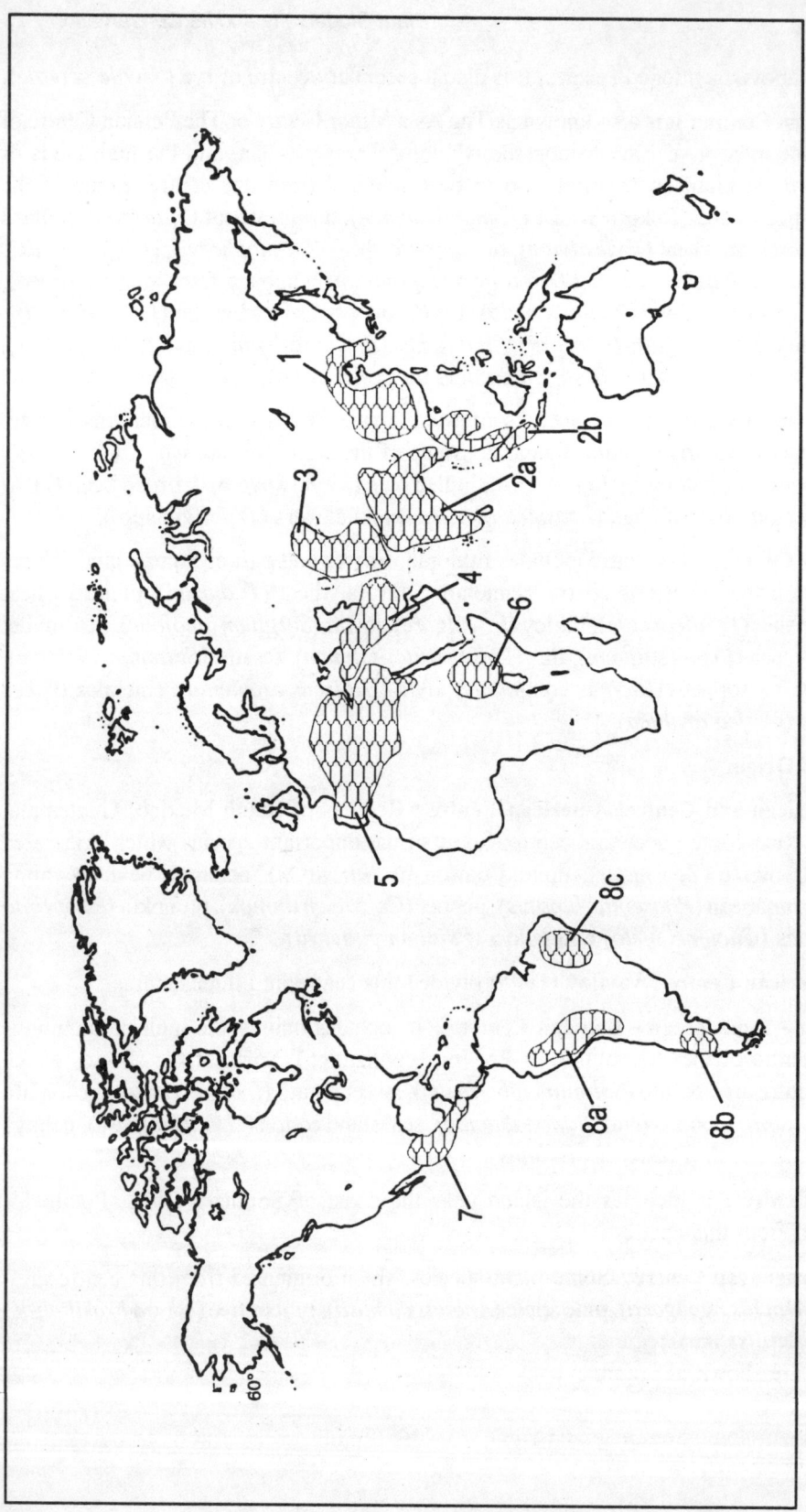

Figure 3.1. Centres of origin of crop plants (shaded areas)

1. Chinese centre
2a. Main centre
2b. Indo Malayan Centre
3. Central Asiatic Centre

4. Near Eastern Cetnre
5. Mediterranean Centre
6. Abyssinian Centre
7. South Mexican and Central American Centre

8a. Peruvian - Equadorean - Bolivian Centre
8b. Chiloe Centre
8c. Brazilian - Paraguayan Centre

Institutional Organization

A number of national and international centres are engaged in plant genetic resources activities.

National Centres

Many countries of the world have established their own research centres/ institutes/departments/ bureaus for collection, evaluation and conservation of plant genetic resources.

National Bureau of Plant Genetic Resources (NBPGR) : With five major divisions (Plant Exploration and Collection, Germplasm Exchange, Plant Quarantine, Germplasm Evaluation and Germplasm Conservation) and a National Facility for Plant Tissue Culture Repository funded by Department of Biotechnology (DBT), the bureau is located in the campus of Indian Agricultural Research Institute (IARI), New Delhi, India. The bureau has three coordinated projects (Medicinal and Aromatic Plants, Under-utilized and under-exploited plants and Cluster bean), 12 base centres (at New Delhi, Shimla, Srinagar, Jodhpur, Shilong, Ranchi, Cuttack, Akola and Hyderabad)/regional centres (at Shimla, Bhowali, Jodhpur, Shilong, Akola/Amrawati and Trichur)/quarantine stations (at New Delhi, Hyderabad and Andamans) representing different agroclimatic zones and an experimental farm at Issapur. The activities of the bureau are : to introduce germplasm from abroad; supply germplasm to its counterparts in other countries : arrange explorations inside and outside the country; inspect newly introduced material; catalogue, evaluate, multiplicate, conserve and distribute the introduced material; disseminate information regarding collection and exchange of material; work in collaboration with international institutes/organizations particularly with International Plant Genetic Resources Institute (IPGRI); and to coordinate activities of national institutes/centres. One of the main objectives of the bureau is *ex situ* conservation of crop genetic resources in the form of national gene bank. Right from its establishment in 1976, the programmes and activities of NBPGR have grown rapidly and the bureau has been doing an excellent job to fulfil its national mandate. The new priority areas of the bureau are : to develop more facilities for conservation, evaluation and documentation of crop genetic resources; increase the capacity of long-term and medium-term storages of base collections and national active collections, respectively; provide more facilities to plant breeders and other researchers by developing a computer-based national database on plant genetic resources; set up crop advisory committees for specific crops; establish a nodal agency in the form of fully developed Indian National Plant Genetic Resources System (INPGRS); work in partnership with leading crop-based institutes, state agricultural universities, national research centres and all India coordinated crop improvement projects; develop more facilities for post-entry quarantine inspection at some specific sites; establish national research centre on biosystematics and evolutionary studies by strengthening national herbarium of cultivated plants; and to bring improvement in the quality of training programmes.

Department of Environment and Forests: In addition to *ex situ* conservation and *in vitro* repository of plant genetic resources, efforts are now being made by NBPGR for *in situ* conservation of landraces, traditionally grown primitive cultivars, wild relatives of crop plants and ecosystem as well as for living collections of threatened and rare plant species in active collaboration with the Department of Environment and Forests. Important agencies like Botanical Survey of India, the Forest Survey of India and the Indian Council of Forest Research and Education are governed by this department. The Botanical Survey of India was engaged right from its establishment (1890) in introducing, testing and maintaining plant materials of botanical and medicinal value (at present, NBPGR is looking after the medicinal plants). Besides this, Forest Research Institute (FRI), located at Dehradun, is engaged in introducing, testing and maintaining germplasm of forest trees.

Central Research Institutes : Sugarcane Research Institute (SRI), Coimbatore; Central Rice Research Institute (CRRI), Cuttack; Central Potato Research Institute (CPRI), Shimla; and central institutes for other crops like tobacco, tea and coffee also introduce, test and maintain germplasm of specific crops. However, the activities of all these institutes are coordinated by NBPGR.

In addition to these Indian centres, a number of national centres in other countries of the world are engaged in collecting, introducing, evaluating and conserving germplasm. Some of such centres are National Biological Institute, Bogor (Indonesia); National Institute of Agricultural Sciences, Kyoto University (Japan); Commonwealth Scientific and Industrial Research Organization (CSIRO) (Australia); Vavilov All-Union Institute of Plant Industry (Russia); Royal Botanic Gardens (United Kingdom); Agricultural Research Service, US Department of Agriculture (United States of America); Institute Nacional de Investigaciones Agricolas (Mexico); and Centro Nacional de Recursos Geneticos (Brazil).

International centres

A brief information about some important international centres is given below :

International Rice Research Institute (IRRI), Manila, Philippines : The institute was founded in 1960 and initiated acquisition, multiplication and preservation of rice germplasm in the form of commercial varieties and breeding lines from various rice growing countries of the world in 1961. In addition to the development of a number of high yielding rice varieties, the scientists of the institute are busy in the collection of germplasm, multiplication, characterization, conservation and distribution of seed and collaboration with national programmes.

International Maize and Wheat Improvement Centre (Centro Internacional de Mejoramiento de Maiz Y Trigo - CIMMYT) Lisboa, Mexico : The centre was founded in 1966 with wheat, maize and triticale as mandate crops and maintains genebanks for maize and wheat.

International Institute of Tropical Agriculture (IITA), Ibadan, Nigeria : The institute was founded in 1967 and its crop improvement programmes were established in 1970. The mandate crops of the institute are maize, rice, cowpea, soybean, cassava, plantain, yam and agroforestry species. The main activities of the institute are collection, conservation, characterization and evaluation of germplasm, conduct of group trainings and research.

International Centre for Tropical Agriculture (Centro Internacional de Agricultura Tropical-CIAT), Cali, Columbia : After its foundation in 1967, the main thrust of the centre has been to alleviate hunger and poverty in developing countries of the tropics. The collection, distribution and enhanced use of germplasm of mandate crops (*Phaseolus bean*, rice, cassava and tropical pastures) are the main activities of the centre.

International Potato Centre (Centro Internacional de la Papa-CIP), Apartado, Lima Peru : After its formulation in 1971, the centre is dedicated to the collection of germplasm of its mandate crops (potato and sweet potato) and their wild relatives from the highlands of South American Andean countries and the tropical zones of Latin America and to facilitate utilization of these collections.

West Africa Rice Development Association (WARDA), Bouke, Cote d' Ivoire : The association was founded in 1971 with a purpose to collect, evaluate, maintain and utilize wide range of rice germplasm found in West Africa (including *Oryza glaberrima*) as well as germplasm introduced from other regions of the world. After a collaborative agreement of WARDA with IITA, IRRI and IRAT (Institut des Recherches Agronomiques Tropicales et des Cultures Vivrieres) in 1977, the association started concentrating on the efficient management of its resources and avoiding duplication of efforts.

International Crops Research Institute for the Semi-Arid Tropics (ICRISAT), Pattancheru P.O., Andhra Pradesh, India : Although ICRISAT was founded in 1972, its Genetic Resources Unit, as a research support programme, was established in 1979 with the objective to collect, evaluate, maintain and utilize germplasm of the institute's mandate crops (sorghum, pearl and finger millets, chickpea, pigeonpea, groundnut and minor millets) and their wild relatives and that of endangered landraces. Training of various types (on-site courses, masters, doctoral and post-doctoral programmes), organization of conferences and workshops on genetic resources of the mandate crops and liaison with national, regional and international organizations are other important activities of the institute.

International Board for Plant Genetic Resources (IBPGR)/International Plant Genetic Resources Institute (IPGRI), Rome, Italy : The board was founded in 1974 with the objective to promote the conservation and use of plant genetic resources so that the erosion of precious genetic wealth is checked and the future of coming generations is safeguarded. The agreement signed by the representatives of five governments in October, 1991 cleared the way for the establishment of IPGRI as the successor to IBPGR. The major objectives of the institute are : (1) assisting countries in meeting their needs for the conservation of plant genetic resources; (2) strengthening international collaboration for the maintenance and utilization of plant genetic resources; (3) bringing improvement in the techniques and strategies used in the conservation of plant genetic resources; and (4) disseminating information on plant genetic resources at international level.

International Centre for Agricultural Research in Dry Areas (ICARDA), Aleppo, Seria : After the foundation and establishment in 1976 and 1977, respectively, the centre has done a commendable job for the people of its mandate area (from Morocco to Pakistan and from Turkey to Ethiopia) by evolving drought tolerant varieties of centre's mandate crops (barley, lentil, faba bean, kabuli chickpea and durum and bread wheats). Other activities of the centre are : (1) collection, conservation and use of landraces and wild relatives of the centre's mandate crops; (2) collaboration with IPGRI and FAO to develop a genetic resources network; (3) descriptive characterization, taxonomic identification and evaluation of the collected germplasm for agronomic traits; and (4) to provide training to staff in national genetic resources programmes.

International Centre for Research in Agroforestry (ICRAF), Nairobi, Kenya : The centre was founded in 1977 with the purpose to conduct and support agroforestry research for improving soil fertility and microclimate of cropping and grazing lands, checking tropical deforestation, controlling soil erosion, safeguarding natural habitats of different plant species and mitigating rural poverty, especially in developing countries.

All these international agricultural research centres are now supported by the Consultative Group on International Agricultural Research (CGIAR) which was established in 1971. The group is a broadly based informal consortium of 40 public and private donors and is aimed at to ensure safeguarding of crop genetic resources both *in situ* and *ex situ* and effective use of the earth's genetic heritage by the present and future generations of research workers and to contribute to sustainable agriculture forestry and fisheries for enhancing nutrition and well-being of people all over the world (especially in developing countries). Also the group has a commitment to promote interdisciplinary approach to strengthen national capabilities in developing countries and to support their problem solving plant genetic resources programmes.

Nevertheless not only there is need of safeguarding natural plant genetic resources, but the conservation of folk knowledge about agriculture is equally important. We must remember that the folk knowledge, like natural plant genetic wealth, is irreversible and irreplaceable and once eroded cannot be rejuvenated. It must be ensured that folk wisdom gets passed down from generation to generation. Agriculture, in fact, is a two-way process, that is, "from lab to land" and "from land to lab". Therefore, the participation of farmers in crop improvement programmes (that is, farmer participating plant breeding) is as important as that of agricultural scientists.

CHAPTER 4

Reproductive Systems in Crop Plants and Methods of Plant Breeding

Reproduction of living beings is necessary for their perpetuation. Further, the requirements of varieties to be evolved and the methods of breeding to be employed in crop plant species are largely determined by their mode of reproduction as the genetic structure of populations in self-fertilized species (autogams) is different from that of populations in cross-fertilized species (allogams) and therefore the two groups show different degrees of response to inbreeding. An autogamous species is usually a group of several homozygous lines and shows a little or no deterioration in general vigour and other attributes when selfed, whereas an allogamous species is consisted of heterozygous individuals and generally exhibits high degree of deterioration in general vigour and other characteristics if subjected to self-fertilization or inbreedng. Wherever vegetative propagation is possible or is a general way of reproduction in a species, it offers a unique opportunity of taking full advantage of heterozygosity for an indefinite period of time as the same genotype irrespective of the degree of heterozygosity can be maintained for indefinite time in this type of reproduction. It means that the mode of reproduction determines the genetic structure of plant populations which in turn largely determines the plant breeding strategy to be followed in a crop.

Floral parts and kind of flowers

A complete flower usually contains two accessory organs, sepals (calyx) and petals (corolla), and two essential (reproductive) organs, stamens (androecium) and pistil (gynaeceum). A flowor lacking one or more of these organs is called incomplete. Further, a perfect flower contains both the essential floral organs and is thus bisexual. A flower lacking one of these organs (pistil or stamens) is called imperfect and is thus unisexual. A flower lacking pistil is called staminate (male) and the one lacking stamens is pistillate (female). Thus a perfect flower may be complete or incomplete (lacking calyx or coralla or both) but an imperfect flower can never be a complete flower. Stamen is the male reproductive part of the flower and commonly consists of a filament (slender stalk) and an anther (pollen developing part). Pistil is the female reproductive organ of the flower and commonly consists of an ovary (an enlarged base), a style (an elongated stalk) and a stigma (pollen receiving part).

Amphimixis

Each of the diploid pollen (microspore) mother cells present in an immature anther undergoes meiosis (two successive nuclear divisions, reductional and equational) and produces a tetrad of four microspores each with haploid number of chromosomes. This production of microspores from the microspore mother cell is called microsporogenesis. Microspore nucleus then divides mitotically to form a vegetative (tube) nucleus and a generative nucleus. Generative nucleus further divides mitotically and produces two male gametes or germ cells. This development of male gametes from a microspore is termed as microgametogenesis.

One or more ovules are found inside the ovary like pollen mother cell, single diploid megaspore mother cell present inside the ovule undergoes meiosis and forms a linear tetrad of four megaspores each with haploid number of chromosomes. This process is called as megasporogenesis. One of the four megaspores (usually farthest from the micropyle) generally undergoes three successive mitotic divisions to produce 8-nucleate embryosac (female gametophyte) and the remaining three megaspores disintegrate. These eight nuclei get organized into three groups, three on the micropylar side (one egg and two synergids), three on the opposite side of the embryo sac (antipodals) and two at the centre (polar nuclei). This formation of female gamete is termed as megagametogenesis.

The transfer of pollen grains from anther to the stigma of the same flower or stigma of another flower of the same or other plant is termed as pollination. Pollination may be anemophilous (by wind), entomophilous (by insects), zoophilous (by animals) or hydrophilous (by water). The pollen grains deposited on the sticky surface of the stigma germinate by showing an outgrowth (pollen tube) which grows into the stylar tissues and enters the ovule through micropyle. The two male gametes are then emptied into the embryo sac and finally double fertilization takes place by the union of one of the two male gametes with the egg cell or female gamete (syngamy) and of the other male gamete with the polar nuclei (triple fusion). The fertilized egg (zygote) divides mitotically and turns into an embryo and the triple fusion nucleus (primary endosperm nucleus) divides to form triploid endosperm. The normal sexual reproduction including micro- and megasporogenesis, micro- and megagametogenesis, pollination, syngamy, triple fusion and formation of embryo, seeds and fruits is called amphimixis (amphi means on both sides; mixis means fusion).

Classification of Plant Species

The classification of plant species is mainly based on two criteria :

1. Distribution of flowers on the plant (s)

Depending upon the type and the way of occurrence of flowers on the plant (s), the plant species may be categorized into following classes :

Hermaphrodite : Both the essential organs, androecium and gynaeceum, are present in the same flower, e.g., wheat, barley, cotton.

Monoecious : Staminate and pistillate flowers are present on the same plant but at different places, e.g., maize, cucurbits. A number of variants of this class are found, for example, **andromonoecious** (presence of staminate and hermaphrodite flowers at the same plant but at different places), **gynomonoecious** (pistillate and hermaphrodite flowers are present on the same plant but at different places) and **trimonoecious** (occurrence of staminate, pistillate and hermaphrodite flowers at different places of the same plant).

Dioecious : In case of typical dioecism, the whole plant is either male (androecious) or female (gynoecious), that is, unisexual plants, e.g., papaya, date palm, hemp. There are a number of variants of this class also, for example, **androdioecious** (when male and hermaphrodite plants are separately found), **gynodioecious** (occurence of female and hermaphrodite plants separately) and **tridioecious** (male, female and hermaphradite plants of the same species are found separately). Development of gynodioecious varieties ensures fruiting on each plant, e.g., papaya.

Monoe-dioecious : Male and monoecious, female and monoecious or male, female and monoecious plants of the same species occur separately. All these variants of monoecious and dioecious plants are found in case of papaya.

It appears that evolution in respect of sex in plants began from hermaphroditism and typical dioecious species are the most recent ones.

2. Mode of reproduction

On the basis of mode of reproduction, the plant species can be classified into two broad groups :

(A) Sexualy reproduced

In this group of plant species, the propagation is by seeds produced through normal amphimixis, that is, embryo develops after the fusion of male and female gametes. This group can be subdivided into following three classes :

Normally self-fertilized : These species are predominantly self-pollinated, that is, the stigma receives pollen from anthers of the same flower or from other flower of the same plant or from the same clone, e.g., wheat (*Triticum aestivum*), barley (*Hordeum vulgare*), rice (*Oryza sativa*), chickpea (*Cicer arietinum*), garden pea (*Pisum sativum*), grass pea or khesari (*Lathyrus sativus*), cowpea (*Vigna anguiculata*), mungbean (*Vigna radiata*), urdbean (*Vigna mungo*), frenchbean or rajma (*Phaseolus vulgaris*), mothbean (*Phaseolus aconitifolius*), hyainthus bean or sem (*Delichos lab-lab*), lentil (*Lens esculanta*), soybean (*Glycine max*), groundnut or peanut (*Arachis hypogea*), guar (*Cymopsis tetragonoloba*), lettuce or salad (*Lactuca sativa*), tomato (*Lycopersicon esculentum*), eggplant or brinjal (*Solanum melongena*), linseed or flax (*Linum usitatissimum*), oat (*Avena sativa*), sesame or til (*Sesamum indicum*), ragi or finger millet (*Eleusine coracana*), foxtail millet (*Setaria italica*), sunhemp (*Crotolaria juncea*).

In a naturally self-fertilizing species, no external agencies like wind, insects, animals or water are required for self-pollination, rather it is enforced by some floral mechanism (s) like cleistogamy, hermaphroditism and homogamy.

Normally Cross-fertilized

These species are predominantly cross-pollinated, that is, the stigma receives pollen from a different plant, e.g., maize (*Zea mays*), pearl millet (*Pennisetum glauccum*), rye (*Secale cereale*), sunflower (*Helianthus annuus*), castor (*Ricinus communis*), sugarbeet (*Beta vulgaris*), sugarcane (*Saccharum* species hybrids), hemp (*Cannabis indica*), berseem (*Trifolium alexandrium*), alfalfa (*Medicago sativa*), sweet clover (*Melilotus officinalis*), white clover (*Trifolium repens*), red clover (*Trifolium pratense*), crimson clover (*Trifolium incarnation*), onion (*Allium cepa*), carrot (*Daucus carota*), garlic (*Allium sativum*), spinach (*Spinacea oleracea*), coriander (*Coriandrum sativum*), cauliflower (*Brassica oleracea* var. botrytis), cabbage (*Brassica oleracea* var. capitata), Brussel's sprout (*Brassica oleracea* var. geminifera), turnip (*Brassica napus*), radish (*Raphanus sativus*), sweet potato (*Ipomoea batatas*), watermelon (*Cucumis melo*), cucumber (*Cucumis sativus*), muskmelon (*Cucurbita moschata*), pumpkin (*Cucurbita maxima*), squash (*Cucurbita melanosperma*), papaya (*Carica papaya*), banana (*Musa paradisiaca*), grapes (*Vitis vinifera*), mango (*Mangifera indica*), date palm (*Phoenix dactilifera*), coconut (*Cocos nucifera*), fig (*Ficus carica*), apple (*Pyrus malus*), pear (*Pyrus cummunis*), cherry (*Prunus avium*), almond (*Prunus amygdalus*), walnut (*Juglans regia*), strawberries (*Fragaria* species).

In cross-fertilized species, cross pollination occurs at its own but some sort of controlled pollination is required for selfing or inbreeding in these species. Therefore, development of hybrids and population improvement are easier in this group of plants.

Often Cross-fertilized (both self and cross fertilized)

This group of plants is in between the self-fertilized and cross-fertilized groups since in these plant species cross fertilization usually exceeds 5 per cent and in some cases may reach upto 50 per cent. Examples of such crop species are cotton (*Gossypium* species), jowar (*Sorghum bicolour*), pigeonpea or arhar (*Cajanus cajan*), okra or lady's finger (*Abelmoschus esculentus*), chillis (*Capsicum annuum*), safflower (*Carthamus tinctorus*), broadbean or faba bean (*Vicia faba*), rai (*Brassica juncea*), toria (*Brassica campestris* variety toria), yellow sarson (*Brassica campestris* variety yellow sarson).

In majority of often cross-fertilized crop species, both selfing and crossing are easy and both homozygous and heterozygous varieties are developed. For example, both the homozygous and hybrid varieties are developed in cotton, jowar and pigeonpea.

(B) Asexually reproduced

In this kind of reproduction, there is no fusion of sexual gametes. This can further be divided into two :

Apomixis (apo means without; mixis means fusion)

It is the development of embryo without the fusion of male and female gametes, that is, here amphimixis is substituted by an asexual reproductive process which does not involve fusion of gametes. Apomicts produce either only apomictic embryos (**Obligate apomicts**) or both normal and apomict embryos (**Facultative apomicts**). The embryo may develop without fertilization from a reduced egg cell (**haploid parthenogenesis**) or from an unreduced egg cell (**diploid parthenogenesis**), from a reduced synergid or a reduced antipodal cell (**haploid apogamy**) or from an unreduced synergid or an unreduced antipodal cell (**diploid apogamy**), from any vegetative cell of the nucellus, integument or chalaza (**adventive embryony**), or from a reduced male gamete (**androgamy**). Diploid parthenogenesis, diploid apogamy and adventive embryony where there is no occurrence of meiosis and thus no genetic segregation and recombination, are put under a broad category, **recurrent apomixis**. The resulting plants could therefore resemble to the mother plant. On the other hand, haploid parthenogenesis, haploid apogamy and androgamy are put under the category **non-recurrent apomixis**. Recurrent apomixis is related to **diploid** (somatic) **apospory** (formation of diploid embryosac) and non-recurrent apomixis to **haploid** *(generative)* (formation of haploid embryo sac). In addition, in some plant species, bulbils, vegetative buds or plantlets are formed from floral primordia without seed formation, that is, floral primordia instead of flowers produce vegetative buds. Such cases come under **vegetative apomixis**. Since there is no formation of seeds in this category, some scientists do not put it under apomixis. Nevertheless, vegetative apomixis has little significance in plant breeding.

According to Poehlman and Sleper (1995), apomixis can be classified into two major categories: **agamospermy** and **vivipary**. In agamospermy, seeds are formed without the union of gametes and thus this class includes both the recurrent and non-recurrent types of apomixis. Vivipary is the vegetative apomixis described above.

Apomicts generally give complex or unpredictable segregation ratios and can have sexually unmaintainable unusual patterns of relationships among chromosome sets. Within species crosses between amphimicts and apomicts suggest that several interacting genes govern the phenomenon of apomixis. However, genetic control of adventive embryony and vegetative reproduction is relatively simple.

Although apomixis may be suspected on the basis of the similarity between the progeny and the mother plant, the presence of the phenomenon can be conclusively confirmed by measuring DNA contents of nuclei using flow cytometer and then the genes governing apomixis may be transferred to desirable plant types. The use of sophisticated biotechnological tools may provide opportunity of cloning of apomictic genes and transformation with these genes.

The recurrent forms of apomixis which produce viable diploid embryos without fertilization and help in perpetuating the genotype over generations may be beneficial to the plant breeder in a number of ways : (i) since apomicts tend to conserve the genetic structure of their carriers, true to type progeny can be produced in autogams (usually homozygous) as well as in allogams (usually heterozygous); (ii) maternal effects can be efficiently exploited; (iii) seed can be saved from year to year without loss of hybrid vigour; and (iv) provides opportunity of the transfer of desirable genes from wild apomicts to their related cultivated species. However, non-recurrent forms of apomixis do not possess these qualities and are thus of academic interest only.

Vegetative propagation

Vegetative parts of the plant (normal or modified stems, roots, leaves and bulbils), rather than the development of sexual gametes and fertilization, are the means of reproduction. For example, tubers are used for vegetative propagation in potato, stem cuttings in case of sugarcane, sweet potato, roses, grapes, etc., layering and grafting in a number of fruit trees and ornamental plants, root cuttings in red raspberry and leaf cuttings in *Bryophyllum*. Tissue culture technique is being used for propagating sugarcane, many horticultural plants and grasses. All the plants that have descended by mitosis from a single plant constitute a clone. All members of a clone are genetically identical and resemble to their parent plant for all characteristics. Since the same genotype can be maintained for indefinite period of time through this kind of reproduction, the system allows best use of outstanding individuals (irrespective of the degree of heterozygosity and the stage of the breeding programme) and artificially induced as well as spontaneously occurring mutations.

Nevertheless, some scientists classify plant species on the basis of mode of reproduction into three categories : (1) sexually reproduced (through amphimixis), (2) asexually reproduced (through apomixis), and (3) vegetatively reproduced (through vegetative propagation). According to them, asexual reproduction should not include vegetative propagation since the former is propagation by seeds developed directly from the mother plant by asexual process. That is, the mode of reproduction in plants may be broadly classified into two : (1) reproduction by seeds produced either through the fusion of sexual gametes (amphimixis) or without fertilization (apomixis), and (2) reproduction by vegetative parts of the plant (vegetative propagation).

In addition to the mode of reproduction, the choice of plant breeding method to be adopted in a crop species would depend on the ease with which selfing and crossing can be done in that species. Therefore, in some cases, different plant breeding methods are to be applied to improve crops falling in the same reproductive group. For example, in case of maize, selfing and crossing both are easy and thus development of inbreds and ultimately the production of commercial hybrids are easy and economical. On the other hand, in sunflower (also a cross-fertilized crop species), some strains show high degree of self-incompatibility and thus sib-mating rather than selfing is to be done for the development of inbreds. The situation is more acute (rather extreme) in case of lucerne where neither selfing nor crossing is easy because of the control of pollination by insects (by rupturing the protective membrane enclosing stigma), self-incompatibility and somatoplastic sterility (where embryos resulting from cross-pollination grow more rapidly and successfully than the embryos resulting from self-pollination, that is, post-fertilization sterility). Similarly, in autogams, emasculation and crossing are quite easy in case of tomato, brinjal, tobacco, etc. with a good seed set. Also, in cotton (an often cross-fertilized crop), selfing emasculation and crossing are so easy that commercial hybrids are being produced by hand emasculation and crossing. Contrary to it, seed set after crossing is very low in many self-fertilized leguminous crops like chickpea, cowpea, mungbean, urdbean and mothbean because of difficulty in emasculation and high degree of sheding of flowers. Production of hybrids is also difficult and costly in case of oats.

Further, all cross-fertilized plant species do not show same degree of deterioration in vigour and other attributes when subjected to inbreeding. Cucurbits, for example, mostly exhibit a little or no deterioration in vigour on selfing probably because of the small population size effect in this group of plants. Hemp and papaya also show similar response to inbreeding. Response to inbreeding by a species also varies manytimes from strain to strain. Rye, onion, sunflower and maize are, in general, moderately tolerant of inbreeding. However, some strains of these crops show less response but some others exhibit more reduction in yield and other morphological characters. A lot of improvement in the inbreds of maize has been obtained. Contrary to a very poor yield of and reduction in height and other attributes in the inbreds of old days, now there are some maize inbreds available which have yield potential and other characters comparable to those of already released varieties. However, lucerne, brown sarson and carrot are the crops where drastic reduction in general vigour and other characteristics can be seen in selfed generations. The deterioration is so high in case of lucerne that after 3-4 selfed generations a very few lines are able to survive.

Whether a species is self-fertilized or cross-fertilized depends on one or more factors discussed below :

Conditions Encouraging/Enforcing Self-fertilization

Three conditions are mainly responsible for encouraging/enforcing self-fertilization : (1) hermaphroditism, (2) homogamy and (3) cleistogamy.

Hermaphroditism : If androecium and gynaeceum both are present in the same flower and no other mechanism is operating against selfing, the species is expected to be self-fertilized, e.g. wheat, barley, pea.

Homogamy : Self-fertilization is generally favoured if both the sexual organs androecium and gynaeceum mature at the same time, e.g. wheat, barley, pea.

Cleistogamy : In strict sense, cleistogamy means that flowers do not open at all. In such a condition, self-pollination is enforced as no foreign pollen can reach the stigma of such flowers. However, complete morphological cleistogamy is rarely found in cultivated species. Of course, partial cleistogamy is a characteristic of almost all the leguminous crops (chickpea, garden pea, mungbean, urdbean, etc.). In this group of plants, the two innermost petals unite to form a keel which completely encloses the stamens and the gynaeceum. Unless this keel is ruptured by insects (as in case of lucerne) or by any other mechanical means, there are no chances of cross-fertilization.

There is another group of crop species (wheat, barley, oats, lettuce, etc.) where the flowers open after the pollination has already taken place. Such a condition is termed as **chasmogeamy**. In this group, therefore, cleistogammy operates upto a specific stage of the development of the flower.

A more interesting situation is found in some crop species like brinjal and tomato. Here the flowers open before the occurrence of pollination but the anthers surround the stigma in such a way that no foreign pollen is allowed to reach the stigma and thus giving no chance for cross fertilization.

Chickpea presents a peculiar situation. In this species, two contrasting mechanisms, partial cleistogammy and slight protogyny, operate simultaneously. The crop is self-fertilized because the stigma does not get foreign pollen and is ultimately pollinated by the pollen of the same flower. However, advantage of protogynous condition can be taken in making crosses between desired genotypes without emasculating flower buds. This can be easily done by simply pollinating flower buds by desired pollen when stigma is receptive but the pollen of the flower bud is immature.

Conditions Encouraging/Enforcing Cross-fertilization

The following conditions check self-fertilization and thus encourage/enforce cross-fertilization.

Dichogamy : Maturation of anthers and stigma at different times will discourage self fertilization. Protogynous condition (where gynaeceum matures earlier than androecium) in pearl millet, walnut, etc. and protandrous condition (where andoecium matures earlier than gynaeceum) in maize, sugarbeet, carrot, etc. enhance chances of cross fertilization in these crop species.

Heterostyly : If the style is longer than the filaments of the anthers, the stigma cannot receive pollen from the same flower and thus enhancing cross fertilization.

Unisexuality of flowers : If staminate and pistillate flowers occur on the same inflorescence (e.g. sunflower) or on the same plant (e.g. cucurbits) but at different places (monoecism), there is more likelihood of cross fertilization than self-fertilization. Of course, occurrence of male and female flowers on the same inflorescence is a weaker barrier for self fertilization than the occurrence of unisexual flowers at different places of the same plant. However, occurrence of staminate and pistillate flowers on different plants is one of the strongest barriers for self fertilization and thus enforces cross fertilization.

Sterility and self-incompatibility : Sterility is usually referred to pollen sterility, that is, pollen is non-functional, while in case of self-incompatibility, the pollen is viable but it is not compatible with the pistil of the same flower. Pollen (male) sterility may be due to nuclear genes (genetic), due to cytoplasm (cytoplasmic) or due to both these factors (cytoplasmic-genetic). The last type has been found more important in crop improvement and is being exploited at a commercial scale in case of bajra, jowar, etc.

Self-incompatibility may depend upon the genotype of the gamete (**gametophytic**) as in case of tobacco, potato and rye or may be controlled by the genotype of the sporophytic parent (**sporophytic**) as in case of cabbage, cauliflower and mango.

A unique type of sterility called as **somatoplastic sterility** is found in lucerne. In this crop species, the embryos resulting from cross fertilization grow more rapidly and successfully than the embryos resulting from self fertilization.

In many cases, two or more mechanisms operate to encourage/enforce self or cross fertilization. For example, effects of both monoecism and protandry in maize, of male sterility and protogyny in pearl millet and of self-incompatibility and monoecism in sunflower, ensure cross-fertilization in these crops. In addition to the mechanisms discussed above, some genetic and environmental factors (crop variety, insect population, temperature, humidity, wind velocity and direction, etc.) may play an important role in changing the degree of self or cross fertilization. For example, cotton, chillies and okra show great variations with respect to their amount of natural crossing under different environmental conditions. Even male sterility in some cases has been found dependent on temperature (TGMS) and photoperiod (PGMS).

Determining Mode of Reproduction in a Species

The made of reproduction in a species can be determined by the following three methods :

1. Growing single plants in isolation : If a plant grown in isolation shows seed setting, it is generally self-fertilized and if there is no seed setting it is cross-certilized. But seed setting in a plant grown in isolation does not always indicate that the species is self-fertilized. There are examples (like maize) where single plants of cross-fertilized species grown in isolation show fairly good degree of seed setting.

2. On the basis of floral morphology : As has been discussed earlier, the bisexuality of flowers in a species encourages self fertilization and unisexuality generally enforces cross fertilization. But there are numerous examples where two species having same floral morphology show great differences in their natural crossing, one may be highly self-fertilized and the other highly cross-fertilized. Although all leguminous species have almost the same floral morphology with the occurrence of bisexual flowers but chickpea, mungbean,, urdbean and several others are highly self-fertilized; pigeonpea is often cross-fertilized and lucerne and berseem are highly cross-fertilized. Similarly, wheat (highly self-fertilized) and rye (highly cross-fertilized) have similar floral morphology.

3. On the basis of deterioration in vigour shown after inbreeding : The degree of deterioration in vigour and other characters on selfing or inbreeding also indicates the mode of reproduction in a species. The cross-fertilized species, in general, exhibit high deterioration in general vigour than the self-fertilized species. But great variations are found in this regard even in the same group of plant species. For example, cucurbits, maize and lucerne all belong to the same (cross-fertilized) group but they show very little, moderate and very high deterioration in the general vigour and other attributes, respectively, when subjected to inbreeding.

Therefore, all these methods, can just give a general idea about the mode of reproduction in a species but these are not foolproof methods.

Determining the Amount of Natural Crossing in a Species

Barring the effect of environment, the amount of natural crossing in a species can be determined by growing two strains of the species one with recessive marker and other with its dominant phenotype in alternate rows or mixed in a plot. The seeds from the recessive strain are harvested and grown in the next season. The percentage of plants with dominant phenotype will indicate the percentage of natural crossing.

Classification of Plant Breeding Methods

The ultimate goal of every plant breeding programme is the evolution of new varieties meeting the requirements of farmers. Likewise, the success of every breeding programme depends mainly on three factors :
1. Objective (s) of the breeder;
2. the selection of base material to start from; and
3. the choice of breeding method.

Although the decisions about all three kinds of factors are based on specific considerations, however, decisions about the designation of a promising complex of objectives (including its flexible pursuing) and the selection of appropriate source material to start with are related to particular crops, whereas the decision about the choice of an efficient breeding method is more related to the type of variety to be bred and the mode of reproduction rather than to particular crop.

The assessment of the general methodology of plant breeding which forms the central part of any general theory of this discipline, is based on several criteria like breeding category, operative processes involved, quantiative variability utilized, possible help from quantitative genetics and efficiency of selection enhanced. Likewise, breeding methods have been classified based on several criteria like natural reproductive system (Baur, 1921; Hayes *et al.*, 1955; Poehlman, 1979, 1987; Poehlman and Sleper, 1995; and others), origination of initial variability (Roemer and Rudorf, 1941; Hoffmann *et al.*, 1971) and propagational category of resulting varieties (Schnell, 1969, 1978, 1981, 1982; Simmonds, 1979). Furthermore, a number of modifications and refinements have been suggested in conventional breeding procedures from time to time and not only the theoretical grounds of plant breeding have undergone fundamental change but also the practical aspect of this subject has changed noticeably.

Based on natural reproductive system, plant breeding methods can be classified into following three categories :

1. Breeding methods for self-fertilizing crops : Mass selection, pure line selection, pedigree method, bulk population breeding, backcross method and their variants.

2. Breeding methods for cross-fertilizing crops : Recurrent selection, development of synthetics and composites, production of single, three way and double cross hybrids.

3. Breeding methods for asexually reproduced crops : Development of clones and land varieties of vegetatively propagated crops

Notwithstanding a significant advancement in crop improvement, this old classification has been most widely used by the authors of textbooks and the teachers teaching the subject. However, after the development of hybrids in self-fertilizing and often cross-fertilzing crops like rice, cotton and pigeonpea, it is difficult to draw a sharp line of demarcation between the methods employed in predominantly self-fertilizing and those employed in predominantly cross-fertilizing crop species. Further, in often cross-fertilized crops, both kinds of methods are being used since long period of time.

Breeding methods can be classified according to origination of initial variability into following four categories :

1. Breeding by selection, that is, exploiting only existing variability through selection methods and no new variability to be produced - mass selection and pure line selection.

2. Breeding by crossing, that is, exploiting existing variability as well as the variability produced by recombination - pedigree method, bulk population breeding, backcross method, variants of these three conventional methods,

3. Polyploidy breeding, that is, exploiting variability produced by polyploidy - development of polyploids,

4. Mutation breeding, that is, exploiting variability created by mutations - development of mutant varieties.

This classification suffers from several weaknesses (Schnell, 1982). Polyploidy breeding and mutation breeding, in fact, are breeding techniques and not breeding methods. Further, most of the breeding methods used for self- and cross-fertilized crops come under the category breeding by crossing.

A more justifiable grouping of breeding methods can be done according to propagational category of resulting varieties. Details of this classification are as follows :

(A) Line breeding, that is, self fertilization with lineal varieties - land varieties, pure line varieties and multiline varieties of self-fertilizing crops :

1. Selection breeding :
     Mass selection
     Pure line selection
2. Recombination breeding
     Pedigree method and its variants
     Bulk method and its varients
     Backcross method and its variants
     Single seed descent method and its variants
     Diallel selective mating system

(B) Hybrid breeding, that is, controlled crossing between parents to evolve hybrid varieties :
     Development of single cross hybrids
     Development of three-way cross hybrids
     Development of double cross hybrids

(C) Population breeding, that is, panmictic cross fertilization to evolve population varieties - open pollinated varieties, synthetics and composites :

        Mass selection and its variants
        Recurrent selection and its variants
        Development of synthetics and composites

(D) Clone breeding, that is, asexual reproduction to evolve clone varieties - clone varieties and land varieties of vegetatively propagated crops :

        Clonal selection

Based on a set of queries, these four categories of breeding can be differentiated as given in Table 4.1.

Table 4.1. Differences between four categories of breeding based on a set of queries

Query	Categories of breeding			
	Line	Hybrid	Population	Clone
1. Is resulting variety reproduced by means of seeds ?	Yes	Yes	Yes	No
2. Is evolution of genetically homogeneous variety feasible ?	Yes	Yes	No	Yes
3. Is reproduction of the variety by its own plants feasible ?	Yes	No	Yes	Yes
4. Is the resulting variety heterozygous ?	No	Yes	Yes	Yes
5. Is exploitation of all kinds of gene effects feasible ?	No	Yes	Yes	Yes
6. Is it advisable to use seed of the variety for a long period ?	Yes	No	Yes	Yes
7. Is it feasible to evolve one-genotype variety ?	Yes	Yes	No	Yes
8. Is the category used in food grain crops?	Yes	Yes	Yes	No

Breeding methods can also be categorised on the basis of the type of quantitative variability exploited. Here, two major categories would be :
(1) Breeding for exploiting only fixable quantitative variability - line breeding
(2) Breeding for exploiting fixable as well as unfixable quantitative variability - population breeding, hybrid breeding and clone breeding.

Nevertheless, this classification is an oversimplification of the problem.

Before we proceed for a detailed description of different plant breeding methods in the following chapters, it is important to discuss differences between the terms breeding category, breeding method, breeding technique, breeding phase and breeding step. Breeding category refers to a class of breeding methods based on a specific criterion. A breeding method can be defined as the total plan to evolve a variety or create an improved population (in the special case of recurrent selection) which can be used either as a new variety or as initial variation in another breeding programme. A breeding technique (mutation breeding, polyploidy breeding, haploidy breeding, etc.) is used to create new genetic variability, increase or decrease ploidy level, produce completely homozygous lines or to perform some other such function. However, it does not imply all the operative processes necessary for establishing a new variety. Breeding phase comprises a functionally different part of the whole process of breeding. A breeding method consists of three distinct breeding phases : (i) procuring initial variation, (ii) forming experimental varieties, and (iii) evaluating experimental varieties. Each of these phases serves a definite functional purpose, but would vary according to the breeding category depending upon the kind and length of the operative processes involved. Breeding step means a single operation (choice of source material, growing the base material, making crosses, testing for aspects like disease resistance and combining ability, final selection among experimental varieties, etc. of a breeding method.

In some literature, heterosis breeding has been described as hybrid breeding. However, heterosis breeding is a very wide term meaning exploitation of hybrid vigour in heterozygous or homozygous condition (when heterosis can be fixed) and thus includes all the four breeding categories (line, population, hybrid and clone breeding), whereas hybrid breeding is restricted to the development of hybrids only. Thus, hybrid breeding is a part of heterosis breeding and the two terms should not be used interchangeably.

SECTION II

Genetic Basis of Breeding Crops

Principles of Breeding Self-Fertilizing Crops

CHAPTER 5

Selection in Self -fertilizing Crops

Selection is the oldest method of plant breeding probably as old as is the agriculture and is used as a complete method (pure line selection and mass selection) or as a part of any plant breeding method to handle segregating generations. It is mainly responsible for shifting plant populations for superior types by changing the array of gene frequencies and is the basis of all the crop improvement techniques. The method is equally applicable to both self- and cross-fertilized crops and is thus an important part of breeding programmes. The rapid advancement in the understanding of the genetic basis of selection in the recent years indicates that selection is a very wide term including a number of procedures and therefore the plant breeder should be very cautious while choosing an appropriate selection procedure for his plant material. The effectiveness of a selection procedure would depend upon several factors like mating system and the type of gene effects governing the charater. However, since selection is incapable of creating new variability and since its action is confined to the heritable differences only (the two main characteristics of selection), the success of any selection programme would mainly depend upon two factors :

(1) The extent of genetic variability already present in the material being handled; and

(2) heritable versus environmental contribution towards the control of character under consideration.

The effectiveness of selection will be high if there is enough genetic variability in the base material and the character under consideration has high narrow sense heritability.

Since selection does not create any new variability, there must be some process which could expose total genetic variability present in the species concerned so that selection may act upon it effectively. In self-fertilizing species, this job is performed by natural selfing. The consequences of selfing/inbreeding are :

(i) The proportion of homozygotes increases at the expense of heterozygotes;

(ii) the population splits into small groups, that is, the concealed genetic variability in the heterozygotes is uncovered;

(iii) the population variance increases in the absence of directional selection (within group variance, however, decreases);

(iv) genetic correlation between close relatives increases irrespective of selection; and

(v) the effect of selfing on population structure is independent of the change in environment and the presence of dominance and epistasis.

Selection is of two types : (1) natural and (2) artificial. Natural selection is the main force behind evolution whereas artificial selection is the main force behind plant breeding. Nature selects those types which possess characteristics (hardiness, resistance to insect pests and diseases, etc.) that are necessary for the survival and perpetuation of the species. These types may be very low yielding and thus may not be suitable for the plant breeder as commercial varieties. Further, the natural selection will depend upon the conditions of that region where the plants are growing and thus the direction of natural selection may be different in different regions. On the other hand, under artificial selection, those types are selected which are best suited to man. Therefore, there will be a specific direction of this selection since it will be according to the will of man. The plant types selected by man may not perform better under natural conditions. For example, if *Saccharum officinarum*, a highly protected species by man and *Saccharum spontaneum* (kans), a wild species which has been facing consequences of natural selection from times immemorial, are planted in a field and left to grow under natural conditions, *Saccharum officinarum* will prove to be a poor competitor.

Early Work on Selection

Although sorting out of individual plants or their groups with desired features from mixed populations is an ancient practice, Van Mons in Belgium, Knight in England and Cooper in America in the last part of the eighteenth century for the first time made systematic efforts to know the extent of the effectiveness of selection. They were able to demonstrate that selection can play an important role in the improvement of crop plants. Patrick Sheriff developed some varieties of wheat and oats through selection. These varieties were grown on large areas. However, John Le Couteur (a farmer), a little later, more clearly demonstrated the importance of individual plant selection in case of cereal crops. He critically observed the diversity present among plants in his wheat field and raised single plant progenies. While publishing his work in 1843, he concluded that the progenies of single plants showed high degree of uniformity, whereas there were large differences for agriculturally important characters between the progenies obtained from different plants. About the same time, Hallett applied single plant selection in small grains by growing the material under best environmental conditions and selecting best plants from the lot in each generation. In spite of his wrong belief that acquired characters were heritable, he was able to isolate some good varieties.

Louis de Vilmorin who became incharge of the Vilmorin seed firm (France) in 1843, probably made the most significant contribution towards the understanding of the effects of individual plant selection in self- as well as cross-fertilized crops. His procedure which was later known as **Vilmorin method** or **Vilmorin Isolation Principle** was based on the progeny testing of single plant selections. He applied his method to four varieties of wheat (a highly self-fertilized crop) and sugarbeet (a cross-fertilized crop) and found that while the selection of plants year after year in wheat was totally ineffective (since the varieties were homozygous and genetically homogeneous), the selection for more sugar content in sugarbeets resulted in a considerable increase in the sugar percentage in the crop. It was for the first time that a clearcut difference in the effectiveness of line selection in self- and cross-fertilized species was demonstrated. The progeny test proposed by Vilmorin helps in knowing the breeding behaviour of a plant and heritability of the character under selection, that is, the method helps in determining the real worth of a plant. This method got a remarkable support from Hays (1888). However, the Swedish Seed Association (established in 1886 at Svalof) followed the method in a most systematic manner and at a very large scale and by making some refinements brought it to the level of pure line selection method of today. Because of its notable contribution towards the understanding of some methods of plant breeding in general and that of Vilmorin's method in particular, the Association became internationally famous.

Definition and Genetic Basis of Pure Lines

Although many of the details about the pure line selection method were known by the end of the nineteenth century, the genetic basis of pure lines was firstly described by Johannsen in 1903. He defined a pure line as **the progeny of a single self-fertilized homozygous individual**. His classical experiment on a common variety of beans (*Phaseolus vulgaris*) named Princess for seed weight provided a well founded scientific basis for selection in self-fertilized plant species. *Phaseolus vulgaris* is a highly self-fertilized species. The commercial seed of the variety Princess showed great variations in size indicating that this variety was consisted of a group of pure lines as regards the character seed weight. Therefore, it can be considered as good example of a land variety (regularly existing in an area since long period of time, is a mixture of types and is well adapted to the environmental conditions of that area) of beans.

Johannsen observed that when seeds of different sizes were grown separately, the progenies derived from smaller seeds had smaller seeds than those derived from larger seeds. This indicated that selection for seed weight was effective and that variability for seed size had a genetic basis.

Johannsen selected 19 seeds of different sizes from the Princess variety, grew their progeneis and ultimately developed different pure lines from these seeds. The mean seed weight of these 19 lines ranged from 350 mg (line 19) to 640 mg (line 1), each line having its own characteristic seed weight. Within pure line variation was wholly environmental since the plants grown from the selected seeds of different weights from the same line gave plants having similar seed weight. The example of line 2 may be cited here. The seeds of this line having 400, 500, 600 and 700 mg weight when grown, produced progenies with mean seed weight of 572, 549, 565 and 555 mg, respectively. Further, extreme selections made in each line (two sublines from each line possessing heaviest and lightest seeds) for six years proved to be ineffective, that is, the mean seed weight of each line remained almost the same. This indicated that each pure line was homozygous and genetically homogeneous.

Another proof in support of the points mentioned above came from parent-offspring correlations. The correlations based on the parents and their progenies of the same line (that is, within line parent-offspring correlations) were found non-significant. On the other hand, parent-offspring correlations calculated from the mixture of all 19 lines were significent. That is, whereas the variability for seed weight within each pure line was non-genetic, the variability between lines had a genetic basis.

The following conclusions can safely be drawn from Johannsen's experiment :

(1) The Princess variety of bean (and similarly any land variety of any other self-fertlized crop) was a mixture of homozygous lines which Johannsen named as pure lines.

(2) Each line had its own characteristic seed weight which could not be changed even by making extreme selections in each line in several successive generations.

(3) Within line variability was attributable to environmental effect only.

(4) Between line variability gave evidence of genetic differences.

Although the credit of describing the genetic basis of pure lines goes to Johannsen, the mathematical proof of this basis had already been given by Mendel. Self fertilization increases homozygosity at the cost of heterozygosity. The 50 per cent of the heterozygosity present at a locus turns into homozygosity after each generation of selfing. For example, if all the plants of a population have A_1a_1 genotype at A_1 locus, the first generation of selfing will split the population into ¼ A_1A_1 : ½ A_1a_1 : ¼ a_1a_1 in respect of this locus, that is, 50 per cent of the individuals will become homozygous for this locus. Since A_1A_1 and a_1a_1 will breed true, 50 per cent of the A_1a_1 individuals will become homozygous after second generation of selfing, that is, there will be 75 per cent homozygosity at this locus after second generation of selfing. After 10 generations of selfing, almost all individuals of the population will become homozygous for this locus.

The term percentage of homozygosity refers to the degree of homozygosity at individual locus, whereas the percentage of homozygous individuals means percentage of individuals that are homozygous for all the loci under consideration. For example, if the plants having $A_1a_1 A_2a_2$ genotypic constitution are selfed, the percentage of homozygosity at each of the two loci will be 50 per cent but the percentage of completely homozygous individuals for both the loci in this case will be 25 per cent, that is, four ($A_1A_1 A_2A_2$, $A_1A_1 a_2a_2$, $a_1a_1 A_2A_2$ and $a_1a_1 a_2a_2$) out of total 16 individuals. If s is the number of selfed generations and l the number of segregating loci, the percentage of homozygosity and the percentage of completely homozygous individuals can be calculated by the formulae $(2^s-1)/2^s$ and $(2^s-1/2^s)^l$, respectively, provided all the loci segregate independently and all the genotypes have equal survival value. Thus, the percentage of homozygous individuals in any generation of selfing will decrease with an increase in the number of segregating loci. If there is presence of linkage between the genes, it will affect the percentage of homozygous individuals irrespective of the genes in coupling or repulsion phase but not the percentage of homozygosity. In the presence of both kinds of linkages (coupling and repulsion), there will be a substantial increase in the proportion of parental types, that is, the types of abundant homozygous combinations will be different in two types of linkages. For example, if the genotype $A_1a_1 A_2a_2$ with 80 per cent linkage between the two loci is selfed, it will produce 34 per cent individuals homozygous for both the loci (16/100 A_1A_1 A_2A_2 : 1/100 A_1A_1 a_2a_2 : 1/100 a_1a_1 A_2A_2 : 16/100 a_1a_1 a_2a_2 in case of coupling phase linkage and 1/100 A_1A_1 A_2A_2 : 16/100 A_1A_1 a_2a_2 : 16/100 a_1a_1 A_2A_2 : 1/100 a_1a_1 a_2a_2 in case of repulsion phase linkage) instead of 25 per cent such individuals when there is no linkage between the genes.

Another point which emerges from the effect of self fertilization on homozygosity is that the increase in homozyzosity is very rapid in first few generations of selfing and later on it becomes very gradual.

The work of Johannsen, therefore, provided a sound genetic basis of selection (its main characteristics, effect and consequences) and greatly helped in understanding the relation between genotype and phenotype.

Genetic Variation in Pure Lines

In self-fertilized species like beans, the pure lines are expected to be homozygous at all the loci, that is, they are supposed to be genetically fixed units. According to Mather (1943), such species become adapted to homozygosity and possess a specific genetic organization called **homozygous balance.** Once this balance is established in a line, it does not exhibit any deterioration in general vigour and other attributes due to inbreeding or selfing. But the pure lines of a self-fertilized species also maintain a **heterozygous balance**, that is, some level of heterozygosity is maintained by them. These ideas are based on the fact that the self-fertilized species have descended from cross-fertilized species and during the course of their evolution they develop a homozygous-heterozygous balance. This hidden or residual heterozygosity in these species helps them in adaptation and allows some degree of natural hybridization and recombination and thus development of some new types. These things are difficult to prove on the basis of short term experiments like that of Johannsen, but by conducting long term experiments it can be easily proved that most of the self-fertilized species have some degree of cross fertilization.

The most important source of genetic variation in pure lines are the spontaneously occurring mutations. Since such mutations usually occur at a very low rate, it is again difficult to prove their effect on pure lines in short term experiments. Mutations are sudden heritable changes and, in broad sense, include point (gene) mutations, structural changes in chromosomes (deficiencies, duplications, inversions and translocations) and changes in the entire chromosomes or sets of chromosomes (aneuploidy, haploidy and polyploidy). Nevertheless, point mutations have played the major role in the evolutionary process since they usually do not affect the survival of the individual in which they occur and thus allow gene recombination and provide raw material for further heritable changes.

As regards the effects of natural mutations on the phenotype, their effect may be specific causing a change in a single character or in more than one character. A gene can change into different allelic forms causing a multiple allelic situation at some loci. However, mutations are not usually favourable to the organism irrespective of their unifold or multifold effects and their single or more possibilities at any locus. Furthermore, mutations are recurrent and thus there is every likelihood that any mutations which seems new today might have already occurred before and because of its less competitive ability might have been eliminated from the species by nature. It is, therefore, safe to conclude that almost all kinds of possible mutations in cultivated species had already occurred before and only those could survive which proved superior to others in facing consequences of natural selection. It is the reason why the newly occurring mutations are not usually favourable to the organism.

In case of quantitative characters, it is manytimes difficult to detect the effect of mutations since the genes controlling these characters have usually similar or supplementing effects. Thus effects of individual genes are not distinctly identifiable. However, since a quantitative character is usually controlled by many genes and there is no reason that these genes should have low mutation frequencies than the major genes, the quantitative characters have a higher frequency of mutations. There are both observational and experimental evidences in support of this statement. One such evidence comes from the detailed studies carried out on the Atlas variety of barley. Long back, this variety was developed from a single plant selected from a land variety and thus was expected to be completely homozygous and genetically homogeneous. Subsequent selection in the variety and careful examination of the progenies of individual plants indicated that though none of these new progenies was better than the original variety, there were differences among them for quantitative and other minor characters. But these observations were subjected to a serious criticism that during such a long time there could be natural hybridization and mechanical mixture in the variety and the differences among the new progenies may therefore be attributed to these two factors.

East (1936) provided a better evidence of mutation rate in pure lines on the basis of his studies in *Nicotiana*. His evidence was based on parthenogenetically developed diploid plants which were expected to be completely homozygous. But after a few generations of selfing the progenies of these plants exhibited variability comparable to that found in ordinary inbreds. This variability in progenies cannot be ascribed to natural hybridization or mechanical mixture since no scope for these two factors was left in the experiment. The experiment of East, therefore, proved beyond doubt the possibility of the occurrence of mutations in pure lines.

However, the most conclusive work in this direction was done by Stadler (1942). He was able to obtain precise information about muation rates of eight genes governing endosperm characters in maize by testing very large number of gametes.

Since the frequency of mutations is generally very low and since mutant alleles are usually inferior to the old alleles, the mutation technique does not seem to have much importance in plant breeding. The methods of plant breeding can bring faster improvement in the crop species. However, in some cases, the rates of mutations for desirable genes are quite high and thus advantage of such opportunities should be taken.

CHAPTER 6

Hybridization and its Genetic Consequences

The crossing of genetically dissimilar individuals is called hybridization. Hybridization results in recombination (both between the genes present on the same chromosome and between genes of different chromosomes) which in turn widens the spectrum of variation among individuals. Therefore, the basic factor responsible for providing genetic diversity in progenies of crosses is the formation of new combinations of genes mainly due to Mendelian recombination. This newly created variability is then channelized into development of better varieties.

The importance of pure line selection as a method of breeding started decreasing when the plant breeders felt that the genetic variability existing in the land varieties of self-fertilized crops had been almost completely exploited and there was little chance for further improvement in these crops by using pure line breeding. Therefore, no option was left for plant breeders except to cross diverse individuals and combine together desirable characters in a single genotype from two or more parents. The planned hybridization which could avoid self pollination and chance cross pollination, soon became the main feature of almost all plant breeding programmes, so much so that some have started considering the hybridization and the plant breeding as synonymous terms.

History of Hybridization

Like plant breeding, it is not precisely known that when and by whom hybridization in plants was attempted for the first time. However, one can safely generalize that selection is an older method than hybridization. Although Assyrians and Babylonians did artificial pollination in date palms in 700 B.C. to know the effect of pollen on the maternal tissue of the fruit (that is, metagenic effect of pollen), a systematic effort in this direction was made by Camerarius in 1694 when he demonstrated that plants are sexual organisms. However, Thomas Fairchild in 1717 provided the first authentic proof of hybridization in plants by crossing the sweetwilliam with the carnation. This first plant hybrid was called as Fairchild's mule. The work of Fairchild created so much interest among the plant scientists that many of them started making crosses in plant species for crop improvement as well as for basic studies. One such scientist was Joseph Koelreuter (1760-1766) who firstly reported sterility in the hybrids of *Nicotiana particulata* and *N. rustica* and suggested that crossing was successful between related plants only. Thomas Andrew Knight (1823) made a notable contribution towards the understanding of plant hybridization. He made reciprocal crosses in plants and established that the female and male parents make equal contribution to the F_1 hybrid and that the segregation starts from the second generation. Gartner made thousands of crosses in numerous species and genera of plant kingdom. On the basis of his extensive work on plant hybridization, he described relationship between the parents, F_1 and F_2 generations and published his work in 1849. A number of other scientists, about the same time, did similar work on plant hybridization and by mid-nineteenth century several facts like the dominance in F_1, uniformity in the performance of F_1 plants of a cross, equal contribution of the female and male parents to their F_1 offspring and segregation in F_2, were known. However, it was Gregor Mendel who followed the most systematic approach towards the understanding of plant hybridization by very carefully selecting his experimental material (garden pea), applying mathematics to his data and very clearly explaining his results. On the basis of his work from 1856-1865 (see Mendel 1866), he put forward two laws of inheritance, the law of segregation and the law of independent assortment of factors which later on provided the scientific basis of hybridization and the scientists started realising that plant breeding is not only an art but it is also a science. The work of Mendel along with that of many others at the end of the nineteenth century and in the beginning of the twentieth century helped the plant breeders in predicting the performance of the progenies of crosses on the basis of the type of parents involved in a particular cross. Now the crosses are made with advance planning and clearcut objectives in mind.

Types of Hybridization

Depending upon the degree of relationship between the individuals used in a particular cross, hybridization may be of the following four types :

1. Intra-varietal hybridization : Making crosses between the plants of the same variety is known as intra-varietal hybridization. In self-fertilized crops, however, this type of hybridization has a very limited use since the modern varieties of this group of plants species are homozygous and genetically homogeneous. This type of hybridization may have its own use in case of land varieties.

2. Inter-varietal hybridization : The crosses between the varieties of the same species come under this category. It is the most common type of hybridization being followed by the plant breeders in the improvement of crops.

3. Inter-specific hybridization : It is the crossing of two species belonging to the same genus. For example, crosses between *Hordeum vulgare* and *H. bulbosum* and between *Gossypium hirsutum* and *G. barbedense* will be inter-specific crosses.

4. Inter-generic hybridization : Making crosses between two different genera is termed as inter-generic hybridization. The crosses between *Raphanus sativus* and *Brassica oleracea* and between *Triticum* species and *Secale cereale* (giving rise to *Triticale*) are two most striking examples of this type.

The types 3 and 4 mentioned above come under a common term **distant hybridization** (or wide hybridization). Although distant hybridization is one of the main patterns of evolution in cultivated plants (in addition to Mendelian variation and polyploidy) leading to a large number of gene differences appearing in the progenies, many of the new gene combinations prove to be disharmonious under natural conditions as well as in the experiments of the breeder. Therefore, the success achieved by plant breeders and geneticists in this direction has been very limited. However, the difficulty in producing interspecific and intergeneric hybrids greatly varies from case to case. In some cases, such hybrids can be produced very easily, while in others it is not at all possible to make successful crosses.

Objectives of Hybridization

A well planned hybridization between carefully chosen parents has the following objectives :

1. Creation of new gene combinations : When the existing gene combinations in a crop have been completely exploited by the breeder, hybridization between diverse individuals or lines and segregation of genes in F_2 and subsequent generations provide opportunities for obtaining gene combinations of different types, that is, more genetic variability becomes available to the breeder. The spectrum of variation can be further widened by random intermating in F_2 or any other segregating generation.

2. Combining desirable characters in a single genotype from two or more different genotypes : The chief objective of hybridization is to evolve a variety which has desirable characters of both its parents. This accumulation of desirable genes in a single variety is necessary because these characters are manytimes found scattered in different varieties, species or genera and thus in such a situation a single genotype or line will be deficient for one or the other good character (yielding ability, adaptability, disease resistance, etc.). Sometimes, a variety possessing almost all the desirable characters starts deteriorating after a few years of its release in respect of some important character like resistance to a popular disease. For eliminating such defects (that is, to remove bottleneck genes) from a variety which is otherwise very good, hybridization is used to transfer a character from one variety to another. The purpose of hybridization is not only to combine desired characters of the parents in a single variety but individuals falling outside the parental range in respect of some character (that is, transgressine segregants) can also be exploited.

3. Understanding and exploiting heterosis : Hybridization helps in both the basic type of studies about the understanding of heterosis and the practical use of heterosis in the improvement of crops. However, the extent of heterosis in the self-fertilized crops is much lower than its degree in the cross-fertilized crop species. But careful handling of transgressive segregants (which are superior to both the parents) may result in the development of valuable varieties in the self-fertilized group of crops.

Consequences of Hybridization

As has been discussed earlier in this chapter that hybridization has now become a necessity rather than an alternative for the plant breeder to achieve desired improvement in his crops. But one thing should be kept in mind that the consequences of hybridization may be favourable or unfavourable and the successes and failures of hybridization as a method of breeding depend upon a number of phenomena of Mendelian genetics (gene segregation and recombination, pleiotropy, modifiers, penetrance and expressivity, xenia, epistasis, linkage, threshold effect, etc.).

Before going into the details of various phenomena of Mendelian genetics, it is necessary to know the relationship between genes and characters. It is now an established fact that though the qualitative characters are governed by relatively simple genetic mechanisms (as compared to the genetic mechanisms involved in the control of quantitative characters), the relation between the basic genetic material (genes) and the final phenotype (characters) is mostly complex and depends upon a number of factors like physico-chemical reactions, gene interaction and linkage and the interaction between genotype and environment.

The following factors affect the expression of a character in segregating generations :

Gene segregation and recombination : When genetically diverse parents are crossed, the F_1 plants will be heterozygous at the loci for which the two parents differ. The selfing of F_1 will give rise to a huge number of recombinants in the F_2 generation. According to Mendelian principles, the kinds of gametes possible in F_1 (also the number of genotypes occurring in backcrossing, the number of homozygous genotypes and kinds of phenotypes in F_2 with full dominance), kinds of genotypes possible in F_2 (also the kinds of phenotypes in F_2 with no dominance and no epistasis) and the smallest perfect population in F_2 will be 2^l, 3^l and 4^l, respectively, where l is the number of segregating loci. Therefore, even with a moderate number of genes segregating, the kinds of genotypes possible in F_2 will be very large. For example, if ten genes are segregating, the kinds of possible genotypes in the F_2 will be 59049. This number increases so drastically with an increase in the number of segregating loci that with 20 genes segregating the kinds of genotypes in F_2 will be 3486784401. The situation becomes more complicated in case of quantitative characters where many genes govern the same character, the effects of genes cannot be identified individually, and the environment has a considerable effect on gene expression. Under such a situation, it is not practically possible to carry out gene-by-gene analyses of differences between parents.

Selection of heterozygous plants in the early segregating generations manytimes gives a confusing picture about the worth of the selected plants. Such plants segregate in subsequent generations and their progenies may be quite different. Further, in case of overdominance, the heterozygotes may be superior than the homozygotes in performance as well as in adaptability and thus the theroretically expected performance as well as the proportion of homozygotes may not be found in the later generations. However, linkages between desirable genes will make the job of the plant breeder easy. Also, the predominance of addive gene effects for a character enhances the effectiveness of selection in early segregating generations.

Nevertheless, the composition of populations derived from hybrids can be known with the help of the formula -

$$\{1+(2^s-1)\}^l$$

where s is the number of selfed generations and l the number of segregating loci. The expansion of this binomial will tell us the composition of the population in terms of individuals heterozygous/homozygous for different number of loci. For example, if we want to know the composition of a population segregating for four loci and selfed for three generations, the picture will be as follows :

$$\{1+(2^3-1)\}^4$$

$$= \{1+7\}^4$$

$$= (1)^4 + 4\,(1)^3\,(7) + 6\,(1)^2\,(7)^2 + 4\,(1)\,(7)^3 + (7)^4$$

$$= 1+28+294+1372+2401$$

Here, the first exponent indicates the number of heterozygous loci and second the number of homozygous loci. Therefore, in the F_4 generation, the number of plants

heterozygous for all the four loci	=	1
heterozygous for three loci and homozygous for one locus	=	28
heterozygous for two loci and homozygous for two loci	=	294
heterozygous for one locus and homozygous for three loci	=	1372
and homozygous for all the four loci	=	2401

Total   4096

The coefficient outside the bracket in each term indicates the ways for which a plant can be heterozygous/homozygous for loci. For example, there are four ways a plant can be heterozygous for three loci and homozygous for one locus. Similarly, there are six ways a plant can be heterozygous for two loci and homozygous for two loci and four ways that the plant can be heterozygous for one locus and homozygous for three loci. The proportion of homozygotes increases with an increase in the number of selfed generations and thus selection of homozygotes will be more effective in later generations.

In addition to the complexity created by gene segregation and recombination, a number of factors tend to disturb the relationship between the genotype and phenotype and thus create difficulty in the identification of superior types during selection. These factors are discussed hereafter.

Pleiotropy : A number of morphological and physiological characters (shape and colour of flower in some ornamental plants, determinate versus indeterminate growth habit in pigeonpea and some other leguminous crops, pollen characteristics in wheat, large versus small nectaries in muskmelon, seed colour in numerous field crops, resistance to a specific race of a disease, etc.) are little affected by internal (genetic) and external (physical) environments and thus show simple inheritance. The effect of the genes controlling such characters is stable, large, well recognizable and easily interpretable on the basis of standard Mendelian techniques. However, all characters governed by major genes do not have such simple inheritance. In some cases single major genes have been found to control more than one character. Such a phenomenon is called as pleiotropy. Examples of pleiotropic genes in crop plants are few. Of course, one interesting example of pleiotropy is found in wheat. The gene controlling presence versus absence of awn in this cereal also influences the yield and the market quality of the produce. The awned version of the variety **Onas** had higher yield and kernel weight than the original awnless **Onas** (Suneson *et al.*, 1948). The Tu-tu locus in chromosome IV of maize which has played a significant role in the evolution of domesticated maize, is also considered to have multiple effects on plant phenotype.

Pleiotropy, however, should not be confused with the tight linkage between the genes. Very tightly linked genes also behave in a similar fashion unless the linkage between them is broken by some means. If the characters governed by tightly linked genes or by a single major gene are desirable, the condition is favourable for the plant breeder and the breeder will not be interested to know whether it is due to pleiotropy or due to tight linkage. But if one desirable character is associated with some undesirable character, than it is necessary to know the cause of this association and if it is due to linkage, all out efforts should be made to break such undesirable linkages. However, if both the desirable and undesirable characters controlled by the same major gene (a case of pleiotropy), one is helpless to break this association. The development of isogenic lines (either by selecting individual plants heterozygous at a particular locus in the F_1 and in the succeeding generations of selfing till the two plants become completely homozygous except for the locus in question or by recurrent backcrossing) can help to identify whether the side effects of major genes are due to pleiotropy or due to tight linkage.

Modifiers : Modifiers are a group of genes which have small effects but their main function is to modify the effect of major gene(s), that is, they may cause an increase or decrease in the expression of major genes. The effect of modifiers is thus an example of alteration in the gene-character relationship by genetic environment. Many of the agronomic characters in crop plants are under the control of both the major genes and the minor genes and, therefore, modifying genes are playing important role in the crop improvement. However, the most important example where clearcut effect of modifiers can be seen comes from mice. In mice, the presence of spotting is under the control of a dominant major allele S. But the degree of spotting is controlled by a group of modifying genes.

Dwarfing recessive genes in sorghum (dw_1, dw_2, dw_3 and dw_4) situated at four loci have shown dramatic effect for reducing height. However, considerable variation in the height of the same genotype is found indicating presence of modifying genes. Similarly, major genes (along with polygenes) controlling maturity in sorghum are found on three loci. The recessive allele ma in homozygous condition (ma ma) masks the effect of dominant alleles Ma_2 and Ma_3 found at other two different loci and thus makes the plant early maturing. The plants with Ma Ma ma_2 ma_2 ma_3 ma_3 and Ma ma ma_2 ma_2 ma_3 ma_3 genotypes have intermediate maturity. Similarly, plants with Ma Ma ma_2 ma_2 ma_3– genotypes show late maturity and plant with Ma– Ma_2– Ma_3– and Ma– Ma_2– ma_3 ma_3 genotypes are very late. However, the expression of these genotypes also depends on the day length.

Penetrance and expressivity : Other phenomena which create confusion in the relation between the genotype and phenotype are penetrance and expressivity particularly when the degree of penetrance and the manner of expression changes according to environmental conditions under which the plant material has been grown. Penetrance is the proportion (or percentage) of individuals in which a gene is able to express itself. Depending upon the environmental conditions and/or the type of modifying genes, a gene may not express itself at all (that is, zero penetrance) or it may express itself in some individuals but not in others though they also carry that gene (that is, incomplete penetrance) or it may express itself in all the individuals which carry this gene (that is, complete penetrance). The expressivity, on the other hand, is the way in which the gene expresses itself, that is, a gene may show uniform expression in all the individuals which carry it or it may have variable expression (in some individuals it expresses in some manner, while in others in a different mananer). The example of chlorophyll deficiency in Lima beans and that of stem fasciation (fusion of branches with stem) in peas will make these terms more clear. In the Ventura variety of Lima beans, a dominant allele causes partial chlorophyll deficiency in the tips and margins of leaves at the seedling stage. This chlorophyll deficiency is, however, usually confined to only 10 per cent of plants though all the plants carry this allele in homozygous condition. Therefore, penetrance in this case is 10 per cent (incomplete). But this percentage of penetrance may come down to zero or may reach to 100 (complete penetrance) under specific set of environmental conditions. Further, the chlorophyll deficiency in the plants carrying the same allele may be restricted to tips of the leaves or to the margins of leaves or the entire leaves become deficient and ultimately get abscised or the entire leaves show deficiency in the beginning but after some time turn into normal green leaves. So, expressivity here is variable.

In peas, the stem fasciation is dependent upon a recessive allele fa. The crosses between normal and fasciated types indicate incomplete penetrance in the F_1, F_2 and test cross progenies. However, the penetrance varies from cross to cross. Further, the fasciation is either so acute that the plants become almost abnormal, or so small that it is difficult to know whether fasciation has actually occurred or not, or with intermediate effect. That is, the expressivity is variable. However, lines with complete penetrance and with remarkably uniform expressivity could be developed by manipulating favourable modifiers.

These two phenomena, therefore, can create problems for the plant breeders in selecting desirable genotypes since the ability of the gene to express itself and its way of expression are not certain.

Threshold effect : According to Mayr (1963) threshold is a point or stage at which if any genotype reaches due to its sensitivity to environmental conditions like temperature, its expression is changed. Such a damage in the phenotype due to the sensitivity to temperature, etc. can thus obscure gene-character relatinship. A well documented example of threshold is found in Drosophila. The phenotype cross-veined changes to cross-veinless under extreme temperatures and once the flies have become cross-veinless, they will not turn into corss-veined even under the normal temperature conditions.

In crop plants, such an example is found in barley. Albinism in a particular type of barley is a temperature sensitive phenomenon. If plants are raised at temperature below 45°F, they are albino. But the expression is different at high temperatures and all plants are green if raised at temperature above 65°F. However, if small green sector are present on the plants at low temperatures, such plants usually fully recover when temperature is increased. The Atlas variety of barley also shows temperature-sensitive albinism. But it seems that in addition to the sensitivity of this homozygous variety to temperature, some genetically controlled threshold also plays role in the production of albinism.

Xenia : Xenia is defined as the immediate effect (or direct influence) of pollen upon the endosperm of the female. This phenomenon is common in maize and tends to obscure the correspondence between gene and character. In maize, if two varieties differing in endosperm characteristic are crossed, the endosperm characteristic of the pollen parent usually appear in some kernels of the female. This situation creates difficulty particularly when one is interested to produce pure seed of a particular type. Suppose one is interested to produce sweet corn, but if some stray pollen from some other type (say starchy) comes to the stigmas of the sweet corn, some kernels with stray type endosperm will appear in the cobs.

Non-allelic interactions (epistasis) : As the name indicates, non-allelic interactions are the interactions between the alleles belonging to different loci, that is, when the alleles situated at different loci are not independent of each other for their action. Here also (like modifying genes), the genetic environment alters the relation between the genotype and phenotype. The dominance (intra-allelic interaction or the interaction between the alleles of the same locus) is a different kind of phenomenon. The relative importance of these two phenomena (epistasis and dominance) in the improvement of crops, however, depends upon the requirements of the varieties to be evolved. In crops where homozygous varieties are evolved for commercial cultivation, homozygote × homozygote type epistasis is more important but in crops where the commercial varieties are hybrids, dominance may play more important role than epistasis.

If epistasis is confined to two loci only, the expected phenotypic ratios may be 9:7 (complementary epistasis), 9:3:4 (supplementary epistasis), 13:3 (inhibitory epistasis), 15:1 (duplicate epistasis) and 12:3:1 (masking effect of the allele). If alleles of three or more than three loci are interacting then it may complicate the situation.

There are numerous examples of epistasis in crop plants. For example, aleurone colour in maize is controlled by complementary genes. However, a dominant allele 1 inhibits the effects of colour genes and when a cross is made between a variety having this inhibitory allele in homozygous condition with allele responsible for colour and another variety with recessive alleles at both the loci, the F_2 gives a 13:3 phenotypic ratio. Interestingly, in some cases, one gene produces the raw material for the other gene to act upon it. Such a situation is found in Ladino clover for HCN content.

Linkage: Linkage is the tendency of two or more genes to remain together from one generation to another, that is, when two or more loci are not independent in their transmission. However, though linkage does not influence the percentage of homozygosity, it can alter the proportion of homozygous individuals, that is, it can increase or decrease the percentage of specific gene combinations. Since linkage causes an increase in the parental types and a corresponding reduction in the recombined types, the proportion of any gene combination will depend on whether the linkage is in the coupling phase (both the dominant alleles come from the same parent and both the recessive alleles from other parent) or in the repulsion phase (one dominant and one recessive alleles come from one parent and another dominant and recessive alleles from the other parent). For example, the expected proportion of AB/AB genotype (dominant homozygote) in F_2 from the double heterozygote with 75 per cent linkage will be 14.06 if linkage is in coupling phase (AB/ ab) and 1.56 per cent if linkage is in the repulsion phase (Ab/aB). The proportion of the individuals carrying this genotype will increase in coupling phase linkage with an increase in the percentage of linkage (or with a decrease in the percentage or recombination since linkage is equal to 1-recombination fraction) and with 99 per cent linkage in coupling phase, the percentae of this genotype will reach to 24.5 per cent. On the other hand, the proportion of AB/AB individuals will decrease with an increase in the percentage of repulsion phase linkage and with 99 per cent linkage, the proportion of AB/AB genotype will come down to almost zero (.0025 per cent). The percentage of these dominant homozygotes in coupling and repulsion phase linkages can be calculated by $\frac{1}{4}(1-p)^2$ and $\frac{1}{4}p^2$ formulae, respectively, where p is the recombination fraction. However, if the genes assort independently (that is, 50 per cent recombination), various combinations will occur equally frequently.

Depending upon whether the linkage is between two desirable characters or between a desirable and an undesirable character linkage can enhance or can delay the improvement in a crop. The linkage between two desirable genes is a favourable situation for the breeder since selection in favour of one desirable character will automatically result in the selection of the other desirable character. Contrarily, in case of linkage between one desirable and other undesirable character, the selection in favour of the desirable character will automatically reject the other desirable character. Such undesirable linkages sometimes pose serious problems before the breeders and unless they are broken by some means and desirable combinations are obtained, the progress of breeding programme will be delayed. The desirable recombinants can be obtained by raising large F_2 populations or by intermating and selection in segregating generations. However, the size of population for recovering desirable combinations will depend upon the tightness of linkage between the genes.

Examples of both kinds of linkages (favourable and unfavourable) are available in crop plants. In wheat, the two favourable genes Rio and Turkey that are resistant to different races of bunt disease, are found linked with each other. Similarly, in barley, the genes resistant to loose smut and stem rust show close linkage. As regards the undesirable linkages, in some strains of wheat, stem rust resistance is found linked with late maturity. There are numerous other examples of this type in wheat and other crops. Linkages between high number of grains and low grain weight are common in most of the cereal crops.

Environmental effect : As has been discussed earlier (penetrance, expressivity and thresholds), the external environment sometimes tends to confuse the relation between genotype and phenotype. However, in case of qualitative variability, the effect of physical environment is much less as compared to its effect on quantitative variability since in most of the cases the genetic ratios given by major genes remain undisturbed under different environmental conditions. Therefore, the results obtained about qualitative variability under a particular set of environmental conditions may be generalized for different sets of environmental conditions.

If the above mentioned factors which can affect the expression of characters in segregating generations and thus can alter gene-character relationship are not operating, the genes will segregate and recombine according to Mendel's laws.

Selection of Parents for Hybridization

Depending upon the objective of the breeding programme, the parents to be involved in crossing must be very carefully chosen since the success or failure of the hybridization programme will depend upon the choice of the parents. While selecting parents for hybridization, consideration of the following points may help in attaining relatively more improvement :

(1) In general, one of the parents should possess high productivity and adaptability (thus, it is usually a well adapted commercially grown variety of the locality) and the other parent or parents should have all those desirable characters which are not presnet in the first present, that is, all the desirable attributes of a species should be present in the parents selected.

(2) The parents should have a high combining ability.

(3) The desirable features which are absent in the first parent and are present in other parent or parents should have a high intensity so that the specific weaknesses of the first parent are complemented in the best possible way.

(4) More genetic diversity with dispersion of desirable genes among the parents would enhance the possibility of obtaining recombinants that are superior to both the parents.

As regards the number of parents to be included in the hybridization programme, no particular limit can be fixed since it will depend upon how the desirable features are scattered among the genotypes available to the plant breeder.

Aids to Selection in Hybridization Programme

The hybridization programme and particularly the pedigree method of breeding involves two main steps : (a) The selection of parents to the included in the programme which determines the potential of the breeding programme and (b) the selection of superior types in the segregating generations which determines the degree to which this potential has been exploited. Say in other words, the degree of success in any hybridization programme will depend upon the degree of ease with which the parents have been chosen and the skill used during the selection in segregating generations. Therefore, chance hybridization and unaccurately practiced selection in segregating generations have little importance in modern plant breeding. The choice of parents has already been discussed in detail. The following factors would determine the accuracy of judgement of the worth of plants selected by the breeder in segregating generations :

Breeder's knowledge about the crop : Good knowledge about the crop (that is, what should be the characteristic features, both morphological and physiological, of a newly evolved variety ? how the selected plants respond to different environmental conditions ? nature and the magnitude of gene effects controlling important characters, expected market value of the produce, etc.) is a pre-requisite for the plant breeder. The breeder has to make selection usually from a very large number of plants and that too on the basis of quick visual observations since it is not usually possible to take precise measurements of each plant for all the characters. If the breeder is not thoroughly acquainted with the crop concerned, he will not be able to judge the real worth of the plants during selection which may lead to wrong decisions.

Artificial inoculation : At present, disease resistance is considered to be a basic requirement of a new variety. Therefore, screening of genotypes for this character should be done under the conditions which are congineal for the spread of disease. However, sometimes when the environmental conditions are not favourable for the proper development of the pathogen, particularly when there is drought, the susceptible plants may also behave like resistant ones. Under such a situation, artificial inoculation helps the scientist in the proper screening of genotypes.

Growing the material in off-season nurseries : Raising of material in off-season nurseries helps the breeder in several ways. For example,

(a) the places where the off-season nurseries are raised usually have the most congineal environment for the spread of diseases and thus many of the genotypes which seem to be resistant, show susceptibility to diseases and thus it helps in the proper screening of genotypes;

(b) it provides the opportunity of the use of desired race or races without contamination; and

(c) the time required (usually 10-12 years) to evolve a variety can be reduced by advancing generations in off-season nurseries.

Laboratory tests : A number of laboratory tests can be carried out to help the breeder in identifying desirable genotypes and in speeding up his breeding programme. These tests may be for quality and resistance to lodging, salinity, thermo- and photosensitivity, drought, post-harvest sprouting, etc. The breeder is not only interested to increase the yield potential of the crop but the improvement in quality is also equally important for him. Sometimes, it is difficult to test the lodging resistance in the field as the conditions may not be favourable for this test (no heavy rains, no fast winds, etc.). However, the breaking strength of straw can be measured in laboratory with the help of machines. Similarly, the resistance to drought can be tested in the laboratory. Additionally, there are some morphological and physiological characters (presence of awns in wheat and barley, colour of leaves, rooting pattern, CO_2 assimilation, etc.) which are associated with drought. Thus, drought resistant genotypes can also be identified on the basis of these characters.

Early generation testing : Early generation testing helps in an early elimination of undesirable types and thus enhances the chances of proper identification of better segregants. It also helps the breeder in reducing the size of his experimental material. Estimation of general and specific combining ability effects and variances would help in rejecting disappointing progeny.

Intermating : Whereas random intermating of individuals in segregating generations checks early fixation of genotypes, increases chances of new gene combinations and helps in breaking linkage blocks, selective intermating provides opportunity for developing desired gene combinations.

CHAPTER 7

Basic Quantitative Genetics: Concepts

Variability in plants is generally defined as the differences among plants for various characters. No two plants are exactly alike for characters like yield, height and maturity and thus differences from plant to plant, row to row, plot to plot, etc. can be easily observed. Variability is the basic requirement of plant breeding programmes. It is the nature and the extent of variability on which the success of hybridization (which can produce new gene combinations but cannot create new genes) and selection (neither creates new genes nor produces new gene combinations) and ultimately of the plant breeding programmes depends. The plant breeding, therefore, has been rightly called as purposeful management of variability.

Variability can be broadly classified into two types : (1) **genetic (heritable)** variability and (2) **environmental (non-heritable)** variability. However, since genotype and environment do not usually act independently of each other, genotype × enviroment (G×E) interaction also plays an important role in the development of phenotype in many cases. The variability in plants can also be divided into **qualitative** variability and **quantitative** variability. Qualitative and quantitative types of variability though have the same Mendelian genetic basis, differ in a number of ways :

(i) Qualitative variability is based on the control of characters by one or a few genes with large distinct effects (oligo genes), whereas the quantitative variability is based on the control of characters by many genes with small effects (polygenes). Thus, in case of former type of variability, usually simple genetic systems are involved in the control of characsters. But in case of the latter type, the characters are generally governed by complex genetic systems.

(ii) The qualitative variability can be easily grouped into discrete phenotypic classes (that is, specific ratios can be obtained), but such a grouping is not possible in case of quantitative variability because of the simultaneous segregation of many genes and the effect of the environment. These two factors (many genes and considerable effect of environment), in fact, translate the effects of individual genes into continuous variation.

(iii) The environment has less effect on qualitative variability and thus genetic ratios are not usually disturbed due to its effect, whereas quantatitive variability usually show high sensitivity to environmental changes and thus a due respect has to be given to the environment while studying this kind of variability.

(iv) Transgressive segregants (individuals falling outside the parent range for one or more characters) may appear in the segregating generations in case of quantitative variability but such segregants are not obtained in case of qualitative variability.

However, there are a number of cases in crop plants where a character is controlled by oligo genes as well as by polygenes. For example, plant height in wheat though is a continuously variable character, different varieties of wheat have been put in four distinct height groups (tall, single dwarf, double dwarf and triple dwarf varieties) as if this character in wheat is governed by three major genes only. Similarly, maturity in jowar is controlled by polygenes and major genes, but different varieties of this crop can be put in four maturity groups–early, intermediate, late and very late.

The characters showing qualitative variability were the basis of genetics when Mendel proposed his laws of inheritance and at the time of the rediscovery of Mendelism and its subsequent development. Not much statistics is involved in the study of qualitative variability. Mostly one simple statistical method the chi-square (χ^2) test, is applied to test the goodness of fit of different genetic ratios and to measure the difference between populations put in contingency tables.

Several characters of economic importance in crop plants are under the control of qualitative genes. These characters (colour and shape of flower, colour of seed, chlorophyll development, pollen and endosperm characteristics, etc.) have been studied in detail. For example, the recessive alleles like fl (floury endosperm, that is, kernels containing soft starch), wx (waxy endosperm, that is, most of the endosperm made of amylopectin), ae (amylose endosperm, that is, endosperm high in amylose) and su (sugary endosperm, that is, kernels have high percentage of sugar) in maize have been studied in detail and their use in home, manufacture of adhesives, manufacture of plastics, etc. has been identified.

Quantitative Variability

In case of quantitative variability, the phenotypes cannot be classified into discrete classes since there is continuity in the variation from one extreme to another. As has been discussed earlier, this continuity in the variation is ascribable to simultaneous segregation of several genes with small similar effects and the unignorable effect of the environment. Further, since single individuals are uninformative, the study of quantitative variability is done at population level rather than at single individual level by estimating some population parameters (mean, standard deviation, variance, correlation, etc.) rather than testing the goodness of fit of some genetic ratios. In this way, a lot of statistics gets involved in the study of quantitative variability.

Now a potent question arises that whether the continuous variation displayed by quantitative characters and discontinuous variation displayed by qualitative characters have the same or different genetic bases. Francis Galton (in last part of the nineteenth century) must be credited for initiating work on continuous variation. He and his followers were able to show that the continuous variation was, at least, partly heritable. Of course, they could not explain that how this variability was transmitted from one generation to the next.

The rediscovery of Mendel's laws in 1900 resulted in a severe controversy between Galtanians or biometricians (particularly Carl Pearson) and Mendelians or non-believers of biometry (Bateson, de Vries and others). The former group tried to explain all kinds of variation on the basis of quantitative inheritance. Contrarily, the latter group regarded continuous variation as a non-heritable part of phenotypic variation. Therefore, the most important issue at that time was whether the classical Mendelian genetics could be reconciled with the continuous variation observed for most of the important characters of living beings. The first attempt towards this reconciliation was made by Yule (1906) who suggested that both the kinds of inheritances (qualitative and quantitative) have the same Mendelian basis (For details see Singh and Pawar, 2005).

The Concept of Multiple Factors

The multiple factor hypothesis was firstly advanced by Nilsson-Ehle (a plant breeder from Sweden) in 1909 on the basis of his studies on kernel colour in wheat and oats. According to him, three loci controlled kernel colour in wheat (between white and various shades of red). It was found that the three loci had a cumulative effect and segregating in a predictable Mendelian way. When one gene was segregating, it gave a 3 (red) : 1 (white) ratio; when two genes were segregating, they gave a 15 (red); 1 (white) ratio; and when all the three genes were segregating, the phenotypic ratio was 63 (red) : 1 (white). In fact, the red phenotypes (3 in the first, 15 in the second and 63 in the third case) were not of the same intensity. A careful study of different intensities of the red colour indicated that segregation at one locus gave a 1:2:1 ratio, segregation at two loci gave a 1:4:6:4:1 ratio and segregation at all the three loci resulted in a 1:6:15:20:15:6:1 ratio, that is, the intensity of red colour depended on the number of alleles (rather than entirely on the type of combination) present in the individual. For example, all the genotypes $A_1A_1a_2a_2a_3a_3$, $A_1a_1A_2a_2a_3a_3$, $A_1 a_1 a_2a_2A_3a_3$, $a_1a_1A_2A_2a_3a_3$, $a_1a_1A_2a_2A_3a_3$ and $a_1a_1a_2a_2A_3A_3$ will have same intensity of colour. The colour of the kernel was white when all the six recessive alleles were present. Thus, kernel colour in wheat is a continuously variable character having a Mendelian basis.

However, the most conclusive proof that continuous variation is Mendelian in nature was provided by East (1916). He conducted a classical experiment in *Nicotiana longiflora* and demonstrated a parallelism between the variation observed for corolla length (a continuously varying trait) and the variation expected on Mendelian basis in different non-segregating and segregating generations. He crossed two homozygous varieties of *N. longiflora* having considerable differences between them for corolla length and obtained F_1, F_2, F_3, F_4 and F_5 generations of the cross. Based on the Mendelian genetics, in an experiment like this, both the parental and F_1 generations are expected to show low magnitude of variability (attributable to environmental effect only) since the parents were homozygous and genetically homogenous and the F_1 plants were though heterozygous but genetically homogeneous, that is, none of these generations should be genetically variable; the F_2 generation should show more variability than that within the parental groups and the F_1 generation because of the segregation and recombination of genes in F_2 plus the environmental effect; F_3 families produced from different F_2 plants should show marked differences in mean corolla length and variability, but in no case an F_3 family or a family in any subsequent generation should show greater variability than that of the family from which it was produced.

The results obtained by East were in complete agreement with these Mendelian expectancies. His observations were as follows :

(1) The parents were genetically diverse for corolla length since they exhibited wide differences for the character when raised under similar environmental conditions. The corolla length of parent 1 ranged between 34 and 43 mm and that of parent 2 between 88 and 100 mm.

(2) In spite of the fact that the parents were homozygous or nearly homozygous, some differences within the parental groups were observed which could be attributed to environmental fluctuations within the experimental area.

(3) The F_1 plants showed similar magnitude of variability (between 58 and 70 mm) as was observed within the parental groups.

(4) The F_2 generation exhibited much greater variability (ranging from 52 to 88 mm) than both the parental and F_1 generations.

(5) The means of the F_3 families differed significantly and these differences depended upon the size of the F_2 plant from which the family was produced.

(6) The frequency distribution of various F_3 families and their coefficients of variability indicated that the differences between these families were wide. One F_3 family showed variability similar to that of F_2. Of course, none of the F_3 families exhibited variability comparable to that of the either parent probably because of a small number of F_3 families grown. However, the variability observed for two out of four F_4 families was comparable to that of the smaller parent.

(7) None of the F_3 or subsequent generation families showed greater variability than that of the F_2 generation or lesser variability than that of either parent (Also see Allard, 1960, 1999).

As expected, the coefficient of variability decreased with an increase in the homozygosity within families. The coefficient of variability was highest in case of F_2. Its value was less in F_3 than in F_2. Similarly, F_4 families had lower coefficient of variability than that of F_3 and similar was the case with F_4 and F_5 families.

It can be roughly concluded from the F_2 distribution that at least five loci were involved in the control of corolla length in *N. longiflora* since out of 444 F_2 plants studied, only one plant reached near to the size of larger parent but none was similar to the smaller parent. Had the character been under the control of four segregating loci, one plant out of 256 F_2 plants would have resembled the smaller parent and one plant out of them resembled the larger parent.

The results of East's experiment thus provided a confirmatory evidence that the corolla length in *N. longiflora* though is a continuously variable character, its inheritance has a Mendelian basis. The explanation of such characters was given the name of *multiple factor hypothesis*. According to this hypothesis, the operation of each gene is considered to be more or less independent. However, Mather (1943) introducing the term *polygenes* (the genes with small effects), believed that the polygenes do not operate independently, rather the action of each polygene depends on the other genes present. But the final shape to the concept of polygenes was given by Lerner (1958). According to him, the polygenic inheritance is an extremely puzzling subject. Since the segregation of a large number of genes controlling the same character takes place simultaneously and because of a considerable effect of environment on the phenotypic expression of polygenically controlled characters, it is not possible to trace the effect of each individual gene in a segregating population.

Under most circumstances, the polygenes behave like isoalleles (alleles having similar effect on ·a character) and thus substitution of one allele for another at a different locus may not be easily detected, that is, a phenotypically identical homozygous population may be consisted of different genotypes even when there is no dominance, no epistasis and no environmental effect. Such genotypically heterogeneous populations thus have great reserves of genetic variability. In many cases, polygenes form gene blocks of specific significance under a set of environmental conditions. Also, they act as modifiers for various characters in almost all the crop plants.

Poehlman and Sleper (1995) have suggested that the number of genes governing a quantitative character can be known by the formula -

$$\frac{(\bar{p}_1 - \bar{p}_2)^2}{8\,(\sigma F_2)^2 - 8\,(\sigma F_1)^2}$$

Where, $\bar{p}_1$ and $\bar{p}_2$ are the means of two homozygous parents and σF_1 and σF_2 are the standard deviations of F_1 and F_2 generations, respectively. The assumptions of the method are absence of dominance, epistasis and gene correlation and that the genes have equal effect. However, since these assumptions are not realistic and can rarely be fulfilled by any plant material, the method cannot be treated as a standard one. The reliability of the method decreases with the increase in the number of genes controlling a quantitative character. The method generally underestimates the number of genes. Further, the estimation is confined to the number of genes for which the two parents differ. Nevertheless, no reliable method for this purpose is presently available.

The complexity in the genetic system of quantitative characters is not only due to multiple genes located at different loci but the sets of multiple alleles at the same locus may also play role in the genetic control of these characters. In case of multiple allelism, the number of allelic combinations will be determined by the number of different alleles present in the population and will be equal to n (n+1)/2 combinations, where n is the number of alleles. Even with a moderate number of alleles at a locus, the number of genotype combinations becomes quite large. For example, with five multiple alleles at a locus, the number of genotype combinations will be 15. So, the number of genotype combinations increases drastically with an increase in the number of alleles.

Role of Environment in Quantitative Variability

Two factors affect the development of an individual, its genotype and the environment in which it has been grown. However, there should not be any controversy about the relative importance of these two factors in the control of a character since the genotype needs some environment for the development of a phenotype and similarly the environment needs some genotype with which it can interact. The phenotypes of different individuals are usually different in the same environment and the phenotype of the same individual may be different in different environments, that is, different genotypes behave differently in the same environment and the same genotype behaves differently in different environments. Thus, for determining genetic differences, different genotypes should be raised under same environmental conditions and for determining environmental differences, the same genotype should be raised under different environments. However, since genotype may be relatively more important in the control of some characters but environment for other characters, the quantitative measurement of the effect of the genotype and that of the environment for agriculturally important traits like yield, plant height and maturity may help the plant breeder a great deal in his selection experiments and deciding breeding methodology for improving a crop species.

Let us first discuss the effect of the number of genes on variability by keeping the enviroment constant. If the character is under the control of one locus only, there will be 3:1 ratio in F_2 if dominance is complete and 1:2:1 ratio if there is no dominance or there is partial dominance. With two segregating loci, with complete dominance and no dominance (or presence of partial dominance), the F_2 ratios will be 9:6:1 and 1:4:6:4:1, respectively, and with three segregating loci these ratios will be 27:27:9:1 and 1:6:15:20:15:6:1, respectively, and so on if the genes have similar effect on the character. This indicates that the number of F_2 classes will go on increasing with an increase in the number of segregating genes (the number of classes is more when there is no dominance or there is partial dominance) resulting in the reduction in the frequency intervals of phenotypic classes. When the number of loci governing a character is quite large, the F_2 distribution will take a form of normal distribution in case of no diminance. In case of complete dominance, the F_2 distribution may be somewhat skewed. However, these F_2 distributions have been calculated when the genes have similar and equal effect, there is no epistasis, linkage, thresholds, etc., no effect of environment, allelic frequencies at each locus are equal and the dominance is isodirectional (that is, all loci show dominance in one direction). In plant breeding experiments, it is very difficult to fulfil all these conditions and thus skewness of the distribution even in case of complete dominance may vanish because of the unfulfilment of one or more of these conditions. Therefore, even though there is no effect of environment, the F_2 distribution of a quantitative character controlled by a sufficiently large number of genes may be symmetrical.

Similarly,theoretical distribution in F_2 for a character controlled by only one locus under different conditions of environment may be predicted. If environmental contribution towards total variability is zero, that is, 100 per cent heritability (heritable portion of variability), the F_2 will give a 3:1 ratio in case of complete dominance and 1:2:1 ratio if there is no dominance or there is partial dominance. But as soon as the environmental variation becomes a part of the total variability, these specific ratio start vanishing and thus the correspondence between the genotype and phenotype is disturbed. This disturbance in the correspondence will increase with the increasing proportion of the total variability due to environment. If the contribution of the environment becomes more than 60 per cent, one can expect the F_2 plants showing almost a normal distribution.

In case of a character like yield which is controlled by a large number of genes and whose heritability is usually low (sometimes even not significantly different from zero), that is, both the factors responsible for changing the discontinuities of variability into continuity are operating simultaneously it is impossible to classify the F_2 population into distinct classes. Thus, simple statistical procedures cannot be applied to study this kind of variability. Furthermore, the heritability of agronomically important characters is different under different environmental conditions and so is the performance of a gentype or a variety. Depending upon the situation, some varieties may be best suited for a specific locality or region, some may perform better over a range of environments but not best in any environment, some may be good under rich environments and show stability over a range of poor environments, etc. The breeder is interested to evolve a variety which is highly responsive to controllable environments (fertilizer dose, irrigation, etc.) and has stable performance in uncontrollable environments (rainfall and other climatic conditions). For this, measurement of genotype × environment interaction by raising the variety over a range of environments becomes necessary. However, the text of this book does not allow detailed study of this subject and thus a simple procedure based on ordinary analysis of variance will be discussed here (For details see Singh and Pawar, 2005).

The following linear equation can be given to demonstrate the control of a phenotype :

$$P = \mu + g + e + (g \times e)$$

where P is the phenotype in respect of a character, μ the general population mean, g the genetic effect, e the environmental effect and $(g \times e)$ the interaction between the genotype and the environment. If the performance of genotype is independent of the changes in the environment, the $(g \times e)$ interaction will be zero. But such a situation has very less likelihood. The studies done on genotype×environment interaction indicate that occurrence of this interaction in plant breeding experiments is very common and thus it should not be ignored. The linear relationship given above may be expanded as follows if the genotypes have been grown in a replicated (r) experiment at various locations (l) for a number of seaons (s) :

$$P = \mu + g + r + l + s + (g \times l) + (g \times s) + (g \times l \times s) + e$$

where $(g \times l)$, $(g \times s)$ and $(g \times l \times s)$ are, respectively, the interactions between the genotypes and locations, genotypes and seasons and genotypes, locations and seasons and e is an error due to differences in the plots of the same replication (within replicate), sampling among plants in the same family (within family) and error due to measurement. The relevant items of the analysis, their degrees of freedom (df) and expected mean squares (EMS) will be as given in Table 7.1.

Table 7.1 : Genotype × environment interaction

Item	df	EMS
Genotypes	g-1	$\sigma^2_w + rk\,\sigma^2_{gls} + rks\,\sigma^2_{gl} + rkl\sigma^2_{gs} + rkls\,\sigma^2_g$
Genotypes × locations	(g-1)(l-1)	$\sigma^2_w + rk\,\sigma^2_{gls} + rks\,\sigma^2_{gl}$
Genotypes × seasons	(g-1)(s-1)	$\sigma^2_w + rk\,\sigma^2_{gls} + rkl\,\sigma^2_{gs}$
Genotypes × locations × seasons	(g-1)(l-1)(s-1)	$\sigma^2_w + rk\,\sigma^2_{gls}$
Error	glsr (k-1)	σ^2_w

Here, k is the number of plants scored in each family.

The significance of various items may be tested and it can be known that which type of genotype × environment interaction is more important. Also, variances due to different items of analysis ($\sigma^2_w + \sigma^2_{gls} + \sigma^2_{gs} + \sigma^2_{gl}$ and σ^2_g) can be estimated on the basis of expected mean squares. Once these observational components of variance are estimated, we can estimate phenotypic variance

$$\left(\sigma^2_g + \frac{\sigma^2_{gs}}{s} + \frac{\sigma^2_{gl}}{l} + \frac{\sigma^2_{gls}}{sl} + \frac{\sigma^2_w}{rls} = \sigma^2_p \right)$$

degree of heritability ($\sigma^2_g / \sigma^2_p = h^2$) and the magnitude of expected genetic advance ($s \times \sigma P \times h^2 = G_S$). Of course, it is better to estimate genetic advance as percentage of mean rather than in absolute units to make it comparable between different characters (For details see Singh and Pawar, 2005).

Partitioning of Quantitative Variability

The two major components of quantitative variability are genetic variance and environmental variance. The genetic variance can further be partitioned into : (a) Additive genetic variance (the variance arising from the additive effects of the genes, that is, from the difference between homozygotes at the same locus and is the chief cause of resemblance between parents and their offspring); (b) dominance variance (it arises due to the interaction between the alleles of the same locus, that is, due to intra-allelic interaction); and (c) epistatic variance (arising from the interaction between the alleles belonging to different loci, that is, non-allelic or interallelic interaction). The epistatic variance can again be divided into three subcomponents : (i) additive × additive (homozygote × homozygote) epistasis, (ii) additive × dominance (homozygote × heterozygote) epistasis, and (iii) dominance × dominance (heterozygote × heterozygote) epistasis.

·The genetic variance can be partitioned in another way into two components : (1) **fixable component** which includes additive and additive × additive components and has a unique importance in selection experiments since more is this component, more will be the effectiveness of selection, and (2) **unfixable component** which includes dominance, additive × dominance and dominance × dominance components.

However, most of the genetic procedures which are used to estimate the components of variation in plant populations, assume the absence of epistasis and are thus based on simple additive-dominance model. But this assumption may not be realistic for all the populations for the same character and for all the characters of the same population.

Estimation of Components of Quantitative Variability

The expected genetic architecture of populations in various generations can be known by considering single locus with two alleles and then generalizing the same for many genes. If A_1- is a locus with two alleles, A_1 (allele with positive effect) and a_1 (allele with negative effect), the values of three genotypes A_1A_1, A_1a_1 and a_1a_1 produced by this locus in terms of additive (d) and dominance (h) effects of genes as deviations from the mid-parent (m) may be represented as shown in Figure 7.1.

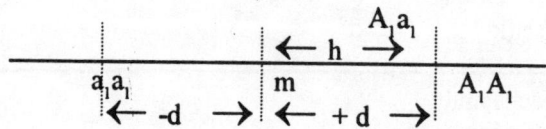

Figure 7.1. Increments of additive (d) and dominance (h) gene effects
as deviations from mid-parent for a locus

Depending upon the position occupied by the heterozygote on the scale, there may be different degrees of dominance. If A_1a_1 lies

(i) at mid point between the two homozygotes (m), there is no dominance, that is, presence of additive gene effects only;

(ii) between m and A_1A_1, there is positive partial dominance, as in the present case;

(iii) between m and a_1a_1, there is negative partial dominance;

(iv) at A_1A_1 point, there is positive full dominanc;

(v) at a_1a_1 point, it is a case of negative full dominance;

(vi) beyond A_1A_1, there is positive overdominance; and

(vii) beyond a_1a_1, it is a case of negative overdominance.

An F_2 population produced by crossing A_1A_1 and a_1a_1 homozygotes will have $1A_1A_1 : 2 A_1a_1 : 1 a_1a_1$ ratio. By putting the proportions of these three genotypes and their genotypic values, we can calculate the variance of the F_2 population.

Genotype	Proportion	Genotypic value
A_1A_1	$\dfrac{1}{4}$	d
A_1a_1	$\dfrac{1}{2}$	h
a_1a_1	$\dfrac{1}{4}$	-d

Mean = ¼d + ½h + ¼ (-d) = ½ h

Variance = ¼(d)² + ½ (h)² + ¼ (-d)² – (½ h)²

$\qquad$ = ½d² + ¼ h²

Considering many genes governing the character under consideration (so that $\Sigma d^2 = D$ and $\Sigma h^2 = H$) and including non-heritable (environmental) portion of the variance, the total variance of an F_2 population will be

$$V_{F_2} = \tfrac{1}{2}D + \tfrac{1}{4}H + E$$

The variances of first backcross generations ($F_1 \times P_1 = B_1$ and $F_1 \times P_2 = B_2$) can also be calculated in the same way. In B_1, the genotypes, proportions and genotypic values will be as under :

Genotype	Proportion	Genotypic value
A_1A_1	$\dfrac{1}{2}$	d
A_1a_1	$\dfrac{1}{2}$	h

Mean = ½d + ½h

Variance = ½(d)² + ½ (h)² – (½ d+½h)²

$\qquad$ = ¼ d² + ¼ h² –½dh

$\qquad$ = ¼(d– h)²

Similarly, in B_2, the genotypes, proportions and genotypic values will be as follows :

Genotype	Proportion	Genotypic value
A_1a_1	$\dfrac{1}{2}$	h
a_1a_1	$\dfrac{1}{2}$	-d

Mean = ½h + ½ (-d) = ½ h − ½ d

Variance = ½(h)² + ½ (-d)² − (½ h−½d)²

$\quad\quad$ = ¼ d² + ¼ h² +½dh

$\quad\quad$ = ¼(d+h)²

Therefore, in case of V_{B_1} and V_{B_2}, it is not possible to completely separate d and h effects and usually these two variances are combined together so that

$$V_{B_1} + V_{B_2} = ½ D + ½ H + 2 E$$

However, if one more parameter F (=Σdh) which is weighted sum of h's (the weights are the corresponding d's) is included,

$$V_{B_1} = ¼ D + ¼ H − ½ F + E$$
$$V_{B_2} = ¼ D + ¼ H + ½ F + E$$

The environmental variance can be calculated from the variances of non-segregating generations (P_1, P_2 and F_1) since these generations are genetically homogeneous. If the F_1 variance does not significantly differ from the variance of P_1 and that of P_2,

$$E = 1/3 (V_{P_1} + V_{P_2} + V_{F_1})$$

If V_{F_1} is different from V_{P_1} and V_{P_2},

$$E = 1/4 (V_{P_1} + V_{P_2} + 2 V_{F_1})$$

With the help of these generation variances, the magnitudes of D, H and F and their some important proportions like heritability in broad as well as in narrow sense and genetic advance can be estimated. For example, $V_{B_2} - V_{B_1} = F$ and $2 V_{F_2} - (V_{B_1} + V_{B_2}) = ½ D$. By putting the values of ½ D and E in the formula of V_{F_2}, the value of H can be known. Let us take a numerical example of grain yield (grammes) per plant where

V_{P_1} = 4.20	V_{F_2} = 25.60
V_{P_2} = 3.76	V_{B_1} = 16.40
V_{F_1} = 6.92	V_{B_2} = 21.80

and F_2 population mean = 22.50

Therefore, the estimates of the components will be

$\quad$ E $\quad$ = $\quad$ ¼ (4.20 + 3.76 + 2 × 6.92)

$\quad\quad\quad$ = $\quad$ 5.45

$\quad$ F $\quad$ = $\quad$ 21.80 − 16.40

$\quad\quad\quad$ = $\quad$ 5.40

$\quad$ ½ D= $\quad$ 2×25.60 − (16.40 + 21.80)

$\quad\quad\quad$ = $\quad$ 13.00

$$\frac{1}{4}H = 25.60 - 13.00 - 5.45$$
$$= 7.15$$

Accordingly,

$$\text{Heritability in narrow sense } (h^2_n) \quad = \quad \frac{13.00}{25.60}$$
$$= \quad 0.507$$
$$= \quad 50.70 \text{ per cent}$$

$$\text{Heritability in broad sense } (h^2_b) \quad = \quad \frac{13.00+7.15}{25.60}$$
$$= \quad 0.787$$
$$= \quad 78.70 \text{ per cent}$$

$$\text{Genetic advance} = s\sigma ph^2 \text{ (at 5\% selection intensity)} = 2.06 \times \sqrt{25.60} \times 0.507$$
$$= 5.28$$

(2.06 is the standardized selection differential at 5 per cent selection intensity and σp is the phenotypic standard deviation).

Therefore, if 5 per cent best yielding plants are selected from this F_2, their F_3 progeny should have mean equal to $22.50+5.28 = 27.78$ grammes per plant. However, genetic advance in per cent of mean provides more reliable information since different populations can be compared on this basis. In the present case, the genetic advance in per cent of mean will be $5.28\times100/22.50=23.47$.

If a random sample of genotypes has been raised in a randomized block design, the values of genotypic, phenotypic and environmental variances can be estimated through simple analysis of variance as follows (Table 7.2).

Table 7.2 : Analysis of variance (random effect model)

Item	df	Expected mean square
Replications	r-1	$\sigma^2 + g\,\sigma^2_r$
Genotypes	g-1	$\sigma^2 + r\,\sigma^2_g$
Error	(r-1)(g-1)	σ^2

Here, σ^2 provides direct estimate of environmental variance, σ^2_g, that is, MS genotypes – MS error/r, provides the estimate of genotypic variance and $\sigma^2 + \sigma^2 g$ = phenotypic variance (σ^2_p)

These estimates of components of variance can be used to obtain estimates of heritability in broad sense $(\sigma^2_g/\sigma^2_p$ if genotypes are heterozygous), heritability in narrow sense $(\sigma^2_g/\sigma^2_p$ if the genotypes are homozygous where heritability in broad sense = heritability in narrow sense) and genetic advance $(s\times\sigma_p \times (\sigma^2_g/\sigma^2_p))$.

The heritability in narrow sense, since it is based on the fixable portion of genetic variance, is more important for the plant breeder than the heritability in broad sense. Thus, for obtaining a reliable value of genetic advance, one should use the value of narrow sense heritability rather than that of broad sense heritability.

B

Principles of Breeding Cross-Fertilizing Crops

CHAPTER 8

Selection in Cross-fertilizing Crops

Basic Population Genetics Concepts

As discussed in Chapter 4, populations of cross-fertilizing crops are generally highly heterozygous and genetically heterogeneous as they are made up of freely interbreeding plants. Since all plants of a population share in a common gene pool and the transmission of genes among these plants obeys the laws of Mendel, such a freely reproductive plant community can be called as a Mendelian population. A cross-fertilizing species is an example of the most extensive Mendelian population which is usually divided and subdivided into several such (Mendelian) populations. Based on the reproductive system of the crop species and the previous evolutionary history of the population, each of such populations has a specific integrated genetic structure in terms of systems of gene frequencies and thus can be described on the basis of simple population genetics concepts. But the members of a clone, pure line or an inbred line do not constitute a Mendelian population.

Genetical study at population level become necessary because (1) the life of an individual is limited in length of time, whereas a population is practically immortal, (2) the genetic make up of an individual remains the same (ignoring mutations) throughout its life, while the genetic composition of a population may change from generation to generation which is of great evolutionary significance, and (3) most of the important characters in crop plants (grain yield, straw yield, test weight, etc.) are quantitatively inherited and since specific genetic ratios cannot be obtained for these metric traits, single individuals are usually uninformative. These points thus indicate that though the genes are carried by individuals, the fate of the different alleles is decided in populations.

Gene and Genotype Frequences

Frequencies of genes and genotypes provide basis for the genetic structure of populations. Gene frequencies may be defined as the proportions of alternative forms (different alleles) of a gene in a population. Similarly, proportions of different genotypes among the individuals of a population are termed as genotype frequencies. Let us consider an autosomal locus with two alleles A_1 and a_1.

There will be, therefore, three genotypes A_1A_1, A_1a_1 and a_1a_1. If the frequencies of the two alleles are denoted by p (A_1) and q (a_1), those of the three genotypes by d (A_1A_1), that is, dominant homozygote, h (A_1a_1), that is, heterozygote and r (a_1a_1), that is, recessive homozygote, and the total number of individuals (d+h+r) by n, the number of alleles in this population will be 2n and thus

$$p = (2d + h)/2n = (d+\tfrac{1}{2}h)/n$$

and $q = (h+2r)/2n = (\tfrac{1}{2}h+r)/n$, so that p+q =1.

For example, in a population of 40 d, 32h, 8r, the values of p and q will be .7 and .3, respectively.

If the three genotypes are given in terms of percentage (that is, d+h+r=1), the frequencies of alleles will simply be

$$p = d+ \tfrac{1}{2}h \quad \text{and}$$
$$q = \tfrac{1}{2} h + r$$

Thus, the above population will now be in the form of 5d, 4h, 1r. The gene frequencies are again the same, that is, p =.5+.2=.7 and q=.2+.1=.3.

Hardy-Weinberg Law

Hardy in England and Weinberg in Germany in 1908 independently putforward an important population genetics principle called Hardy-Weinberg law. They independently demonstrated that (1) in a large panmictic (random mating) population, the gene and genotype frequencies remain constant from generation to generation in the absence of selection, differential migration and differential mutation, (2) genotype frequencies are determined solely by gene frequencies, and (3) equilibrium frequencies ($p^2A_1A_1$, $2 pq A_1a_1$, $q^2a_1a_1$, that is, an expansion of a binomial) are attained after a single cycle of random mating irrespective of the genotype frequencies in the parent population. Therefore, there can be two or more than two populations with different genotype frequencies but with the same gene frequency. For example, populations 40d : 32h : 8r, 60d : 20h : 20r, 50d : 40h : 10r, 55d : 30 h: 15r, 48 d : 44h : 8r and 42 d : 56h : 2r all have gene frequencies .7 (p) and .3 (q) and after one cycle of random mating all (populations) will become 49 d : 42h : 9r (the equilibrium frequencies). Let us consider numerical example of a population 60 d : 20 h : 20 r (Table 8.1).

Table 8.1. Relative frequenceis of different kinds of offspring produced by the population 60d : 20h : 20r

Parents		Offspring		
Mating type	Frequency	A_1A_1	A_1a_1	a_1a_1
$A_1A_1 \times A_1A_1$	$p^4 = d^2 =0.36$	$p^4=0.36$	–	–
$A_1A_1 \times A_1a_1$	$4p^3q = 2dh =0.24$	$2p^3q=0.12$	$2p^3q=0.12$	–
$A_1A_1 \times a_1a_1$	$2p^2q^2 = 2dr =0.24$	–	$2p^2q^2=0.24$	–
$A_1a_1 \times A_1a_1$	$4p^2q^2 = h^2 =0.04$	$p^2q^2=0.01$	$2 p^2q^2=0.02$	$p^2q^2 = 0.01$
$A_1a_1 \times a_1a_1$	$4pq^3 = 2hr =0.08$	–	$2pq^3=0.04$	$2pq^3=0.04$
$a_1a_1 \times a_1a_1$	$q^4 = r^2 = 0.04$	–	–	$q^4=0.04$
Total	$p^2+2pq+q^2$	$p^2 (p^2+2pq+q^2)$	$2pq(p^2+2pq+q^2)$	$q^2(p^2+2pq+q^2)$
	$= 1.0$	$= 0.49$	$=0.42$	$=0.09$
	$=d+h+r$	$= (d+\tfrac{1}{2}h)^2$	$= 2(d+\tfrac{1}{2}h) (\tfrac{1}{2}h+r)$	$= (\tfrac{1}{2}h+r)^2$

If a population is in equilibrium and random mating occurs, its gene and genotype frequencies will remain the same in all subsequent generations.

The two characteristics of Hardy-Weinberg principle that any population (d:h:r) reaches its equilibrium proportions after a single cycle of random mating and that this eqilibrium persits thereafter until the gene frequencies are changed, can be proved by a much shorter method, that is, random union of gametes, since complete random union of all the gametes produced by the population is equivalent to the total result of random mating among members, and the subsequent random union of gamates produced by the mates.

Males

	p (A$_1$)	q (a$_1$)
p (A$_1$)	p^2 (A$_1$A$_1$)	pq (A$_1$a$_1$)
q (a$_1$)	pq (A$_1$a$_1$)	q^2 (a$_1$a$_1$)

Females

However, single generation of random mating is not enough to bring zygotic equilibrium in a population segregating for two loci. The situation becomes increasingly complex if there are many segregating loci and thus in such populations an equilibrium point is reached only after long continued random mating. Of course, absence of close linkage helps in bringing an equilibrium point much more rapidly in populations segregating for two or more loci.

Broadly, two kinds of processes, namely, **systematic** (directional) and **dispersive** (non-directional), affect gene frequencies and consequently genotype frequencies in populations. Mutation, selection and migration come under the systematic processes where the action in changing the frequencies of alleles is dependent upon the gene frequencies themselves and thus both the magnitude and the direction of change are predictable. The effect of the dispersive process is, however, independent of the existing gene frequences in the population and thus the change in gene frequencies is predictable in magnitude but random in direction.

Effect of Mutation

Mutations affect gene frequencies in populations directly as well as indirectly since they themselves act as a force in changing the gene frequencies and also by enhancing genetic recombination and by increasing genetic variability, they provide the working material for other forces (like selection) to act upon it. The effect of mutation, however, depends on its kind (reversible or non-reversible) and occurrence (recurrent or non-recurrent). If there is a single mutation, the chance of survival of the mutant allele is very small since the probability of elimination of such an allele increases with each generation from the time of the occurrence of mutation. If the mutations are recurrent and non-reversible (say the allele A$_1$ goes on mutating to a$_1$), the frequency of the mutant allele will increase with each generation until it completely replaces the wild allele. The frequency of the wild allele will reduce to $p(1-m)^n$, where p is the initial frequency of the wild allele, m the rate of mutation and n the number of generations. However, it is not necessary that mutations occur in only one direction but reverse mutations can also take place. The net amount of change in the frequency of the mutant allele (Δ_q) will be $m_1p - m_2q$, where p and q are, respectively, the initial frequencies of the wild (A_1) and mutant (a_1) alleles, m_1 the rate of mutation of A_1 to a_1 and m_2 the rate of mutation of a_1 to A_1. This relationship indicates that the value of Δ_q will be larger if the differene between the frequencies of the two alleles is more and the frequency of the abundant allele will decrease rapindly. However, as the difference between two gene frequencies becomes narrower, the value of Δ_q also goes down. A point may come when $m_1p = m_2q$ or $\Delta_q = 0$, that is, a **mutational equilibrium** between the frequencies of A_1 and a_1. However, the mutational equilibrium which is an equilibrium of the gene frequencies themselves under opposing mutation pressures and thus basically more stable, must not be confused with the Hardy-Weinberg equilibrium which refers to a relative stability with regard to the genotype and phenotype frequencies in populations on the basis of assigned gene frequencies.

The effect of mutation on population structure may be summarized as : (1) Mutations with normal rates (that is, 5×10^{-5} or less) may be important from evolutionary point of view but have little effect on the gene frequencies and, therefore, such mutations have a very minor or no importance in plant breeding experiments; (2) backward (reverse) mutations are mostly less frequent than forward mutations (molecular basis); and (3) mutational equilibrium frequency can be altered by inducing a change in the mutation rate through some mutagenic agent unless that agent has a proportionate effect on the two rates (forward and backward) of mutations.

Efect of Selection

The frequency of any allele in a natural population largely depends on the fitness (adaptive value, selectine value, relative reproductive success or relative ability of any genotype to produce surviving offspring) of its carriers. Fitness, in turn, is dependent upon selection which includes a number of simple and complex mechanisms. Different genotypes generally show varying degrees of fitness at the same location and the fitness of the same genotype is different at different locations. The selection coefficient (the force acting against genotype to reduce its fitness) is inversely proportional to fitness, that is, fitness=1 – selection coefficient. The value of fitness usually varies between zero and unity except in case of overdominance where it may be more than unity. Further, fitness is different from dominance and a dominant allele may have even zero fitness.

The effect of selection on population structure is of primary interest to the plant breeder. It may affect the individuals of the population at gametic stage or at zygotic stage. If it affects any genotype at the gametic stage, the recessive allele cannot be shielded by the dominant allele and thus if the fitness of the allele is zero it will be eliminated from the population in a single generation. But if the selection coefficient (s) is less than unity, $\Delta_q = (sq^2 - sq)/(1-sq)$. Same will be the fate of the dominant allele if selection acts against it.

The change in gene frequency at the zygotic stage depends not only on the initial frequency of the allele but also on the condition of dominance, selection and lethality and thus several situations may arise. If there is absence of dominance, selection acts against the allele a_1 and a_1 is not lethal in homozyzous condition, the relative fitness of the three genotypes A_1A_1, A_1a_1 and a_1a_1 will respectively be 1, 1-s and 1-2s. (where s is the selection coefficient) and Δ_q will be $(sq^2 - sq) / (1-2sq)$. Further, if dominance is full, selection acts against a_1 and a_1 is lethal in homozygous condition, the relative fitness of the three genotypes will respectively be 1, 1 and 0 and Δ_q wil be $(-q^2)/(1+q)$. Nevertheless, in case of overdominance (that is, heterozygote is superior to both the homozygotes), the relative fitness of the three genotypes will respectively be $1-s_{A1}$, 1, and $1-s_{a1}$ (where s_{A1} is selection coefficient of A_1 allele and s_{a1} the selection coefficient of a_1) and Δ_q will be $[(s_{A1}p - s_{a1}q)pq]/(1-s_{A1}p^2-s_{a1}q^2)$. The value of Δ_q in this case may become zero if $s_{A1}p = s_{a1}q$, that is, the population achieves equilibrium. At equilibrium, therefore, $p=s_{a1}/(s_{A1}+s_{a1})$ and $q = s_{A1}/(s_{A1} + s_{a1})$. One conclusion can be safely drawn here that p and q will reach a stable equilibrium if the values of s_{A1} and s_{a1} are constant. If the frequency of any allele departs from its equilibrium value, it will be forced back by selection pressure and this is how, the selection preserves genetic variability in natural populations. Such a stability in the values of s_{A1} and s_{a1} and consequently in the frequencies of the two alleles has a great evolutionary significance as it results in a balanced polymorphism. Similarly, several other situations may arise.

It is not always true that selection is most effective in first few generations and its effectiveness decreases in later generations, rather the effect of selection in changing gene frequency is substantial when value of q is intermediate, that is, under constant selection (s fixed) the effectiveness of selection is maximum for characters that are common and minimum for those that are rare in populations. This is why selection is quite effective in reducing the gene frequency of recessive alleles in Mendelian populations but is ineffective to eliminate than completely as recessives are shielded by dominant alleles in heterozygotes.

Most of the characters of plant breeder's interest are quantitative in nature and thus show continuous variation. Since such characters are governed by many genes, they require different methodology for describing and analysing data than do the qualitative characters. Selection intensity for such characters can, however, be assessed by calculating selection differential (mean deviation between the mean of selected plants and the mean of the original population) expressed in standard deviation units. Such selection differential is more helpful for breeder than that with the actual units of measurement as it is easily transformable into percentage of individuals saved for breeding and it also allows comparison of selection intensities between different characters or between different populations.

If a character is governed by additive genes only, improvement in that character is expected to continue till fixation of all desirable alleles. Contarily, if a character is governed by non-additive genes (dominance, epistatic or both), progress due to selection may decrease rapidly, become unpredictable and at a stage, selection may become virtually powerless. In fact, the present knowledge about the effect of epistasis on the progress of selection is not enough to give specific recommendation.

The theoretical limit of progress through selection can be determined by population genetics concepts. However, the probability of this limit is attributable to several reasons and is usually very small. In addition to non-additiveness of genes, the rate of progress towards the theoretical limit is influenced by some practical and theoretical factors (degree of heritability, selection intensity permissible under different reproductive biologies of various species, number of characters simultaneously considered by the breeder, type and degree of correlation between these characters, etc.). The number of characters for which selection is practiced is inversely proportional to the intensity of selection possible for any one character.

Simultaneous Action of Mutation and Selection

Each of the two pressures, namely, mutation and selection, if operates unhindered, will ultimately cause either fixation or complete elimination of the allele in question. In nature, however, it is not generally the case but mutation and selection act simultaneously. The conclusion drawn on the basis of one of these factors only, therefore, will be misleading. The fixation or elmination of an allele will be faster if both the factors exert their pressures in the same direction. However, if they oppose each other, the population may reach a stable equilibrium because of the cancelling effects of the two processes. As far as the relative effectiveness of these two factors is concerned, it depends on the gene frequency but in different ways. Whereas mutation is most effective when the mutant allele has low frequency, selection is most effective when the allele is abundant in the population. A population under joint effects of mutation and selection will attain equilibrium if the allelic gain through incoming mutations equals to the allelic loss on account of selection.

Effect of Migration

The change in the gene frequency due to migration will mainly depend upon : (1) the difference in the gene frequencies between the donor (migrant) and the recepient populations, and (2) the proportion of alleles that is being received by the hybrid population each generation. If there is no change in the population size, the gene frequency in the recepient population and the difference in gene frequency between recepient and donor populations after n generations will respectively be (q_{rn-1}) $(1-m)$ + mq_d and (q_r-q_d) $(1-m)^n = q_{rn}-q_d$, where q_r is the initial gene frequency in the recepient population, q_d the frequency of the same allele in the donor population, m the proportion of newly introduced alleles each generation and n the number of generations.

Effect of Dispersive Process

The consequences of the dispersive process are mainly three :

(1) It tends to divide the original population into smaller groups (populations) because there is a more tendency of mating between individuals which are more close to one another.

(2) Since individuals belonging to the same group become more and more alike, genetic variability within group is reduced.

(3) Homozygotes are increased at the expense of heterozygotes.

To have a better understanding of the dispersive process and its consequences, one must study the process from two viewpoints :

(i) To consider it as a samplong process, that is, the frequencies of different alleles in infinitely large random mating populations remain constant from generation to generation (ignoring effects of directional forces), while the same (gene frequencies) drift drastically in finite populations since such populations will mostly be associated with large standard deviations (chance or random derift), and

(ii) to consider it as an inbreeding process, that is, even though there is random mating among individuals of an effectively small population, more consanguineous matings will take place than in a sufficiently large population.

Random Drift (a restriction of the population size)

Although the idea of random drift was firstly given by Brooks (1899), precise information about the process as random fluctuations in gene frequencies in finite populations was provided especially by the work of Wright (1921, 1931a,b, 1932). Later on, Wright (1948, 1949), Li (1948) and Lerner (1950) discussed the process in more detail. Since random drift can alter the gene frequencies drastically in effectively small populations and can thus disturb the Hardy-Weinberg law, the study of the process becomes more important for the plant breeders as they have to handle such populations more frequently.

Importance of random or genetic drift as a sampling process can be easily proved by taking into account two hypothetical populations, one consisted of 4 parents and the other of 4000 parents. If the locus in question is represented by two alleles A_1 and a_1 with their respective frequencies p and q, the standard deviation of proportion (σ) for diploid parents will be $\sqrt{pq/2n}$, n being the number of actual parents used to start a generation. Further, if p=q=0.5, $\sigma = \sqrt{0.5 \times 0.5/8} = \sqrt{0.03125} = 0.1767$ for the first population. For the second population, $\sigma = \sqrt{0.5 \times 0.5/8000} = \sqrt{0.00003125} = 0.0055$, that is, the gene frequency in the second population will fluctuate mostly between 0.4945 and 0.5055 (0.5±0.0055), while the same will fluctuate between 0.3233 and 0.6767 (0.5±0.1767) in the first population indicating that if the same small size is continued in each generation, the population will ultimately reach a point of no return (since gene frequency cannot go beyond 0 or 1). It means that the next sampling will lead to a further dispersion as each line in the second generation will start from a different gene frequency. However, if the fixation has reached, it does not mean that all the lines will be fixed for the same allele but the fixation will be according to the initial gene frequencies if we consider several such small populations together.

As such drift is usually beyond the control of the breeder. However, the breeder may enhance or reduce the rate of chance fixation by exerting selection pressure and/or by increasing or decreasing the population size.

Inbreeding Effect of a Small Population

Random mating in an effectively small population is nothing but matings between relatives. Thus, inbreeding becomes an important factor affecting the genetic composition of such (finite) random mating populations. If all gametes of the parents unite completely at random (including selfs), the rate of fixation (loss or decay of heterozygosity) per generation because of this inbreeding will be 1/2n, where n is the number of actual parents.

However, if there is presence of self-incompatibility in monoecious plants (that is, no self-fertilization) or the population is consisted of dioecious parents with equal numbers of males and females, the formula for the rate of fixation takes the form 1/2n + 1. In large populations, 2n and 2n+1 will be approximately equal.

One conclusion can be easily drawn from the above discussion that whereas fixation proceeds very slowly in large populations, its rate in small populations is significantly high. This indicates that population size must be an important consideration in plant breeding experiments. However, the effective number of parents is not always equal to the total number of individuals in the population. For example, the numbers of males and females of a population may not be necessarily equal. Such a situation can be found in plant breeding experiments where each of a large number of lines is sometimes crossed with a few male testers or vice versa (that is, a few male sterile parents may be crossed with a large number of pollen parents). Also, in animal breeding, several dams are usually crossed to the same sire. Effective population size under such a situation can be estimated on the basis of the effective population size, the number of females and the number of male parents (For details see Allard, 1960).

Patterns of Response to Selection

As has been discussed earlier, the progress under selection is determined mainly by the number and nature of genes controlling the character, the degree of heritability of the character and in some cases by physiological factors. A character may be governed by a simple genetic system or by a complex system. Further, a character may be controlled mainly by genetic factors (high heritability) or mainly by prevailing environmental conditions (low heritability). Allard (1960) has very lucidly described various patterns of response to selection by giving examples of each type. In fact, these patterns depend upon the type of character under study. Based on the results of a number of experiments carried out in plant and animal species for various characters, the common patterns of response to selection according to him fall in the following five categories :

High response in the first few generations followed by a prolonged period of slow response : This type of response is shown by quasiqualitative characters which are governed by few major genes and some minor genes. In the control of such characters (resistance to certain diseases, plant height, colour, etc.), there is a small number of genes which have quite different magnitudes of effects on the phenotype. Selection causes large increase in the frequencies of these genes resuling in rapid progress in early generations. This rapid progress continues till these genes are fixed almost completely. Nevertheless, due to further small changes in the frequency of major genes and slow increases in the frequencies of minor genes, a slow progress under selection continues to occur for a long period of time.

In case of quasiqualitative characters, selection shifts population mean in the desired direction resulting in higher magnitude of the phenotype (increased resistance, more intense colour, etc. than the unselected base population) and appearance of transgressive segregants. Continuous selection in a particular direction will result in the near fixation of major genes and ultimately in the reduction of genetic variability in the population.

Continued slow response for a long period : The best examples of this type of response come from selection experiments conducted for high oil content and high and low protein content in maize. These characters are governed by complex genetic systems and each gene has a small effect on the phenotype. In such cases, the rate of change in gene frequency is small leading to slow fixation of alleles, gradual shift of the population mean, appearance of transgressive segregants and little or no reduction in genetic variability in the population.

Selection for high oil content in an unselected base population of maize (Burr White stock) caused an increase in the oil in the first selected generation by changing the proportions of differnt genotypes but no transgressive segregants appeared in this generation. The appearance of a few new genotypes began in the second and third selected generations. Further selection resulted in a steady increase in the mean oil content and after ten generations of selection the selected population became distinctly different from the original population. A similar slow, steady and long-continued response until fifty generations of selection was observed for both high and low protein content in maize.

Selection for oil and protein content (chemical composition) in maize also had some side effects and led to changes in the characteristics of the cobs, kernels and plants and with the advancement of selection these changes became progressively more pronounced. The side effects on these characteristics at the advanced stage of selection (by the end of fifty generatins of selection for protein and oil content) were so severe that the selected strains had highly distinctive plant types in terms of cob and kernel type and the yield potential. The selected strains had approximately one half of the yield as compared to that of adapted hybrids.

Slow response ending in a plateau : Selection for low oil content in maize can be taken as an example of this pattern of response. The trend of response upto 25 generations was the same (of course, in opposite direction from selection for high oil content) as for high oil and high and low protein content. But after about 25 generations, the selection for low oil content became almost completely ineffective probably because of a strong association between oil content and the germ size and it is rather impossible to reduce the size of germ considerably. This halt in the reduction in oil content may be attributed to a physiological threshold as it was not apparently due to the fixation of alleles. That is, despite presence of genetic variability for oil content it was not possible to reduce the oil content below a certain point. It is probable that a similar physiological threshold may be operating for high oil and high and low protein content after a stage of 50 generations of selection.

No response to selection : Early maize breeders concluded that mass selection and some of its variants (particularly ear-to-row method) were virtually powerless in improving yield in adapted varieties of this cereal. Of course, it was conceded by the breeders that the usefulness of ear-to-row selection in improving the yield of unadapted maize varieties cannot be completely denied. Probable reasons given for the lack of response to selection for yield in adapted varieties were :

(1) Limited additive genetic variability for yield in maize populations;

(2) only a small portion of the total additive genetic variance can be exploited by ear-to-row method; and

(3) low heritability of yield.

Nevertheless, the maize breeders later on realized that there is no dearth of additive genetic variance in open pollinated varieties of maize. Therefore, the first reason had no sound basis and cannot be as such accepted.

Further, ear-to-row method is inefficient both as a plant breeding scheme for improving productivity because it is essentially a type of mass selection and as a genetic technique since it takes care of only 1/8th portion of total additive genetic variance (equal to half sib covariance). Thus, it is not justified to ascribe the failure of this method to improve yield solely to its genetic limitations.

Low heritability of yield was perhaps one important cause of the failure of ear-to-row method for improving productivity in maize. In fact, for accurate measurement of yielding ability of ear rows and drawing worthwhile conclusions, it is necessary to conduct multilocation and multi-year replicated experiments. But at the time when ear-to-row method was used to evaluate the yielding ability of ear rows, the measurements were based on the yields of single plots resulting in low heritability of yield in such ear-to-row breeding programmes. Hence failure of mass selection in bringing desired improvement in maize productivity is more due to the inability to identify superior genotypes than to the presence of limited additive genetic variability for yield. Rather use of recurrent selection in improving maize populations and identifying outstanding inbreds from heterogeneous source material provided a better explanation (based on population genetics concept) for the failure of mass selection to improve yield in maize. In fact, the frequency of desirable alleles in the samples taken from heterozygous source population was generally very low. However, the proportion of favourable alleles can be effectively increased by interbreeding of selects (selective intermating).

High response-plateau-high response ending in another plateau : The experiment conducted by Mather and Harrison (1949) in *Drosophila melanogaster* for abdominal chaetae number of more than 100 generations provides valuable information regarding the storage and release of genetic variability in Mendelian populations. They described the results of their exepriment on the basis of quantitative charaters governed by many tightly linked genes (with plus and minus effects) scattered along the chromosomes. In such situation, a large part of the variability generally remains hidden in the heterozygotes as potential variability and only a part of this variability is freed by recombination by turning heterozygotes into homozygotes in the next generation. There may be homozygotic potential variability also if the genes have repulsion and opposition effects. For example, if a character is governed by two genes $A_1A_1\ a_2a_2$ and $a_1a_1\ A_2A_2$, non-extreme homozygotes have counterbalancing effects. There is, therefore, presence of potential variability in these two homozygotes. To release such variability, it is necessary to first convert homozygotic variability into heterozygotic variability and then to convert this heterozygotic potential variability into free variability by producing extreme homozygotes. The degree of fitness would depend upon the number of heterozygous loci and the number of crossovers required for total release of the potential variability (that is, the type of the arrangement of genes with positive and negative effects). For example, if a character is governed by six genes, different arrangements of genes may be given (Fig.8.1).

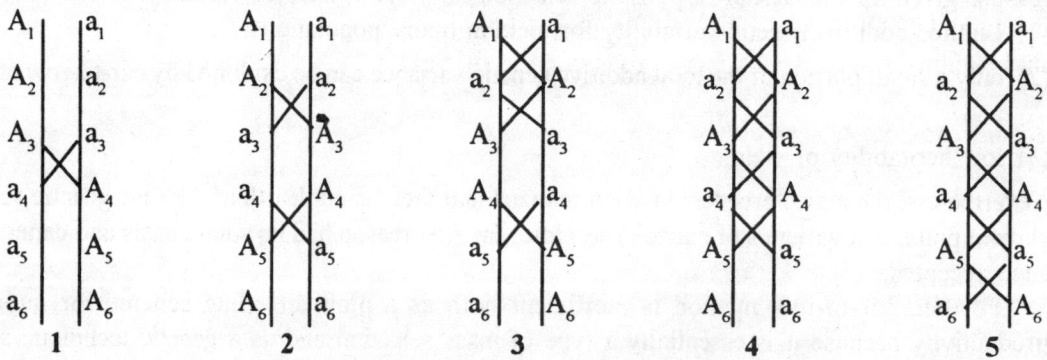

Figure 8.1. Different arrangements of six genes

Whereas, the first arrangement required only one crossover to obtain $A_1A_2A_3A_4A_5A_6$ and $a_1a_2a_3a_4a_5a_6$ extreme combinations (that is, for the total release of potential variability), the second, third, fourth and fifth arrangements would require two, three, four and five crossovers, respectively, to produce these two extreme combinations. Since, the fifth arrangement requires maximum number of crossovers for releasing total potential variability, it will be superior to other arrangements in terms of immediate fitness.

Important features of Mather and Harrison's experiment may be summarized as follows :

1. The base population had enough free and potential variability for abdominal chaetae number.
2. There was high response to selection for about 20 generations because of the free variability present in the population.
3. Continued selection for increased number of chaetae caused severe reduction in reproductive capacity of flies.

4. Suspension of selection restored reproductive capacity but resulted in a plateau for chaetae number. Two types of evidence can be given for this plateau in bristle number :
    (a) Suspension of selection could cause only a slight reduction in chaetae number.
    (b) Selection became virtually ineffective for a long period of time for increasing chaetae number and chaetae number remained almost constant for about 50 generations as the effect of directional selection were counterbalanced by the opposing force of natural selection.
5. After this plateau period, there was again a rapid gain in bristle number presumably because of the occurrence of crossovers between genes, that is, due to the change of hidden variability into free variability.
6. This gain culminated in yet another plateau after the segregation of genes had exhausted.
7. Since many genes governing different characters are situated on the same chromosome, selection for one character would affect the response of other characters to selection. Therefore, response of any character to selection should not always be considered as an independent activity of that character.

Mather and Harrison, like Fisher (1930) and Wright (1951), believed that natural selection tends to maintain a harmonious relationship among all fitness determining characters by favouring individuals that cluster about mean values of all characters and thus counterbalancing effects of directional selection aimed to change specific characters rapidly. The individuals of a population that exhibit such a harmonious combination of all quantitatively varying characters have highest adaptability in the prevailing environment. According to this assumption, natural selection acts against extreme phenotypes and such phenotypes have least chances of survival in population. More the genetic variability in population, more are the chances that such population will be able to face the consequences of long term evolutionary changes. For both immediate fitness and long range flexibility, potential variability plays an important role.

Genetic Structure of Mendelian Populations

Two hypotheses, namely, substitution of more favourable alleles for less favourable ones and genetic homeostasis, have been putforward to explain the genetic structure of Mendelian populations.

Substitution of more favourable for less favourable alleles : According to this classical concept of population genetics, allele is the basic unit for studies of populations and population is a pool of individual genes. This hypothesis also assumes that the substitution of more favourable for less favourable ones has major significance in the evolutionary process and, as a consequence of this substitution, individuals of a Mendelian population should be homozygous at most of the loci and only some loci should be in heterozygous condition since the main function of selection according to this hypothesis is the simple fixation of genes. The variability in populations is stored due to the development of adaptively neutral phenotypes by alleles, shielding of deleterious recessive alleles by dominant alleles in heterozygotes, sequential favour of multiple alleles in fluctuating environment leading to balanced polymorphism, and due to non-displacement of unfavourable alleles by rare desirable mutants. Nevertheless, this hypothesis has been criticized by many investigators who say that though population is composed of individuals but it has some properties like genetic cohesion (governed by self-regulating complex integrated system) which is not a property of its component units.

Genetic Homeostasis

Lerner (1953, 1954) gave the term 'genetic homeostasis' for **the resistance displayed by Mendelian populations to modifications of individual characters by selection.** The conservatism of Mendelian populations is ascribable to the development of organized systems (gene pools) with complex integrative properties. The formation of linked gene complexes plays a great role in the development of these integrated systems. In fact, the fate of a single gene or a gene group in such gene pools under selection is neither simple nor independent rather it depends mainly on the effect of that gene or gene group in combination with other genes or gene groups present in the gene pool of the population. Contrary to the assumption in the first concept of population structure described above, the effect of selection on Mendelian populations according to Lerner is not only confined to the fixation of alleles but it affects the overall pattern of the gene pool of the population. Such systems provide maximum fitness to the population as they have been built up after facing consequences of selection over a long period of time. An individual is less important than the population except where individual superiority makes important contribution towards general fitness of the population. These systems in due course of time become very well balanced, self-regulating and generally show great resistance if any effort is made to destroy their overall balance by modifying individual characters through any type of selection.

It is not wise to consider the two hypotheses discussed above as mutually exclusive nor it is easy to suggest a general genetic system for all kinds of population structures found in different organisms. In fact, the spectrum of population structure is very wide and different genetic systems including the two discussed above may be operating in different organisms. Further, since our knowledge about the reproductive biology of populations is very limited, some modified or quite different genetic model(s) may be needed to explain the whole situation.

CHAPTER 9

Mating Systems

In addition to selection (decision regarding discrimination among individuals to be used as parents and the relative number of offspring produced by the parents in the next generation, that is, relative reproductive success of different individuals used as parents), mating system (the manner in which the crosses are made among parents on the basis of ancestral or phenotypic relationship) is another important thing which the plant breeder can do for improving the genetic pattern of Mendalian populations. Whereas selection can considerably change the genetic value of populations by changing gene frequencies and the frequency of different gene combinations, mating system can be used as a powerful tool for producing homozygous lines, developing an extreme phenotype, maintaining source populations with maximum stability or for progeny testing.

Basically, mating systems can be divided into two types : (1) random mating and (2) non-random mating. The latter can further be divided into four types (a) genetic assortative mating, (b) phenotypic assortative mating, (c) genetic disassortative mating, and (d) phenotypic disassortative mating. In non-random mating, the matings are like-to-like (genetic assortative mating on the basis of ancestry or phenotypic assortative mating on the basis of phenotypic appearance, that is, phenotypic resemblance in this case) or between unlikes (genetic disassortative mating based on the ancestry or phenotypic disassortative mating based on the phenotypic appearance, that is, phenotypic distinction in this case). Phenotypic assorative mating may be positive if the matings are positive × positive or negative if matings are negative × negative.

Random Mating

Random mating, where individuals are mated at random, has two theoretical requirements :
(1) That there is no discrimination among individuals of the population regarding the chance in producing offspring, and,
(2) that every female gamete has equal chance for fertilization by every male gamete.

However, these requirements are hardly fulfilled by plant breeding experiments since selection is almost invariably done in these experiments and since plant materials used in these experiments mostly are not Mendelian populations in true sense due to their effectively small size. If selection is practiced during random mating then this mating system should be called as random mating with selection rather than random mating. Further, every female gamete of the population may not get equal chance to be fertilized by every male gamete due to several factors like dichogamy (protandrous or protogynous condition), genetic incompatibilities, male sterility or somatoplastic sterility and wind direction.

Due to limited resources (land, labour, time, etc.) available to the plant breeder, the population size in plant breeding experiments is mostly effectively small. In such situation, it is imperative to consider two factors, namely, genetic drift and inbreeding. Genetic drift causes drastic fluctuation in gene frequency (ultimately leading to fixation of alleles) and random mating in a small population leads to inbreeding. These two factors and their effects on the composition of populations have been discussed in detail in Chapter 8.

Genetic consequences : If the two theoretical requirements of random mating are fulfilled, no change in population with regard to gene frequency, amount of genetic variability, genetic correlations among individuals and degree of homozygosity and heterozygosity is expected.

If selection is combined with random mating, the expected effect on the genetic composition of the population will be according to the combined effect of random mating and selection as discussed in chapter 8. There will be a shift in the gene frequency and mean of the population in the desired side (that is, direction of selection). But the population variance, homozygosity or genetic correlations among chose relatives are not expected to change.

Furthermore, if the population size is effectively small, genetic drift and inbreeding both start operating and bring changes in the gene frequency and homozygosity of the population.

Genetic Assortative Mating

In genetic assortative mating, the matings among individuals are based on the ancestral relationship, that is, genetically related individuals are mated together. Therefore, this mating system is basically a form of inbreeding (matings between closely related individuals). Selfing (an extreme form of inbreeding), inbreeding or sib-mating (mating between sibs) ultimately fragments the population into small homozygous groups by fixing the genes. But since the fixation of genes under genetic assortative mating is independent of the effects of environment, degree of dominance, epistasis and degree of heritability, the genes governing characters with low heritability can be fixed equally easily as those governing characters with high heritability. Important characters like grain yield, fruit yield and total biomass which are difficult to improve because of low heritability can be brought into homozygous condition under genetic assortative mating. Since genetic assortative mating is the mating of genetically similar individuals, it is the most powerful phenomenon to increase probability of transmission of the same genes from the parents to offspring and thus tends to increase percentage of homozygosity at the cost of reducing heterozygosity.

The expected increase in homozygosity (or decrease in heterozygosity) under various types of genetic assortative mating can be determined by using Wright's Coefficient of Inbreeding or F. This coefficient is based on the genetic distances (number of generations) of the male and female parents from their common ancestor and the inbreeding coefficient of the common ancestor and can be given as follows :

$$F_y = \Sigma \ c(\tfrac{1}{2})^{d_1+d_2+1}(1+F_c)$$

where F_y is the individual whose inbreeding coefficient is to be determined, c the common ancestor, d_1 the genetic distance between the male parent and c, d_2 the genetic distance between the female parent and c and F_c is the inbreeding coefficient of common ancestor. Both d_1 and d_2 are zero in self-fertilization and thus the above formula in this situation reduces to $F = \tfrac{1}{2} (1+F_p)$, where F_p is the inbreeding coefficient of the preceding generation. Self-fertilization causes rapid increase in the value of F and makes it equal to 1 as the degree of heterozygosity reaches zero. But as the closeness of matings decreases, the rate of increase in F also decreases and it reaches zero if there is random mating.

The coefficient of inbreeding is not only a measure of the probable degree of homozygosity but can also provide information about population structure. If F>0, the zygotic proportions of a population under genetic assortative mating will be

$$(p^2+ Fpq) \ A_1A_1 : (2 \ pq- 2 \ Fpq) \ A_1a_1 : (Fpq + q^2) \ a_1a_1 =1$$

If F=1, the above equation becomes

$$(p^2+ pq) \ A_1A_1 : (0) \ A_1a_1 : (pq + q^2) \ a_1a_1 =1$$

that is,

$$p(p+q) \ A_1A_1 : q(p+q) \ a_1a_1 =1$$
$$p \ A_1A_1 : q \ a_1a_1 =1$$

It means, the proportions of two genotypes A_1A_1 and a_1a_1 become according to the gene frequencies of A_1 and a_1, respectively.

If F=0, the proportions of the three genotypes A_1A_1, A_1a_1 and a_1a_1 will be as in case of a random mating, that is,

$$p^2 \ A_1A_1 : 2pq \ A_1a_1 : q^2 \ a_1a_1 =1$$

Genetic consequences : Though the variance within groups (families) sharply decreases and ultimately reaches zero, genetic assortative mating without selection increases total variability of the population. Of course, at an advanced stage of genetic assortative mating, the variability of the population is entirely ascribable to the between families variability. But inbreeding with directional selection tends to decrease variability and in an extreme situation when a single line is retained at the end of the programme, the total genetic variance will become zero.

Genetic correlations are closely and directly related to the individual's prepotency which in turn depends on the degree of homozygosity, dominance, linkage and epistasis. Further, since homozygosity is the most important factor affecting prepotency and homozygotes are highly prepotent, the genetic correlations between relatives will increase with or without selection as genetic assortative mating increases homozygosity.

Phenotypic Assortative Mating

In this mating type, phenotypically similar individuals are mated together. The desirable parents used in this mating system may have a positive value (high yield, more number of days to flower, more number of days to mature, high gossypol content, etc.) or a negative value (less number of days to flower, less number of days to mature, low gossypol content, etc.). Since matings will be between similar individuals, this system may be subdivided into positive (if matings are positive × positive) and negative (if matings are negative × negative) types. The genetic effect of this mating system depends mainly on the number of genes governing the character and the extent to which the character is affected by the prevailing environmental conditions (that is, the degree of heritability). If the character is solely governed by genetic factors (that is, heritability is 100 per cent), the effect of phenotypic assortative mating is almost similar to that of genetic assortative mating and ultimately there will be complete fixation of genes since in this situation there is possibility of perfect assortative mating. If the character is governed by a single major gene or by a few genes and there is no effect of environment (that is, no errors of classification), the fixation of genes occurs very speedily and homozygosity is achieved in a single or few generations. But if the character is governed by many genes as is the case with most of the characters economically important to the plant breeder, the speed of the fixation of genes, especially the fixation of extreme types, is greatly slowed down. However, so long heritability is 100 per cent, there is possibility of perfect assortative mating leading to complete fixation of genes. Further, if the character is governed by many genes and there is considerable effect of environmnent, the rate of fixation of genes becomes extremely slow and population generally comes to an equilibrium before the extreme types are completely fixed.

Dominance is the third factor which can affect the rate of fixation of genes and like environmental effect can obscure hereditary values. For example, in case of complete dominance, the dominant homozygote and the heterozygote are indistinguishable and thus such dominance hinders identification of extreme genotypes. Of course, under such a situation, the extreme genotypes will be fixed but with a slower rate.

Genetic consequences : In the absence of dominance and environmental effect, the proportion of extreme phenotypes increases speedily at the cost of decrease in the proportion of intermediate phenotypes. There is an increase in total variability if extreme phenotypes are maintained but within family variability decreases. Mating of extreme phenotypes in the absence of dominance and environmental effect leads to complete fixation of extreme homozygotes in a single generation if the character is governed by a single pair of alleles or a few genes. Complete absence of environmental effect allows perfect assortative mating which in turn leads to perfect genetic correlations between close relatives since this type of mating is a powerful tool in increasing extreme diversity in a population.

Genetic Disassortative Mating

Here, matings are between genetically dissimilar (unlike) individuals (that is, cross pollination or outbreeding) to produce varietal or species crosses. Whereas crossing of genetically related individuals results in decreased vigor, reduced fecundity, more defectives, general weakening of the progeny and separation of population into distinct lines, crossing of genetically unrelated individuals (genetic disassortative mating) restores vigour and resistance to diseases and insect pests. Hybrid varieties are developed by crossing genetically dissimilar inbreds, clones or other populations. Such varieties can be evolved by hand emasculation and crossing (cotton), detasselling of female parent (maize), genetic emasculation by male sterility (tomato and peppers), emasculation by cytoplasmic-genetic male sterility (pearl millet, sorghum and onions), using self-incompatibility (cabbage), etc. Genetic disassortative mating is the best method for maintaining diversity in population and holding populations together.

Genetic consequences : This system decreases population variance, reduces genetic correlations between relatives and maintains heterozygosity in a population. It may increase heterozygosity if a character is highly heritable and is governed by a few genes.

Phenotypic Disassortative Mating

Mating of phenotypically distinct individuals is called as phenotypic disassortative mating (positive×negative mating). It is the most conservative of the mating systems and is used to reciprocally compensate the weaknesses of the two parents or to develop an intermediate type from two extreme parents in opposing directions. It is also the best method for holding populations together and maintaining diversity in populations. The best example of this mating system is disruptive selection.

Genetic consequences : Genetic consequences of this mating system are similar to those of genetic disassortative mating.

No mating type has absolute superiority over any other type, rather advantages/disadvantages of a mating system chiefly depend upon the goal of plant breeder. For example, if progeny testing is the goal, random mating is more appropriate. Genetic or phenotypic assortative mating types should be preferably applied if the purpose is to develop homozygous lines. Similarly, assortative mating with selection is useful for the development of an extreme phenotype and disassortative mating is more appropriate to develop populations with maximum stability.

CHAPTER 10

Pollination Control Systems and Heterosis

As has been discussed in Chapter 4, there are a number of systems (dioecy, monoecy, protandry, protogyny, male sterility, incompatibility, etc.) which partially or completely control pollination in crop plants. Of these, protandry, protogyny and monoecy though encourage but are not efficient systems for enforcing cross fertilization because the plant species falling under these categories show varying degrees of self-fertilization. Similarly, there are only a few crop plant species which are strictly dioecious. For example, hemp and spinach which are generally considered to be dioecious, present a wide range of proportions of staminate to pistillate flowers on each plant. Further, almost all different forms (dioecious, gyno-dioecious, andro-dioecious, monoecious, hermaphrodite, etc.) occur in papaya, a normally dioecious species. Therefore, in most of the so-called dioecious species there is no complete control on pollination and some degree of self-fertilization does occur.

However, male sterility and self-incompatibility are two efficient systems of pollination control. The basic difference between these two phenomena is that in case of male sterility the pollen is nonfunctional, whereas in case of incompatibility the pollen is functional but due to some biochemical hindrance it fails to effect self fertilization.

Male Sterility

Male sterility is usually defined as the absence or nonfunction of pollen in plants (Allard, 1960). Howeve, in a broader sense, it is a phenomenon in which normally bisexual plants fail to produce or release pollen due to some genetic, physiological or any other factor (s) (Gabelman, 1956; Ohta, 1980) and thus includes a wide range of cases from complete absence of male sex organs to the nondehiscence of normal viable pollen. There may be a complete or partial suppression of male sex

113

due to failure of orderly stamen differentiation (Frankel and Glun, 1977) as in case of maize, rice, cotton, jowar, pea, tomato, tobacco and cucumber or because of severe malformation of stamens or male flowers as petaloidy (production of petals instead of stamens) in case of begonia (*Begonia semperflorens*) and stock (*Mathiola incana*); sex reversion due to faulty differentiation of anthers as carpelloidy (conversion of andoecium into carpels) in case of tobacco, carrot and cabbage; pistilloidy (conversion of androecium into pistil) in case of tobacco, maize and carrot; stigmoidy (conversion of androecium into stigma-like extensions) in case of tobacco and development of external ovules on staminal tubes; pollen sterility due to complete absence or extremely scarce presence of pollen as in case of tomato, lima bean, onion, carrot and beets; or positional, structural or functional sterility due to the nondehiscence of normal viable pollen as in case of tomato, brinjal and cabbage. The suppression of male sex organs and sex reversion may be put under a common term *staminal sterility*. In fact,pollen sterility in which pollen aborts due to some irregularities in microsporogenesis and microgametogenesis, is the true male sterility. It may be under the control of nuclear genes as in case of tomato and lima bean or may be controlled by a specific cytoplasm in the presence of a particular gene (usually recessive) in homozygous condition as in case of onion, carrot and beets or may be under the control of a particular cytoplasm alone as in case of maize, jowar, carrot and onion.

Causes of Male Sterility

The factors causing male sterility may be classified as under :

(A) Genetic factors

    (1) Meiotic abnormalities
        (i) Failure of spindle fibre formation
        (ii) Failure of pairing followed by no crossing over
    (2) Chromosomal aberrations
        (i) Imbalance in chromosome number - haploidy, polyploidy and aneuploidy
        (ii) Structural changes in chromosomes - deficiencies, inversions and reciprocal translocations
    (3) Nuclear genes
        (i) Nuclear genes as well as small structural differences of chromosomes sometimes cause hybrid sterility even though the two parents (varieties or subspecies) have the same chromosome number with normal behaviour of chromosomes, e.g., hybrids of *indica* × *japonica* rice.
        (ii) Recessive genes causing genic male sterility.
    (4) Cytoplasm - A particular type of cytoplasm alone may cause sterility (cytoplasmic sterility)
    (5) Cytoplasm and nuclear genes - A particular type (sterile) of cytoplasm with specific recessive genes can cause male sterility (cytoplasmic-genetic male sterility)
    (6) Viral infections - A specific virus transmitted to species can interact with the cytoplasm of the host species and can cause cytoplasmic male sterility, e.g., tomato ring spot virus in soybean and tobacco ring spot virus in petunia.

(B) Non-genetic factors

   (i) Due to unfavourable environment - Extremely high or low temperature and light, lack of mineral nutrition, etc.

   (ii) Chemicals - chemicals like maleic hydrazide, ethephon and gibberelins can cause male sterility in some crop plants.

(C) Interactions between genetic and environmental factors

The degree of expression as well as the occurrence of genetically controlled male sterility may be modified by the environment in which the plant material has been grown. The occurrence of genetic male sterility may depend upon temperature (TGMS) and photoperiod (PGMS).

Types of Male Sterility

Usually three types of male sterility occur in cultivated plant species : (1) cytoplasmic, (2) genetic, and (3) cytoplasmic-genetic.

Cytoplasmic male sterility : As the name indicates, this kind of sterility is caused by a particular (sterile) type of cytoplasm. The plants carrying sterile cytoplasm possess female fertility and thus produce seeds when crossed with male fertile plants. However, the plants grown from the F_1 seed are also male sterile because the male parent does not contribute cytoplasm at the time of fertilization. According to Pring *et al.* (1979), on the basis of their work in maize and sorghum, the genetic determinants controlling this sterility are present in the mitochondrial DNA rather than in chloroplasts.

In this type of male sterility, the male sterile (A) line can be maintained simply by crossing with the maintainer (fertile) line. The F_1 hybrid is produced by crossing the A line with the pollinator (C line). The whole scheme of the maintenance of male sterile line and the production of F_1 hybrid may be represented as shown in Figure 10.1.

The transfer of the sterile cytoplasm (rather the genotype of the pollinator) is also simple and can be accomplished by repeatedly backcrossing the line (to which the cytoplasm is to be transferred) as pollinator with cytoplasmically sterile line and thus reconstituting the genotype of the pollinator in the background of sterile cytoplasm. This is shown in Figure 10.2.

The occurrence of cytoplasmic male sterility has been noted in a large number of cultivated species. However, the frequency of its spontaneous occurrence may vary from species to species and from variety to variety in the same species. Its spontaneous occurrence has been noted in crops like maize, pearlmillet, sorghum, onion, carrot, red pepper and beets. It has also been artificially induced in crops like pearlmillet and sorghum. However, the progenies of intergeneric, interspecific and distant varietal hybridizations have been the main sources of the frequent occurrence of this kind of sterility (Edwardson, 1970). *Aegilops caudata* × *Triticum aestivum* is an example of intergeneric hybridization. Similarly, *T. timopheevii* × *T. aestivum*, *Oryza glaberrima* × *O. sativa*, *Gossypium arboreum* × *G. hirsutum*, *G. anomalum* × *G. hirsutum*, *Nicotiana bigelovii* × *N. tabacum*, *Linum floccosum* × *L. usitatissimum* and *Capsicum fruitenescens* × *C. annuum* are the examples of interspecific hybridization. Distant varietal hybridizations (or hybridizations at sub-species level) between *Indica* and *Japonica* in rice, *Milo* and *Kafir* in sorghum and *Procumben* and *Tallrace* in linseed have also been the sources of the occurrence of cytoplasmic male sterility. In all these intergeneric and interspecific hybridizations, the sterile cytoplasm from wild species has been

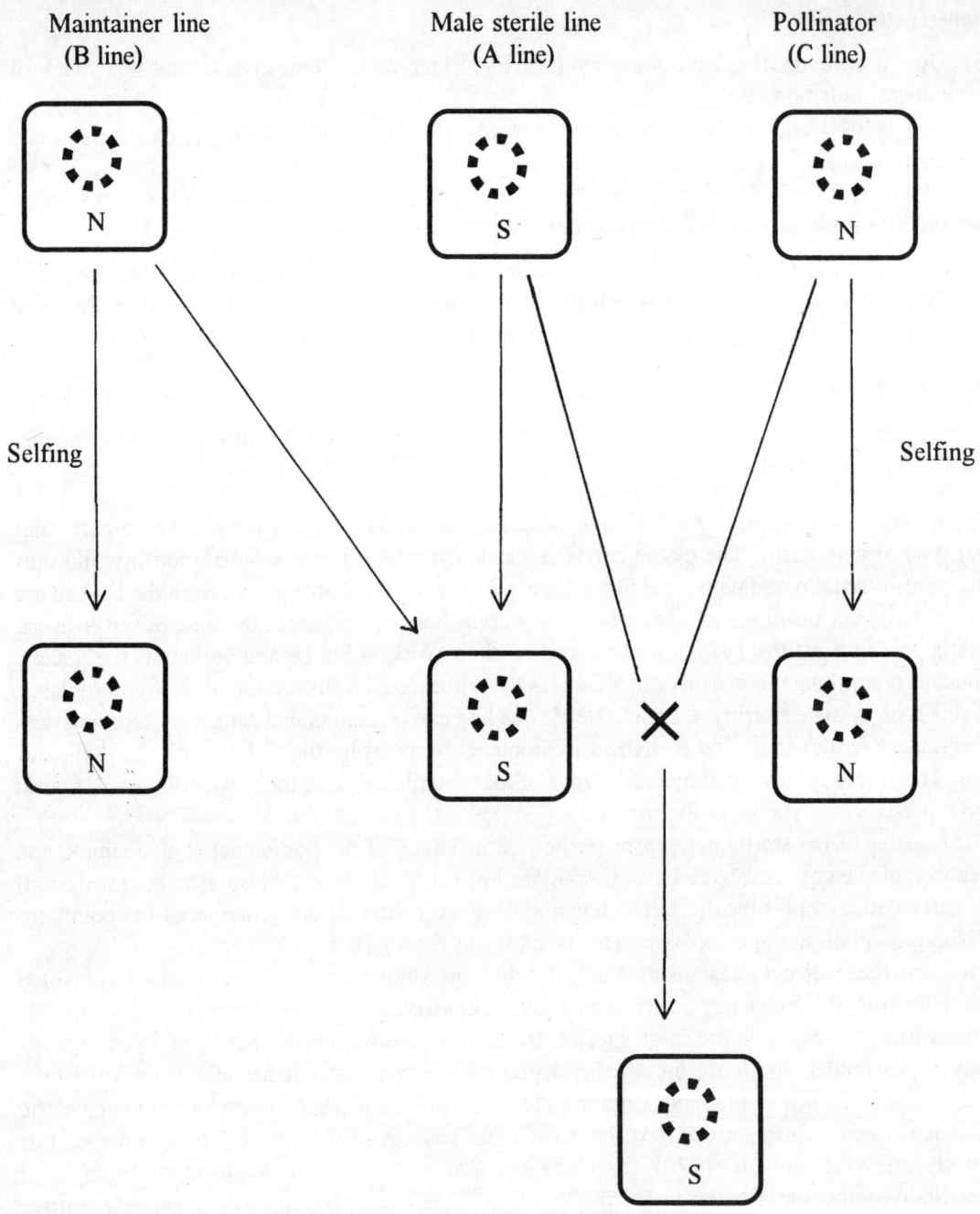

Figure 10.1 : **Constitution and maintenance of A, B and C lines and the production of F₁ hybrid**

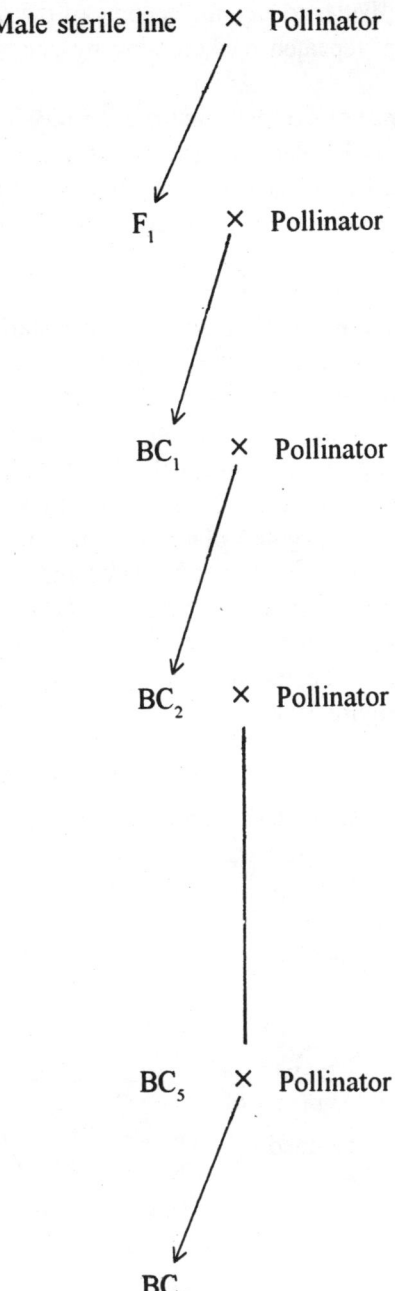

(Sterile cytoplasm with genotype of pollinator).

Figure 10.2 : Transfer of sterile cytoplasm to a desired line

transferred to the cultivated species or, better to say, the genome of the cultivated species has been transferred to the wild species through repeated backcrossing by using the cultivated species as recurrent parent.

In spite of the widespread occurrence of cytoplasmic male sterility in cultivated species, its use in crop species has been restricted since the plants grown from the crossed seed will also be male sterile and this will not produce fruits and seeds unless some pollinator is present. It is advantageous only in those species where fruit and seed are not commercial products. For example, in certain ornamental species, this nonfruitfulness causes longevity in the blooming period and freshness of flowers (desirable characteristics in ornamentals). In those species also where some vegetative part of the plant is a commercial product, advantage of this type of male sterility can be taken.

Genic or genetic male sterility : This type of male sterility is dependent upon nuclear genes. The genes causing sterility are usually recessive (ms) and have a deleterious effect in natural populations. But these genes have proved to be very useful and interesting to the plant breeder. The male sterility is maintained by crossing the A line (carrying recessive male sterile gene in homozygous condition) with B line or maintainer which is heterozygous (Msms) and fertile, thus producing 50 per cent male sterile and 50 per cent male fertile (heterozygous) plants. For the production of hybrid seed, a crossing block is raised with alternate rows of male sterile plants and the pollinator (C line) which carries dominant Ms gene for fertility in homozygous condition. Before anthesis, the male fertile plants from the male sterile rows (which in fact have 1:1 ratio of male sterile and male fertile plants) are to be rogued out. The whole scheme of the maintenance of sterility and the production of hybrid seed may be represented as shown in Figure 10.3.

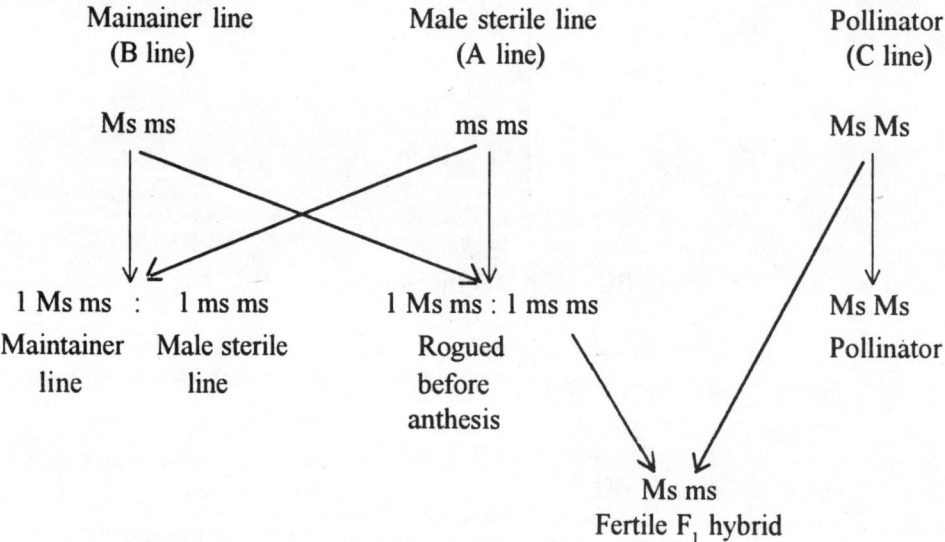

Figure 10.3. Production of hybrid seed using genic male sterility

This type of male sterility suffers from a serious drawback that it is usually under the control of a recessive allele. In some cases, great difficulty is encountered in roguing the male fertile plants from the male sterile rows because it is not easy to identify these fertile plants before flowering. In cases where closely linked marker genes are available or the male sterile allele has identifiable pleiotropic effect, the roguing of male fertile plants and the production of hybrid seed become easy.

In cotton (Weaver, 1968) and maize, this kind of sterility has also been found to be controlled by double recessive genes. There are examples where genetic male sterility is determined by dominant alleles (in potato by Salaman in 1910; in wheat by Sasakuma *et al.* in 1978; in cotton by Bowman and Weaver in 1979). In addition, epistatically controlled genic male sterility has been noticed in case of lettuce (a complementary epistasis reported by Ryder in 1963) and sunflower (a duplicate epistasis noted by Putt and Heiser in 1966).

In addition to the spontaneous occurrence of this type of male sterility in a large number of crop plants, it has also been artificially induced by radiation. This sterility has been used in the hybrid seed production in crops like pigeonpea, barley, tomato, pepper, zinnia and marigold. However, in some cases (barley, tomato and beans), the poor pollen dispersal by the pollinator creates difficulty in producing large quantity of hybrid seed. In such cases, identification of better pollinators and the most conducive environmental conditions become important.

Cytoplasmic-genetic male sterility : Both the cytoplasm and the nuclear genes play role in the control of this type of male sterility. The presence of sterile cytoplasm as well as of the nuclear genes responsible for sterility is necessary for pollen sterility. In the absence of any of the two, the plants will be fertile, that is, only the presence of normal cytoplasm or that of the restorer genes is enough to restore fertility. In some cases, the control of this type of sterility is found quite complicated because of the effect of the modifiers and the environment. However, since it is possible to maintain the male sterile line and the restorer line in homozygous condition and fertile F_1 hybrid can by produced, this kind of male sterility has been found most favourable for commercial hybrid seed production in crop plants like pearlmillet, sorghum, sunflower and maize. The maintenance of male sterility and the production of hybrid seed through cytoplasmic genetic male sterility will be as shown in Figure 10.4.

The genes which can restore fertility in the presence of sterile cytoplasm are known as restorer genes. They are usually dominant alleles. Such genes along with male sterile cytoplasm have been found in both the cross-fertilized and the self-fertilized species. Some examples are given in the Table 10.1. All the cytoplasms in pearlmillet, sunflower and maize are mutant ones. The cytoplasmic-genetic male sterility has been commercially used in case of pearlmillet, sorghum, sunflower and maize (T, C and S cytoplasms, most commonly T cytoplasm which has become susceptible to leaf blight caused by *Helminthosporium*) for hybrid seed production.

The restorer gene may be transferred to another strain or variety by first crossing the strain carrying the restorer gene with the cytoplasmic male sterile line and then crossing the F_1 to the strain to which the restorer gene is to be transferred. All plants of the progeny of this cross will have sterile cytoplasm but only 50 per cent of these plants will possess the restorer gene (in heterozygous condition). The plants carrying restorer gene are repeatedly backcrossed as female parent to the strain to which the restorer gene is to be transferred till all the desirable characteristics are reconstituted. The backcross progeny is then selfed to obtain the restorer gene in homozygous conditon. The whole scheme may be represented as in Figure 10.5.

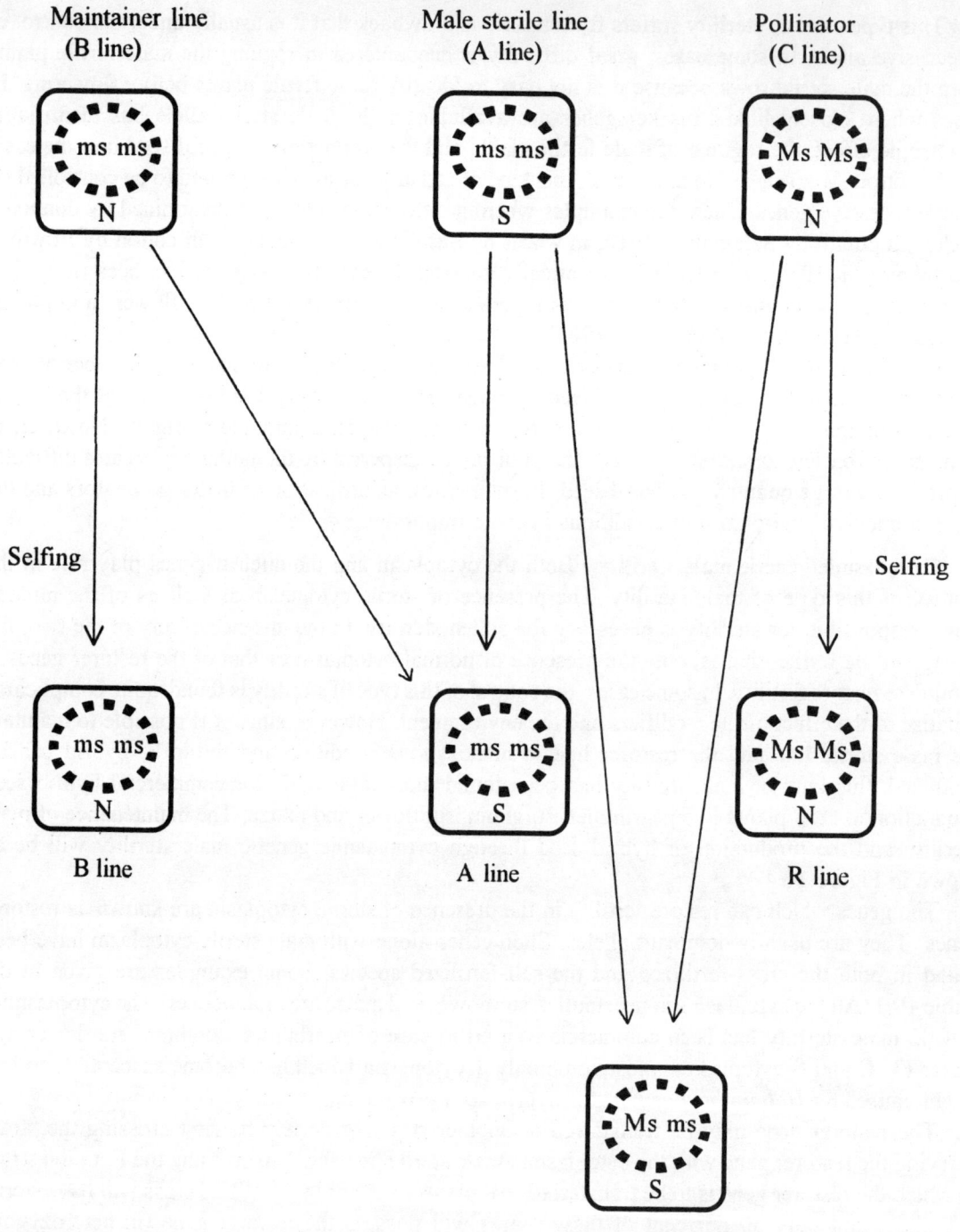

Figure10.4. Constitution and maintenance of A, B and R lines and the production of F₁ hybrid seed

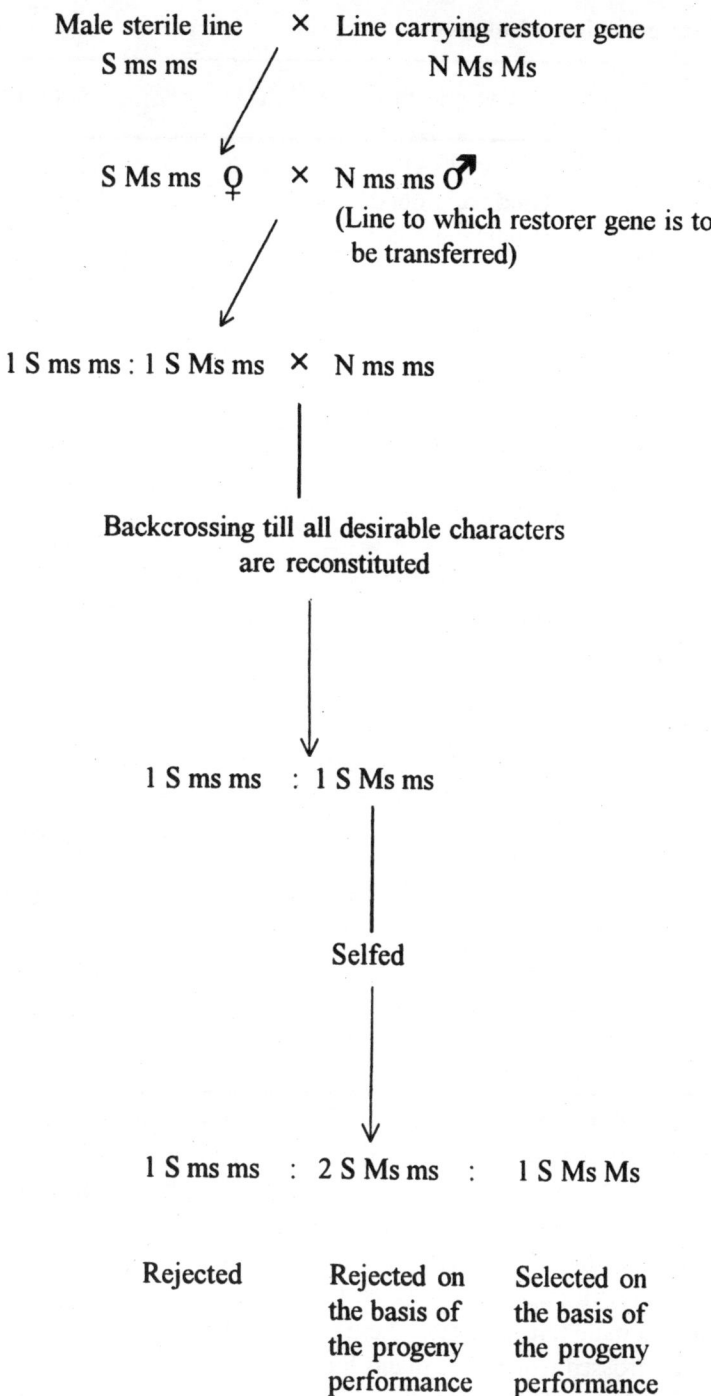

Figure 10.5 : Transfer of restorer gene

Table 10.1. Restorer genes in various crops

Crop	Cytoplasm	Restorer gene (s)
Pearlmillet	A_1 (Tifton 23A)	Ms1
	A_2 (Ludhiana 66A)	Ms2, Ms3
	A_3 (Ludhiana 67A)	Ms1, Ms2,Ms3
Sorghum	Milo	Msc
Sunflower	F_1 of *Helianthus petiolaris* × *H. annuus*	Rf1, Rf2
Maize	T (Texas)	Rf1, Rf2
	C (Charrua)	Rf4, Rf5
	S (USDA)	Rf3
Cotton	*Gossypium anomalum*	Ms3
	G. arboreum	Ms3
	G. harknesii	Ms3
Rice	Chinsurah Boro II (WU 10 for *japonica*) V20A, Zhen Shan 97A ('WA' for *indica*)	Rf1, Rf2, Rf3, Rf4
Wheat	*Triticum timopheevii*	Rf1, Rf2
Wheat	*Triticum timopheevii*	Rf6 (Chinese R-lines)
Wheat	*Aegilops kotschyi*	Rfv
Wheat	*Aegilops crassa* (PCMS)	Rfd1 (Chinese Spring)

The period required to completely transfer the restorer gene in a desirable genotype can be reduced by making selections in segregating generations for all the desirable characters.

Jones and Davis (1944) were the first to use cytoplasmic-genetic male sterility for the production of hybrid seed in onions. A single male sterile plant found in the Italian Red variety (this plant was propagated vegetatively through bulbils) when crossed with other male fertile plants produced three kinds of progenies : (1) All male sterile plants; (2) all male fertile plants; and (3) 50 per cent male sterile and 50 per cent male fertile plants. These results indicate that in the first case the male sterile plant (S ms ms) was crossed with fertile plants having N ms ms constitution, in the second case with plants carrying N Ms Ms constitution, and in the third case it was crossed with plants possessing N Ms ms constitution. However, for hybrid seed production in onion, the pollinator could be N Ms Ms, N Ms ms or N ms ms since vegetative part (bulb) and not the seed is the commercial product in this crop.

Although this kind of male sterility is being used most successfully for hybrid seed production in crop plants, a number of difficulties may have to be faced in maintaining sterility and restoring fertility because of the influence of modifiers, environmental conditions and a small amount of cytoplasm of the male parent which in some cases is transmitted alongwith the male gamete and due to the susceptibility of the sterile cytoplasm to some disease. For example, the most commonly used T type of sterile cytoplasm in maize became susceptible to leaf blight and compelling the maize breeders to depend more on detasselling technique for hybrid seed production. Similarly, Tifton 23A cytoplasm of pearlmillet became susceptible to downy mildew disease.

Utilization of Male Sterility in Plant Breeding

The male sterility has been applied to plant breeding with two main purposes : (1) To produce hybrid seed and (2) to facilitate intermating in self-fertilizing crops for breaking undesirable linkages and obtaining desirable combinations. The first application is mainly confined to cytoplasmic-genetic type of male sterility, whereas genetic male sterility has been mainly used to facilitate intermating in self-ferilized crop species. Nevertheless, environment dependent genetic male sterility (temperature dependent genetic male sterility or TGMS and photoperiod dependent genetic male sterility or PGMS) is as good as cytoplasmic-genetic male sterility for the production of hybrid seed at commercial scale. Such male sterility need not be maintained through crossing with the maintainer line, rather the male sterile stock is grown in isolation when temperature/photoperiod is suitable for fertility. Hence, instead of three lines (A, B and R) in cytoplasmic-genetic male sterility, only two lines (A and R) are required in TGMS and PGMS. Accordingly, hybrid seed production through this environment dependent genetic male sterility is called two-line system of hybrid seed production.

Use of Male Sterility in the Production of Hybrid Seed

The male sterility has been and is being used very successfully for producing hybrid seed in a number of crops. However, various male sterility systems have been used in the production of hybrid seed in different plant species. For example, cytoplasmic-genetic male sterility has been most favourably utilized for hybrid seed production in crops like bajra, jowar, maize and onion. The use of genetic male sterility has been suggested to produce hybrid seed in case of tomato, barley, pepper, etc. The balanced tertiary trisomic (BTT) technique suggested by Ramage (1965, 1975) for producing hybrid seed in barley is the most potential procedure of using genetic male sterility. Driscoll (1972) suggested almost a similar technique (XYZ system) for producing hybrid seed in wheat. But the use of genetic male sterility in the production of hybrid seed in these crops has not been very much encouraging. However, this type of male sterility has been successfully used for producing hybrid seed of castor in USA. According to some scientists, use of functional male sterility in the hybrid seed production in tomato has a good scope since the male sterility can be easily maintained by selfing the male sterile parent by opening the pollen sacs with the help of small forceps or needles and hybrid seed can be produced by simply crossing the male sterile parent with the pollinator. However, the use of this type of male sterility suffers from two main drawbacks : (1) Sensitivity of non-dehiscence to environment and (2) low amount of outcrossing. As has been discussed earlier, the use of cytoplasmic male sterility is confined to only those plant species where fruit and seed are not the commercial products and thus in most of the species where seed is the commercial product, the use of this type of male sterility is very much limited.

Use of Male Sterility to Facilitate Intermating in Self-fertilized Crops

The gap between the plant breeding methods used to improve the two main groups of crop plants, the cross- and the self-fertilized, is now-a-days getting narrowed down. The methods traditionally used in cross-fertilized crops are now being suggested and followed in self-fertilized crops also. However, these methods usually involve large number of crosses to be made which in case of self-fertilized crops is a big problem because of the small and hermaphrodite flowers. The use of genetic male sterility in such cases has proved to be very helpful to the plant breeder.

Genetic male sterility has been used to facilitate crossing for breaking undesirable linkages and obtaining better gene combinations in population improvement schemes like diallel selective mating (Jensen, 1970), recurrent selection (including backcross breeding), biparental matings and composite crosses. It was clearly demonstrated by Suneson (1956) that the genetic variability in composite cross of barley developed by means of male sterile genes was greater than in the composite where there was no use of male sterility. A similar situation has been demonstrated by Ikehashi and Fujimaki (1980) in rice by using monogenic male sterility in the development of a composite population.

If the difficulty of making large number of crosses in self-fertilized crops is overcome, the recurrent selection can be used in this group of plants also with almost the same efficiency as in case of cross-fertilized crops. Gilmore (1964) suggested male sterile facilitated recurrent selection method for improving self-fertilized crops. After that, a number of scientists namely, Doggett and Eberhart (1968) in sorghum, Brim and Stuber (1973) in soybean, Ramage (1977) in wheat and Ikehashi and Fujimaki (1980) in rice, have described application of male sterility in recurrent selection. Similarly, recessive male sterile facilitated backcross breeding procedure has been described in rice by Fujimaki (1978). Sorrells and Fritz (1982), on the other hand, presented a dominant male sterile facilitated backcross breeding scheme, that is, transfer of dominant allele through backcrossing using male sterility.

Self-incompatibility

Self incompatibility may be defined as the inability of a hermaphrodite plant having viable pollen and ovules to produce zygote after selfing. It appears to be a genetically controlled biochemical process and involves recognition of similar gene products in pollen and stigma, style or ovule. The process has the prime function of inhibiting identical genotypes but is totally ineffective when the genes of the male and female gametes are unlike. The occurrence of self-incompatibility in natural populations as well as in cultivated species is very wide. East (1940) made a survey regarding the distribution of self-incompatibility in plant species and found that it occurred in more than 3000 species distributed among 20 families of flowering plants. However, according to Lewis (1954) and many others this number is much higher than that estimated by East.

Types of Self-incompatibility

The self-incompatibility can be classified in various ways : (1) On the basis of the floral morphology– *heteromorphic* (different plants have different floral morphology) and *homomorphic* (no differences in the flower morphology of different plants) and (2) on the basis of time of gene action and the site of gene expression – *sporophytic* (the incompatibility is determined by the genotype of the parent plant, that is, premeiotic and germination of the pollen is inhibited on the stigma) and *gametophytic* (the incompatibility is determined by the genotype of the pollen itself, that is, postmeiotic and inhibition is in the stylar region).

Heteromoprhic self-incompatibility : It is associated with floral polymorphism (occurrence of plants with different floral morphology in the same species). Distyly in many species of *Primula*, that is, short style with long stamens in thrum type of plants and long style with short stamens in pin type of plants and tristyly in *Lythrum* and *Oxalis*, that is, flowers having short styles with mid or long stamens, those having mid styles with short or long stamens and those having long styles with short or mid stamens, are good examples of floral polymorphism exhibiting self-incompatibility. The pollination occurs only between the anthers and stigmas of the same height or to phrase it otherwise, pollinations occur only between morphologically different flowers. For example, in case of *Primula*, pollinations occur between thrum and pin and pin and thrum only. The gene determining floral morphology and incompatibility in *Primula*, also controls the size of pollen and the stigmatic tissues, that is, thrum and pin type of plants also differ for these two latter characteristics. The gene has two alleles, S and s and S is completely dominant over s. The genotypes of the thrum and pin are Ss and ss, respectively. However, the s allele in thrum type of plants also behaves just like S, that is, the incompatibility is determined by the previous sporophytic generation (premeiotic). This incompatibility, therefore, is of heteromorphic and sporophytic type. The compatible and incompatible crosses in *Primula* species may be shown as follows :

Compatible crosses :

   (1) Thrum × pin → 1 thrum : 1 pin

   (2) Pin × thrum → 1 thrum : 1 pin

Incompatible crosses

   (1) Thrum × thrum

   (2) Pin × pin

However, since heteromorphic systems are not common in cultivated plants, these systems do not have any importance in these plants.

Homomorphic self-incompatibility : This type of incompatibility is not associated with floral polymorphism. But it may have a premeiotic control, that is, determined by the genotype of the parent plant as in case of sporophytic type of incompatibility discussed above or it may have a postmeiotic control, that is, determined by the genotype of the pollen itself (gametophytic type).

Gametophytic self incompatibility : The gametophytic system of self-incompatibility is also called as opposition-factor system. East and Mangelsdorf (1925), for the first time, noted it in *Nicotiana sanderae*. After that, it has been noted by a number of workers in several crop plants like rye (*Secale cereale*), beet (*Beta vulgaris*), grasses, clovers (*Trifolium* species), potato, coffee and pineapple. The genotype of the pollen grain determines the incompatibility (a post-meiotic control). The pollen usually germinates on the stigma, but the growth of the pollen tubes is inhibited in the stylar tissues (containing same allele of s) followed by very slow growth or swelling and bursting of the tubes. This system of incompatibility is usually controlled by a single gene s, with a series of multiple alleles. Since this incompatibility is determined by the genotype of the gamete, all plants will be heterozygous at this locus and thus the condition of two alleles with this kind of incompatibility without dominance cannot exist.

The pollination with gametophytic system can be classified into three types :

(1) Fully incompatible – Both plants have the same genotype (it is similar to selfing), that is, $s_1s_2 \times s_1s_2 \to$ no progeny.

(2) Half the pollen is incompatible and half compatible – The two plants differ for one allele, that is, $s_1s_2 \times s_1s_3 \to s_1s_3, s_2s_3$ or $s_1s_3 \times s_1s_2 \to s_1s_2, s_2s_3$.

(3) Fully compatible – The two parents differ for both the alleles, that is, $s_1s_2 \times s_3s_4 \to s_1s_3, s_2s_3, s_1s_4, s_2s_4$ or $s_3s_4 \times s_1s_2, \to s_1s_3, s_1s_4, s_2s_3, s_2s_4$.

In fully incompatible and fully compatible pollinations, the reciprocal crosses give the same breeding behaviour and produce same genotypes. However, in case of half compatible pollinations, reciprocal crosses though give the same breeding behaviour but do not produce the same genotypes. Therefore, with gametophytic incompatibility, no reciprocal differences are found as regards the breeding behaviour.

One self-compatible allele, Sf on S locus in *Nicotiana* was found by East and Yarnell (1929). Plants having this allele in homozygous or heterozygous condition are self-compatible. In 1931, Anderson and de Winton found another allle SF which is able to enforce incompatibility by inhibiting the growth of the pollen tubes carrying Sf allele.

Two independent loci each with multiple alleles have been found to control gametophytic self-incompatibility in case of beet (Owen, 1942), diploid rye (Lundquist, 1956) and some other plant species. A two-locus control of self-incompatibility is highly effective in preventing selfing and enhancing wide cross compatibility. Therefore, such an incompatibility system may be of widespread occurrence in wind-pollinated plant species.

Sporophytic self-incompatibility : This incompatibility has a premeiotic control and was firstly discovered in 1950 by Hughes and Babcock in *Crepis foetida* and by Gerstel in guayule (*Pasthenium argentatum*). The occurrence of this incompatibility has also been reported in several other plant species like radish (*Raphanus sativus*), cabbage, cauliflower, broccoli, kale and mango. However, it is relatively of less occurrence in cultivated species than gametophytic incompatibility. The main characteristics of sporophytic incompatibility are the trinucleate condition (mostly) of the pollen grains and the inhibition of pollen germination on the stigma since the incompatibility proteins which probably are produced by the tapetum are found localized in the outer wall of the microspore, the exine.

Like gametophytic incompatibility, the sporophytic type is usually under the control of a single locus with a series of multiple alleles, but it is different from gametophytic type because here the alleles may show dominance, individual action or competition in either pollen or style. The types of pollinations with sporophytic incompatibility with order of decreasing dominance in the pollen as s_1, s_2, s_3 and s_4 and individual action (no dominance) in the style, may be shown as follows :

(1) Both plants having the same genotype, e.g., $s_1s_2 \times s_1s_2 \to$ no progeny.

(2) Parents having one allele in common, e.g., $s_1s_2 \times s_2s_3 \to$ no progeny or $s_2s_3 \times s_1s_2 \to s_1s_2, s_1s_3, s_2s_2, s_2s_3$ (reciprocal difference).

(3) Parents differing for both the alleles, e.g., $s_1s_2 \times s_3s_4 \to s_1s_3, s_1s_4, s_2s_3, s_2s_4$ or $s_3s_4 \times s_1s_2 \to s_1s_3, s_2s_3, s_1s_4, s_2s_4$.

The results of the second type of pollination above clearly indicate that in this type of incompatibility there are frequent differences between the reciprocal crosses as regards the breeding behaviour and that the homozygotes may normally occur. Further, incompatibility can occur with the female parent, a family can be consisted of three incompatibility groups and an incompatibility group may possess two genotypes. None of these is possible in the gametophytic type.

Physiology of Incompatibility

The self-incompatibility reaction appears to be a simple genetically controlled biochemical process and probably involves some sort of antigen-antibody reaction. As indicated by cytological studies, the incompatibility may be expressed at any stage between pollination and fertilization. The stages at which the incompatibility process can operate may be broadly classified into three :

(1) Either the pollen does not germinate at all or if it does germinate, the pollen tubes are not able to penetrate the stigma. The examples of this type are radish, cabbage, broccoli (*Brassica olecracea*), rye, etc. Since the incompatibility reaction in this type is localized in the stigma, complete or partial removal of stigma or even maceration of its surface is sufficient to change a normally incompatible flower to a compatible one.

(2) The pollen germinates normally but the development of the pollen tubes is inhibited in the stylar tissues (in any part of the style from the begining of the stylar tissues to its bottom). The growth of the pollen tube is so slow that usually it does not reach the ovary in time to effect fertilization. The examples of this type are species of *Nicotiana* and petunia (*Petunia violacea*), etc.

(3) The pollen tubes grow normally and do discharge the gametes in the ovule but there is no development of seed (a post fertilization incompatibility). For example, in *Gasteria*, the germination of pollen and the development of pollen tube are normal but the degeneration occurs after fertilization when embryo is uninucleate and the endosperm two-nucleate. In cacao (*Theobroma cacao*), the degeneration occurs because of the failure of the development of endosperm as the male gamete fails to fuse with the polar nucleus. However, this type of incompatibility occurs very rarely.

Self-incompatibility and Plant Breeding

Self-incompatibility acts as a widespread natural barrier against selfing in plant species and is thus one of the most important mechanisms that enforce cross pollination. However, whether it is a handicap or a benefit to plant breeder depends on the type of the plant part of the crop used as a commercial product (seed or vegetative part) and the kind of reproduction (sexual or vegetative). It is extremely undesirable in case of crops where seed is the commercial product or where crop is dependent upon seeds as in clonally propagated fruit crops. However, it is essential where the crop must not contain seeds as in case of pineapple. It is also desirable in many flowering ornamental plants and several leafy and root crops for prolonging blooming and vegetative stages.

The achievement of self-fertility in cherry (*Prunus avium*) through mutations has resulted in an unexpected advance in the breeding of this important fruit crop. Self-fertile alleles (Sf types) were obtained by Lewis (1954) in this crop by applying mutagens. Self-fertility not only proved helpful in overcoming the difficulty of growing different cross compatible varieties in the same garden but also resulted in overcoming variability through inbreeding and identification of early plants.

One of the main attractions of self-incompatibility to the plant breeder is its probable use in the production of hybrid varieties. However, breeding of F_1 hybrids has two mutually exclusive basic requirements : (1) Repeated selfing must be possible to develop homozygous lines; and (2) the possibility of selfing and sib mating in the lines to be used in the production of hybrid seed must be completely excluded (if self-incompatibility is to be used for hybrid seed production, these lines must be completely self incompatible), that is, two contrasting mechanisms, one enforcing selfing and the other enforcing outbreeding, must operate in the same crop species.

The induction of self-compatibility in a self-incompatible species, in addition to its use in the production of superior inbreds with high combining ability, maybe desirable to maintain high fertility under conditions unsuitable for cross fertilization and to develop and maintain homozygotes with valuable recessive gene combinations. A number of devices like bud pollination, use of pollen mixtures (live incompatible and killed compatible pollen), double pollination, pollination after completely or partially removing the stigma, macerating the stigmatic surface, steel brush pollination, intra-ovarian pollination, test tube pollination by growing ovule and pollen on a nutrient medium in test tubes, increasing CO_2 content, irradiation of styles, hormone treatment of pollen and pistils, high or low temperatures, high relative humidity, treatment of stigmas with ether-soluble pollen-coat material, thermally aided and electrically aided pollination and parasexual hybridization by isolating protoplasts of incompatible genotypes, fusing them and then culturing the hybrid protoplasts, have been suggested for overcoming self-incompatibility or to induce self-compatibility in self-incompatible species. However, none of these devices has been of general use in overcoming self-incompatibility since a number of factors like the type of crop, the genetic background, the age and vigour of plant and flower, the nature of the self-incompatibility alleles and the type of the self-incompatibility, control their effect. The induction of pseudo self-incompatibility may lead to the selection of weak s alleles or unsuitable genetic background. The production of haploids through parthenogenesis or anther culture and doubling of these haploids suggested by Hermsen (1974), use of permanent self-compatible mutations with a gametophytic system proposed by Gastel and Nettancourt (1975) and interaction type V of the one-locus sporophytic system (involving a reverse dominance relationship of s alleles in pollen and style) pointed out by Litzow and Ascher (1977) are though recent and quick methods for producing inbreds, have their own merits and demerits. Further, the development of completely homozygous lines is not manytimes advisable for economical production of hybrid seed. However, the development of near-isogenic lines through optimal number of bud pollinations, is still a more efficient method for this purpose.

The hybrid seed can be produced by interplanting two self-incompatible but cross-compatible lines in the field and collecting seed from both the lines or by interplanting a self-incompatible line with a self-compatible line and collecting seed from the self-incompatible line only. Hybrid seed has been produced in case of Brussels sprouts (*Brassica oleracea*), cabbage, turnip (*Brassica rapa*), alfalfa (*Medicago sativa*), sunflower, etc. The use of self-incompatibility in the hybrid seed production of some self-fertilized crops like tomato, lettuce, barley and some beans where cytoplasmic male sterility restorer system is not yet available, seems to have a good future.

Heterosis

The beneficial effects of crossing and depressing effects associated with inbreeding are widely accepted and well known phenomena. Since crossing leads to heterozygosity and inbreeding to homozygosity, these beneficial and depressing effects are considered to be the manifestations of heterozygosity and homozygosity, respectively. The manifestation of heterozygosity has been termed as heterosis or hybrid vigour and that of homozygosity as inbreeding depression. However, neither heterozygotes are always superior to homozygotes nor inbreeding is always associated with loss in vigour and other characteristics. In fact, whether a heterozygote will be superior to its corresponding homozygote or not depends upon the genetic basis of heterosis, that is, which one of the two important genetic mechanisms of heterosis (over-dominance and dominance of favourable genes or gene dispersion) is predominantly operating (Gallais, 1989). If overdominance is the main contributor towards heterosis, a major part of heterosis cannot be fixed and heterozygote is always expected to be superior to its corresponding homozygote. However, only a few evidences are available in support of this mechanism. But if dominance of favourable genes is the main contributor towards heterosis, it is possible to fix a major part of heterosis and to derive homozygotes equal (in case of full dominance) to or better (in case of partial dominance) than their corresponding heterozygotes (Jayasekara and Jinks, 1976; Pooni and Jinks, 1981), that is, heterozygosity is not a necessary condition for heterosis (Mac Key 1976). In highly self-fertilized plant species (wheat, barley, etc.) which maintain a largely homozogous balance under natural conditions and show little inbreeding depression, homozygotes can be derived which are equal or superior to heterozygotes. On the other hand, in highly cross-fertilized species (maize, sunflower, etc.) which possess a largly heterozygous natural balance and deteriorate drastically on inbreeding, it seems almost impossible to derive homozygotes equal to heterozygotes despite the use of recurrent selection which improves lines faster than hybrids. Recurrent selection definitely decreases difference between lines and hybrids but hybrids remain always better than homozygous lines. Of course, a long term recurrent selection programme may minimize the difference between lines and hybrids to the extent that lines may be almost equal to hybrids. But it is yet to be practically demonstrated. Heterosis and inbreeding depression assume considerable importance in the breeding of crop plants and thus will be discussed separately.

The term 'heterosis' was derived from two Greek words, 'heteros' means different and 'osis' means condition. This term was coined by Shull in 1914 for the superiority of F_1 hybrid over its parents in respect of general vigour, yield, reproductivity ability, etc. It is the converse of depression associated with inbreeding. The term 'hybrid vigour' is mostly used as a synonym of heterosis but some investigators do not agree with this opinion. According to Shull, a sort of physiological stimulus (his physiologic stimulation hypothesis) was responsible for the increased vigour of the F_1. This stimulus was directly related to the diversity of the uniting gametes, that is, more the diversity between the uniting gametes more is the stimulus followed by more vigour of the F_1. Shull believed that whereas heterosis is not entirely Mendelian in nature, hybrid vigour is purely a genetic phenomenon and thus made a clearcut distinction between the two terms. Similarly, according to Whaley (1944, 1952), hybrid vigour is the superiority of the F_1 over the parents whereas heterosis is the process by which this superiority has occurred, that is, heterosis is the mechanism of hybrid vigour. However, the distinctions made between these two terms do not serve any specific purpose and thus the two terms will be used as synonyms hereafter. However, the terms 'luxuriance' used by Dobzhansky (1952) and 'somatic heterosis' by Gustafsson (1952) do not satisfy the definition of true heterosis since these two terms do not take into account the superiority of the hybrid in respect of reproductive ability and adaptation as does the heterosis.

History of Heterosis

Although the term heterosis was firstly given by Shull in 1914, the superiority of the F_1 hybrids over their parents had been reported long back when the first artificial plant hybrid between carnation and sweet william was produced by Thomas Fairchild in the early part of 17th century. Probably, the first authentic report in this regard comes from the work of Koelreuter (1763) who observed hybrid vigour in artifially produced tobacco hybrids. After that, a number of intervarietal, interspecific and intergeneric hybridizations were attempted by various workers. However, the experiments reported by Darwin in 1876 in maize, opened a new door for exploiting heterosis in this crop. Beal (1876-1882) produced many intervarietal artificial hybrids of maize and recorded hybrid vigour upto 40 per cent as compared to their parents. But, more systematic work towards the understanding of heterosis was carried out by East (1908) and Shull (1909) in maize. The causes of heterosis were also explained by Davenport (1908), Bruce (1910), Keeble and Pellew (1910), Jones (1917), Collins (1921) and several others. The speed of producing hybrids in maize was so fast that hybrid maize soon became a commercial proposition and surprisingly by the year 1940, hybrids had occupied about 50 per cent of the world's maize acreage. Various aspects of heterosis were discussed in detail at Iowa State University in 1950 and the proceedings of the discussion were published by Gowen (1952). These proceedings still provide the best resource material on heterosis. Dobzhansky (1952) and Gustafsson (1952) described different kinds of heterosis. The utilization of the effect of heterosis has resulted in the considerable improvement in the productivity of a number of crops and is rightly considered as one of the greatest achievements of the twentieth century in the field of agriculture. The exploitation of heterosis in a number of crops also represents the best example of the application of genetical theory to practical plant breeding. A special Hybrid Research Project launched by the Indian Council of Agricultural Research in 1989 signifies the importance of heterosis in crop improvement.

Measurement and Detection of Heterosis

Heterosis is usually measured in three following ways:

(1) Heterosis over midparent (relative heterosis) : It is calculated as follows :

$$\frac{F_1 \text{ mean} - \text{Midparent mean}}{\text{Midparent mean}}$$

This kind of heterosis does not serve any useful purpose for the plant breeder. However, it is important from the point of view of an evolutionist who can use it as a device to maintain the genetic flexibility and influence gene frequencies.

(2) Heterosis over the better parent (heterobeltiosis) : The formula for calculating this type of heterosis is

$$\frac{F_1 \text{ mean} - \text{Mean of the better parent}}{\text{Mean of the better parent}}$$

This type of heterosis is relatively more useful for the plant breeder than the heterosis over midparent.

(3) Heterosis over the best check variety of the locality (standard heterosis) : It is calculated as

$$\frac{F_1 \text{ mean} - \text{Mean of the best check variety}}{\text{Mean of the best check variety}}$$

It is the most useful type of heterosis for the plant breeder. However, if the better parent is also the best commercial variety of the locality, the heterosis over better parent and over best check variety become the same. For calculating heterosis in percentage, its estimate in each case is multiplied by 100.

Depending on the objective of the plant breeder, heterosis may be calculated for increase in the general vigour, size of flowers (ornamental plants), fruits (many vegetables and fruit trees), leaves (lettuce), stems (sugarcane) and roots (many root crops), yield, reproductive ability, earliness, resistance to diseases and insect pests, quality characteristics, adaptability etc. In addition to these characteristics, heterosis may also be measured for some physiological (increased growth rate, dry matter accumulation, etc.) and biochemical (enzymatic activity) characteristics.

In addition to the techniques based on phenotypic values obtained after field experimentations, some physio-biochemical tests can be used to predict heterosis and choose high combining parents without making large number of crosses and evaluating them in the field. These tests include biotest (to know the effect of yeast growth on the extract of seed mixture of parental lines), serological test (electrophoretic analysis of protein extracts of hybrids and their parents), analysis of DNA and RNA synthesis and enzymetic activity, study of the composition of newly formed RNA and other nucleotide characteristics in seedlings, test of electrokinetic properties of cell nuclei of hybrids and mitochondrial complementation test. However, some workers have indicated their doubts about the validity of mitochondrial complementation test particularly the inefficiency of this test to precisely detect high combining parents. One common drawback of all the laboratory tests is that they fail to discriminate between hybrids which do not indicate yield heterosis for agronomic reasons (disease, lodging, etc.) and genetic reasons (hybrid necrosis, partial restoration of fertility, etc.).

Causes of Heterosis

Although the causes of heterosis are not yet fully known, some hypotheses have been put forward by various investigators to explain the basis of heterosis. However, since inbreeding is the converse of heterosis, the hypotheses given to explain the basis of heterosis are also applicable to explain the deterioration associated with inbreeding. Broadly, heterosis has been explained on three bases : (1) Genetical; (2) physiological; and (3) cytoplasmic.

Genetical Basis of Heterosis

Two hypotheses have been put forward to explain the genetical basis of heterosis.

Dominance of favourable genes : This hypothesis was independently proposed by Davenport in 1908, Bruce in 1910 and by Keeble and Pellew in 1910 and was named as dominance hypothesis. It provides the most widely accepted genetic explanation of heterosis and inbreeding depression. According to this hypothesis, increase in general vigour and other characteristics of the F_1 hybrid produced by crossing two genetically diverse parents is because of the accumulation of favourable dominant alleles dispersed in the parents and the depression associated with inbreeding is due to Mendelian segregation. In the hybrid, the dominant favourable alleles mask the effect of the deleterious alleles, whereas in the progenies obtained after selfing or inbreeding, the alleles with detrimetnal or negative effect become homozygous and thus cause reduction in yield and other characteristics. This can be shown by taking a very simple example where two homozygous parents $A_1A_1\,a_2a_2\,a_3a_3$ A_4A_4 and $a_1a_1\,A_2A_2A_3A_3a_4a_4$ have been crossed to produce the F_1 hybrid (the alleles $A_1A_2\,A_3$ and A_4 being dominant with favourable effect, while $a_1\,a_2\,a_3$ and a_4 being recessive with detrimental effect). The F_1 will be heterozygous carrying all the four dominant favourable alleles. But, when this

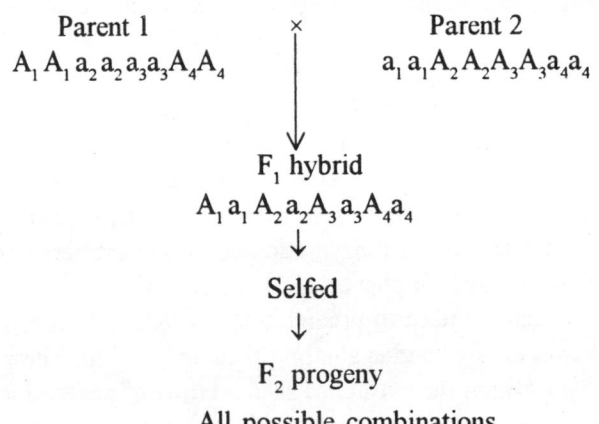

Parent 1
$A_1\,A_1\,a_2\,a_2\,a_3a_3A_4A_4$ × Parent 2
$a_1\,a_1\,A_2\,A_2A_3A_3a_4a_4$

F_1 hybrid
$A_1\,a_1\,A_2\,a_2A_3\,a_3A_4a_4$

Selfed

F_2 progeny

All possible combinations

Phenotypic ratio $81:27:27:27:27:9:9:9:9:9:9:3:3:3:3:1$

hybrid is selfed, it produces all different kinds of possible combinations in F_2 with a phenotypic ratio $81:27:27:27:27:9:9:9:9:9:9:3:3:3:3:1$.

This hypothesis, however, is subjected to the following criticisms :

(1) According to this hypothesis, one should be able to obtain homozygote ($A_1A_1A_2A_2A_3A_3A_4A_4$ in the present case) which must be equal to the F_1 hybrid. But such a situation is not usually found.

(2) With dominance at all the segregating loci, the F_2 distribution must be skewed. But the F_2 population generally shows a normal distribution.

(3) The hypothesis does not provide a proper explanation for multiplicative interaction where the F_1 hybrid exhibits considerable heterosis for a dependent character (say yield) because of large and opposite effects of its component traits. Multiplicative interaction has been observed for plant height in peas (Keeble and Pellew, 1910), fruit yield in tomato (Williams, 1959), grain yield in sorghum (Sinha and Khanna, 1975) and for grain yield in wheat (Gale *et al.*, 1986). This interaction is perhaps a sort of epistatic interaction.

(4) The actual magnitude of heterosis in many crosses cannot be explained on the basis of the replacement of all undesirable recessive alleles by their dominant counterparts (Crow, 1948, 1952).

(5) According to Barabas (1992), dominance should not be considered as the fundamental cause of heterosis as the relation between dominant traits and the level of productivity is not always direct. Had this relationship been direct, wild species which possess more dominant traits than cultivars should have been higher yielder.

Jones (1917) and Collins (1921) tried to remove the first two objections. According to Jones, the dominant favourable genes present in the parents are linked and thus when a character is governed by several genes (as is usually the case with most of the quantitative traits), it is very difficult to obtain a homozygote with all the favourable dominant alleles. Therefore, according to Jones, this hypothesis may be called as **dominant favourable linked genes hypothesis** or **dominance of linked genes hypothesis**. Collins has interpreted that in case of quantitative characters, skewed F_2 distribution will not be generally observed even in the absence of linkage because of the simultaneous segregation at several loci and the effect of the environment.

The best experimental evidence in favour of this hypothesis comes from the work of Jayasekara and Jinks (1976), Pooni and Jinks (1981), Jinks (1983) and Jinks and Pooni (1986) in *Nicotiana rustica* and from the conclusions drawn by Sprague (1983) from the experiments carried out in maize by several workers. The results of all these experiments indicated that the dispersion of favourable dominant genes in parents was the major cause of heterosis.

Overdominance or heterozygosity hypothesis : This hypothesis has also been called as superdominance, single-gene heterosis, stimulation of divergent alleles or cumulating action of divergent alleles and was suggested in 1908 by Shull and East independently. According to it, the interaction between two divergent alleles of the same locus provides a stimulus to development, that is, the **hybridity** or **the state of being heterozygous** itself is responsible for the superiority shown by the hybrid over its parents. Therefore, if A_1 and A_2 are the two divergent alleles, A_1A_2 combination would be superior to both the homozygotes A_1A_1 and A_2A_2. It has further been emphasized by East (1936) that more the divergence between alleles, more would be the stimulus and that the degree of stimulus would depend on the extent of divergence between the alleles. Thus, if there is a series of alleles A_1, A_2, A_3, ... with increasing divergence of function, A_1A_3 combination would be superior to A_1A_2.

Although clearcut cases of single gene heterosis are rare, some examples available in the literature give evidence of this hypothesis. Probably, the most direct evidence comes from the work of Gustafsson in barley who found heterozygotes for chlorophyll mutants superior to homozygous normal plants as regards the number and size of seeds. Similarly, heterozygotes for R locus in maize show more pigmentation than either homozygote.

These two hypotheses (dominance and overdominance) lead to the following two similar expectations :

(1) According to both the hypotheses, the decrease in heterozygosity causes loss in vigour and fertility.

(2) Outcrossing causes restoration of vigour and other characteristics and thus more the genetic diversity between the parents, more vigorous should be the F_1 hybrid.

However, the two hypotheses have one major difference. According to dominance hypothesis, one can obtain an inbred as vigorous as the F_1 hybrid, but it is impossible according to overdominance hypothesis. It seems more appropriate that both the mechanisms (dominance of favourable genes and overdominance) simultaneously cause heterosis. However, manytimes, the apparent overdominance may not be a true overdominance but may be due to epistasis or repulsion phase linkages or both. But the possibility of some contribution of true overdominance towards hybrid vigour cannot be completely ruled out.

In addition to the accumulation of favourable dominants and overdominance, epistasis and multiplicative interaction may also be the genetical causes of heterosis.

Physilogical basis of heterosis : There are evidences where hybrids and homozygotes show basic difference in their metabolic activities. According to Whaley (1952), the hybrids possess more efficient enzyme system than homozygotes and thus can mobilize stored food materials at an earlier stage resulting in an initial advantage in the growth of the hybrid seedlings. In some cases, this increased enzyme efficiency of hybrids has been attributed to larger embryo and endosperm size of the hybrid seeds in comparison to those of homozygotes. Of course, with the present understanding about the mechanism of heterosis, it is not possible to generalize the cause of heterozygote superiority as different mechanisms are operative in different cases. Production of metabolically superior protein by the two different alleles of heterozygote, production of optimum amount of p-amino benzoic acid by the heterozygote in *Neurospora crasa*, production of polymorphic haemoglobin by the heterozygote in man controlling sickle cell anaemia disease and optimum activity of enzyme alcohol dehydrogenase by heterozygote in maize are some such mechanisms showing heterozygote advantage. However, based on the results of some biochemical studies (Hageman *et al.* 1967, Warner *et al.* 1969, Schwartz 1973, Scholl 1974), it can be generalized that hybrid advantage over their inbred parents lies in more rapid unfolding of balanced metabolic processes rather than in greater metabolic efficiency of hybrids over their parents. Therefore, the **balanced metabolism theory** proposed by Hageman *et al.* (1967) perhaps provides a better explanation regarding hybrid advantage over their inbred parents.

The physiological basis of heterosis described above, in fact, should be called as physio-biochemical basis.

Cytoplasmic basis of heterosis : A number of investigators have emphasized the role of cytoplasm in the manifestation of heterosis. Dhawan *et al.* (1966), on the basis of the results obtained from reciprocal inter-racial crosses in maize, found that the cytoplasm exercised varying degrees of control (ranging from complete masking to little or no masking) on the expression of heterosis. They postulated that both the nuclear genes and the plasma genes underwent a process of divergence during the evolutionary differentiation of the races of maize and that the inhibitory cytoplasm might have acted as an isolating mechanism.

Mitochondrial complementation has also been suggested as a cause of heterosis by some investigators (McDaniel and Sarkissian 1966, Sarkissian and Srivastava 1971, 1973, Sage and Hobson 1973, and others). According to them, the mitochondria of heterotic hybrids absorb more oxygen and have a higher phosphorylation/oxidation ratio as compared to those of non-heterotic hybrids and the inbreds used as parents in the production of hybrids and thus such complementation between mitochondria may be used as an indicator of heterosis for yield and other quantitative characters in crops like maize, wheat and barley. Srivastava (1983) emphasized that an additive effect as well as genomic complementation in the hybrid are responsible for a higher mitochondrial activity of heterotic hybrids. Furthermore, the presenc of reciprocal differences in the mitochondrial activity in several materials indicate the possibility of unequal contribution of mitochondria by the parents to the hybrid. Of course, greater phenotypic stability of hybrids under stress environments may not be solely due to the intracellular homeostasis maintained by mitochondria of the hybrid.

In addition, the interaction between nuclear genes and plasmagenes and their interactions with environment may also contribute towards heterosis.

In view of the above discussion on the causes of heterosis, it is clear that none of the presently available hypotheses satisfactorily explains all cases of hybrid vigour, that is, the nature of heterotic loci is not yet fully known.

Maximization of heterosis : There are two important components of F_1 heterosis, h (directional dominance) and d^2 (squared deviations in gene frequency between parents). The degree of heterosis depends upon the product of these two components. For a quantitative character, F_1 heterosis (HF_1) = $\Sigma h\ d^2$ (For details see Falconer 1981). Since the direction of dominance in parents cannot be changed, h may be used here as a constant. The component d, however, can be considerably increased by involving genetically diverse parents in the cross. The choice of diverse parents may be based on d^2 values, pedigree relationships, specific combining ability of cross combinations or geographical diversity. However, a large number of studies indicate that geographical diversity may or may not be associated with genetical diversity.

It is true that genetic divergence between parents is a prerequisite for manifestation of heterosis, but it is also true that the F_1 of two divergent parents may or may not show heterosis. For example, the F_1 of $A_1A_1A_2A_2A_3A_3 \times a_1a_1a_2a_2a_3a_3$ is not expected to be superior to $A_1A_1A_2A_2A_3A_3$ parent in the absence of overdominance. Further, the results of a number of studies indicate that F_1s obtained by crossing extremely divergent parents generally do not exhibit maximum heterosis. Combining ability analysis may perhaps help the plant breeder to a great deal in the choice of desirable parents (Also see Singh and Pawar, 2005).

Fixation of heterosis : The greatest drawback associated with F_1 heterosis is the breakdown of hybridity in the subsequent generations resulting in loss of vigour and other characteristics and thus compelling the farmers to use fresh F_1 seed in every crop season. This drawback may be overcome by fixing heterosis in homozygous or heterozygous condition. If dominance of favourable genes is the major cause of heterosis, a large part of heterosis can be fixed in homozygous condition. But this advantage can be taken only in those crops where the ultimate variety to be evolved is homozygous as in case of most of the self fertilizing crops.

Heterosis can also be fixed in heterozygous condition by omitting meiosis or through chromosomal manipulation. The omission of meiosis leads to permanent heterozygote advantage as there will be no segregation and no recombination. Meiosis can be omitted by changing the mode of reproduction of the crop from seed to vegetative propagation or by stable apomictic reproduction. However, different crop species have their own inherent mode of reproduction and mostly it is impossible to change their mode of reproduction. But there is a need to screen the germplasm of different crop species thoroughly and to explore every possibility of transferring gene or gene complexes conditioning vegetative propagation to the species presently being reproduced by seed. Most of the cereals, pulses and oilseed crops do not respond to vegetative propagation, but there are reports that rice can be vegetatively propagated (Athwal and Virmani, 1972) in addition to its ratooning capacity. Of course, vegetative propagation is of much greater importance in those crops where vegetative parts of the plant are of commercial value (sugarcane, potato, sweet potato, etc.). The change of reproduction of a crop from seed to vegetative propagation can save lot of money, labour and time, can overcome problems faced by the plant breeder in creating an effective male sterility-fertility restoration system and can help in making successful crosses between distantly related species as hybrid sterility and hybrid breakdown would become unimportant.

Apomictic development of embryo from a cell having somatic chromosome number (diploid parthenogenesis, diploid apogamy and adventive embryony) allows heterozygote advantage for indefinite period of time just like vegetative propagation. Unfortunately, such a development of embryo is of rare occurrence and thus transfer of genes governing this mechanism to a desired crop species is not easily possible. A thorough screening of germplasm may perhaps help in this matter.

Chromosomal manipulations may be helpful in the fixation of heterosis in two ways : (i) Transfer of heterozygosity to polyploid level (creating intergenomic heterozygosity) and (ii) development of permanent structural heterozygotes. Diploidized polyploids like tetraploid and hexaploid wheats possess heterozygosity between genomes and homozygosity within genomes. Conversion of diploid heterozygotes into true breeding tetra- or hexaploid ones has also been suggested as a method to maintain heterozygosity. But this method is not without problems as it needs some device for diploidization of the polyploids.

Another method of fixing hybridity is the development of permanent structural heterozygotes. Such a mechanism is naturally operating in evening primrose (*Oenothera lamarckiana*) where homozygotes do not survive due to a balanced lethal system and only translocation heterozygotes survive fixing hybridity forever. Development of such a mechanism in crop plants would be a step further in the pragmatic exploitation of heterosis.

Commercial exploitation of heterosis : In fact, there are numberous plant species where considerable degree of heterosis has been observed. However, commercial exploitation of heterosis has so far been mainly confined to vegetatively propagated crops like sugarcane and almost all fruit trees, cross-fertilized crops like maize, pearl millet, sunflower, sugarbeet and forage crops and often cross-fertilized crops like sorghum, cotton and pigeonpea. Its commercial application in self-fertilized crops has been very limited. Of course, hybrid rice has now become a commercial proposition in China.

Inbreeding Depression

Inbreeding depression is usually defined as the loss of general vigour and other characteristics of the progeny as compared to its parents due to inbreeding. In fact, the real cause of depression in the performance of progeny is not the process of inbreeding itself, but it is a consequence of Mendelian segregation. The degree of depression depends on the number and kind of Mendelian characters for which the parental population is heterozygous. Inbreeding in case of crop plants usually relates to controlled selfing (the extreme form of inbreeding) or sib mating (brother-sister matings).

History of inbreeding : The effects of inbreeding have been recognized since ancient times. Its both the beneficial and harmful effects have been advocated. For example, in ancient times, some Nordic tribes, Egyptians, Greeks and Hebrews preferred matings between the individuals related by descent for maintaining specific family characteristics and preventing dilution of superior bloodlines. A similar practice is still being followed in many castes. A more widespread, direct and deliberate example of the practice of inbreeding comes from the long-continued inbreeding in case of livestocks for maintaining specific breed characteristics. Matings between the bull and his daughters and grand daughters became a common practice by the turn of the seventeenth century into eighteenth century. However, though this practice proved fruitful in maintaining breed characteristics, sooner or later, there was a reduction in the reproductive ability of the progeny and it became difficult to maintain the lines through inbreeding.

In plants, the first recorded proof of the recognition of beneficial effect of outcrossing came from the experiments of plant hybridists of the eighteenth century, particularly from the work of Koelreuter (1763, see Koelreuter 1766) and Sprengel (1793). They concluded that hybrids, in many cases, were unusually vigorous as compared to their parents (Koelreuter) and that nature also favours outcrossing (Sprengel). In 1868, Darwin categorically said that if free crossing is dangerous, too close inbreeding is also not free of danger. He reviewed the voluminous literature available on this subject including results of his own experiments in the form of a book titled 'Cross and Self Fertilization in the Vegetable Kingdom' which appeared in 1876 and concluded that inbreeding is an unnatural process with harmful effects. Darwin was the first to give precise measurements of responses to outbreeding and inbreeding in corn. Later on, East in 1908 and Shull in 1909 independently gave detailed information on inbreeding in maize. Now it has become an established fact that in most of the cross-fertilized and vegetatively propagated plant species, the process of inbreeding results in great deterioration in vigour and other characteristics.

Effects of inbreeding : The main effects of inbreeding are :
1. Inbreeding increases homozygosity.
2. Uncovers genetic variability hidden in the heterozygotes.
3. Ultimately splits the original population into genetically distinct homozygous groups, each group being uniform within itself.
4. Leads to the appearance of lethal and sublethal alleles.
5. Results in the reduction of reproductive ability of the progeny.
6. Causes general weakness in the stock.
7. Causes reduction in yield and other several characteristics.

Response to inbreeding : The self fertilized species do not respond to inbreeding, that is, they do not show any inbreeding depression. On the other hand, many of the cross-fertilized species respond highly to inbreeding. However, in the cross-fertilized group also, there are great differences as regards the degree of inbreeding depression shown by different species ranging from little or no inbreeding depression to extreme loss in vigour and other characteristics due to which most of the lines even after a few generations of selfing or sibmating fail to survive. Even in the same species, different strains may show different degrees of inbreeding depression. Although there is a continuous range of degree of inbreeding depression exhibited by the cross-fertilized plant species, these species may be grouped in three major classes on the basis of their response to inbreeding.

1. Species showing high inbreeding depression : The highest degree of inbreeding depression is probably exhibited by hayfield tarweed. It is self-incompatible and the closest form of inbreeding possible in this species is sibmating. Many lines are not able to tolerate even two generations of sib mating because of extreme reduction in vigour and reproductive ability. Next in order comes alfalfa. In this species also, the rate of deterioration in general vigour and productivity, is very high and, as a result, very few lines survive after the third selfed generation. Corrot and some strains of brown sarson (*Brassica campestris* variety brown sarson)also show drastic reduction upon inbreeding.

2. Species showing moderate inbreeding depression : There is a long list of species falling in this group. Maize, jowar, bajra, rye, sunflower, onion and many others are fairly tolerant of inbreeding. Among these, maize is probably least tolerant of inbreeding.

3. Species showing little inbreeding depression : Cucurbits (monoecious), hemp (dioecious) and papaya (dioecious) though belong to the cross-fertilized group, show very little or no inbreeding depression. These species behave like self-fertilized species as regards the response to inbreeding.

Inbreeding Depression in Relation to Homozygous and Heterozygous Balance

All the self-fertilized plant species and some species belonging to cross-fertilized group, particularly cucurbits, do not respond to inbreeding, some cross-fertilized species (rye, sunflower, onion, etc.) respond moderately and some others (carrot, alfalfa, hayfield tarweed, etc.) respond very highly to this process, that is, different species generally show different degrees of response to inbreeding. The differences in the degree of inbreeding depression shown by different species, according to some investigators may be attributed to the type of mating system occurring in a particular species and how it (mating system) affects the kind, magnitude and the manner of regulation of genetic variability in populations.

In self-fertilized species, the plants are homozygous at most of the loci and thus a population is a group of homozygous lines. Due to self fertilization, therefore, the lethal and sublethal genes become homozygous and individuals carrying such genes in homozygous condition are eliminated from the population. This is happening since times immemorial and, as a result, a **homozygous balance** (a term given by Mather in 1943) has established in the autogamous species. The self-fertilized species though do not show inbreeding depression, they may show heterosis if crosses are made between genetically diverse lines. However, along with homozygous balance these species also possess some hidden variability which leads to the presence of some **heterozygous balance** in these species.

On the other hand, the outbreeders possess a different kind of genetic organization called as **heterozygous balance** since these species are highly heterozygous. These species usually have a **genetic load** (the sum total of all recessive alleles with deleterious effect) since many deleterious recessive alleles are generally retained in populations because of the masking effect of the dominant favourable alleles. However, like self-fertilized, the outbreers also possess a good homozygous balance, particularly those species where the plant size is large and effectively small populations (in kitchen gardens or small plots) are grown like cucurbits and papaya. There is forced inbreeding in such species due to small population size. This is probably the main reason why cucurbits and papaya behave like autogams as regards their response to inbreeding.

It can, therefore, be concluded that, in general, higher the heterozygous balance, higher the inbreeding depression and vice versa.

SECTION III

Plant Breeding Methods

Line Breeding

Selection Breeding

CHAPTER 11

Mass Selection and Pure Line Selection

Although selection is the oldest method of crop improvement, it was put on sound footing in the beginning of the twentieth century only when Johannsen gave the scientific basis of pure lines. Spontaneous mutations, natural hybridization and recombination though slowly but continuously create variability in pure lines. But because of continued selfing in self-fertilized crop species, the heterozygosity created by these factors goes on turning into homozygosity. These two processes go on occurring in nature simultaneously and form the basis of further improvement in the self-fertilized crops. The populations or land varieties in this group of plants are, in general, consisted of several homozygous lines.

Generally, two selection methods are used to improve or purify the self-fertilized crops. These are mass selection and pure line selection. The basic difference between the two methods is that in mass selection a number of homozygous lines are retained at the end of the programme and mixed to develop a variety, while in case of pure line selection the progeny of a single line constitute the variety. These two methods (mass selection and pure line selection) come under selection breeding to develop line varieties.

Mass Selection

Mass selection is the simplest, safest and most rapid method of plant breeding. It may be defined as selecting individually a large number of plants for desirable characters (earliness, resistance to diseases, insect pests and lodging, tillering ability, better grain size and colour, etc.) and bulking them together to evolve a new variety or purify an already existing variety. Thus, the variety evolved through mass selection in a self-fertilized crop will be homozygous but genetically heterogeneous, that is, a group of homozygous lines. The basic idea of mass selection in self-fertilized crops is to eliminate undesirable types from the population in which the selection is being practiced. If the objective of the plant breeder is to purify an already existing variety, the try is made to maintain all the important characteristics of that variety and to reject all the plants with obvious shortcomings.

The Procedure of Mass Selection

Originally, the method was confined to the selection of individual plants with desirable characteristics from the field, tagging and harvesting them separately and bulking them together or roguing out the undesirable types from the field and bulking the seed of the remaining plants together. This selection of desirable types or elimination of undesirable types may be repeated for some seasons or years so that the new variety can perform better under different environmental conditions. However, the selection will be more effective for the characters which have high heritability.

Allard (1960) has suggested improvement in method. According to him, the progenies of the initially selected plants should be raised and these progenies should be tested for all desirable characteristics under different environmental conditions, particularly in the sick plots or hot spots of the diseases prevalent in that particular crop. The mass selection method suggested by Allard therefore involves two distinct steps :

Step 1 : A large number of plants are selected from the genetically heterogeneous population so that maximum genetic variability present in the base population can be exploited. These initial large selections are necessary since all plants of the base population are homozygous and if some desirable plants are ignored at this stage, the loss caused due to such plants cannot be compensated at any later stage of the programme. However, the number of plants to be selected will depend on (1) the type of crop species, (2) the objective of the plant breeder and (3) the facilities available to the breeder in respect of funds, land, manpower, etc. The number of initial plant selections, therefore, may vary from few hundreds to few thousands (generally, from 200-2000).

Step 2 :The progenies of all the individual plants selected in step 1 are raised for evaluation. Observations on these progenies are taken and undesirable types are immediately rejected. However, drastic elimination of plants at this stage must be avoided and the rejection should be confined to a maximum of 25 per cent. The selected progenies are then grown at different locations in different seasons or years for testing their adaptability under different environmental conditions. Again, the progenies with low adaptability and other obvious defects are eliminated. This evaluation of the progenies of single plants usually takes 3-4 years. In the end, all the selected lines are bulked together.

Since a large number of lines are retained at the end of step 2 of mass selection and since the new variety is not usually very much different from the old variety (except that the defective plants have been eliminated), the varieties evolved or purified through mass selection are not tested in replicated yield trials to compare them with standard check varieties. However, depending upon the objective of the plant breeder, appropriate modifications can be made in the method. For example, while purifying an already exiting pure line variety, one can avoid the progeny test of individual plants for several years but can restrict it to one year only. On the other hand, if the genetic variability present in the base variety is large, evaluation of progenies of selected plants for 3-4 years becomes an essential part of the programme.

Functions of Mass Selection

The mass selection method performs two important functions in the improvement of crop plants: (1) Purification of existing varieties and (2) improvement in land varieties.

Purification of existing varieties : The pure line varieties which are 100 per cent pure at the time of their release (since a pure line variety is the progeny of a single self-fertilized homozygous plant), become variable after a few years due to the occurrence of mutations, natural hybridization and mechanical mixture. Therefore, after every few years, such varieties need purification. Generally, few hundred plants possessing characteristics of the original variety are selected in the beginning from the base variety. The progenies of the single selected plants are raised in separate rows and off types and diseased progenies are rejected. The selected progenies are harvested in bulk and the seed thus produced becomes a source of large quantity of the seed to be distributed among farmers. The use of mass selection method is not confined to the plant breeder, but the seed producing agencies use the method quite frequently for producing pure seed of the existing varieties of self-fertilized crops.

Improvement in land varieties : A land variety of a self-fertilized crop is a commercially grown mixture of homozygous genotypes since long period of time and is well adapted to its area of growing. These varieties are being grown in large areas of the world particularly in the countries where agriculture is not well advanced. The different genotypes of the same variety usually differ for several important characters like general vigour, height, disease resistance and maturity and thus possess different agronomic worths. Some of them may be very poor in yield and other attributes, while others may be good enough in respect of these characteristics. Mass selection is the safest and most rapid method of plant breeding for improving the land varieties since defective and unproductive genotypes may be eliminated from the variety without drastically affecting its adaptability which is the most important characteristic of a variety.

As regards the procedure of mass selection for improving land varieties, it involves the initial large individual plant selections and vigorous testing of the progenies of these plants under different environmental conditions. The whole plan may take 4-5 years or more. However, the majority of the genotypes of the original variety are retained and bulked to evolve the new variety so that the new and the old varieties do not show much difference in adaptability.

Advantages of Mass Selection

1. It is the simplest, safest and most rapid method of plant breeding and one does not need much technical knowledge to follow the method. In fact, it is being applied in one form or the other, knowingly and unknowingly, by the plant breeders, seed producing agencies and the farmers since immemorial times.

2. Since the variety evolved through this method is a mixture of many related genotyeps, it has wider adaptability than a pure line variety. The different lines constituting a variety may have different resistant genes for the same disease and thus if one gene becomes susceptible to a particular race the other genes save the crop from big losses due to diseases.

3. Since majority of the lines of the old variety are retained and bulked together at the end of the programme, there is usually no need of testing the new variety in replicated yield trials. This cuts short the time required for evolving a new variety.

4. The adaptability of the old variety is not usually altered.

5. While following mass selection, the breeder can run other breeding programme (s) simultaneously since mass selection method does not need much attention and efforts to be made.

6. Presence of considerable genetic variability in the variety evolved by mass selection and repetition of selection in such a variety after few years provides scope for further improvement in the variety.

Disadvantages of Mass Selection

1. Since the varieties evolved through mass selection are genetically heterogeneous, there may be differences in seed size, seed colour, maturity period, etc. in the plants of the same variety. Seed producing agencies generally hesitate to accept such varieties since these varieties have lesser marketing value than the completely uniform varieties developed through pure line selection. Therefore, the method must be mainly confined to the purification of existing pure line varieties rather than developing new varieties from the land varieties.

2. The retention of majority of the lines of the original variety for maintaining adaptability makes the progress of the improvement slow since the new variety is not very much different from the old variety.

3. Sometimes it becomes difficult to decide identifying criteria for the varieties evolved through mass selection and thus such varieties may create problems at the time of seed certification.

4. Mass selection is effective when the characters under consideration have high heritability. If the heritability of a character is low (which is true for the most important character, the yield), the selection on the basis of phenotypes without growing the progenies of individual plants may not include the really good genotypes in the variety. Therefore, mass selection without progeny test should not be applied for evolving new varieties.

Pure Line Selection

A pure line variety is homozygous and genetically homogenous since it is made up of the descendants of a single self-fertilized homozygous plant. Whatever variation is found within a pure line, it may be solely attributed to environmental effect. Therefore, whereas the variability found in a variety developed by mass selection is ascribable to both genetic and non-genetic factors, the variability exhibited by a variety developed by pure line selection is entirely non-heritable.

The pure line selection differs from mass selection in the following ways :

1. The most basic difference between the two methods in self-fertilized crops is the number of lines used to constitute a variety (progeny of a single best performing line in case of pure line selection and many homozygous lines in case of mass selection). Thus, a pure line variety will be 100 per cent genetically pure, while a variety evolved through mass selection will be a mixture of different homozygous lines.

2. In pure line selection, a drastic elimination of the progenies of lines is done during the step 2 (evaluation of the progenies of single plants on the basis of eye observation) of the method to retain only a few promising lines that are finally evaluated in replicated yield trials with each other and with standard check varieties (step 3 of the method).On the other hand, in mass selection,

majority of the lines of the old variety are retained and all the progenies are bulked together.

3. In mass selection, the newly evolved variety is not usually evaluated in the replicated yield trials to compare it with the local check variety. But, in pure line selection, progenies of the individual single plants (few in number) are evaluated with each other and with standard check varieties in extensive replicated yield trials and thus the method takes usually 3-4 years more than the mass selection method.

4. The variability witin a pure line is solely due to the effect of the environment, while variability within a variety evolved through mass selection is due to both genetic and non-genetic factors.

5. The application of pure line selection method is confined to the self-fertilized crops, whereas the mass selection method is equally applicable to both self- and cross-fertilized crops.

6. The produce of the pure line variety is uniform for all the characters and usually has a high market value than the produce of a variety evolved through mass selection. Therefore, identification of pure line varieties in seed certification programmes is easier than that of the variety developed by mass selection.

7. In pure line selection, the best performing line is evolved as a variety and thus improvement over the old variety is more, while in mass selection several lines which are inferior to the best line are mixed to evolve the variety and the improvement over the base variety is less.

8. The pure line variety usually has less adaptability than a variety developed through mass selection.

The Procedure of Pure Line Selection

Although depending upon the rate of progress at a particular stage, appropriate modifications can be made in the pure line breeding programme, the method generally involves the following three steps:

1. Initial single plant selections : This step of pure line selection is exactly similar to the step 1 of mass selection. Here also, a large number of plants possessing desirable characteristics are selected to exploit maximum portion of the genetic variability present in the base material (a land variety or any other group of homozygous genotypes). The care is taken that no desirable plant is left during these original selections, otherwise it will be left forever. All these single plant selections (generally 200-2000 in number) are separately harvested. Spaced planting of the base material will help in selecting superior types and rejecting inferior and defective types. The selection for characters that are controlled by simple genetic systems and have high narrow sense heritability will be more effective than for characters which are governed by complex genetic systems and have low heritability. No effort should be made to make within lines selections since in self-fertilized crops the total genetic variability exists between lines and within lines variability is due to the environmental effect only. The plant breeder, however, should have a clear objective in mind while making these initial selections.

2. Visual evaluation of the progenies of individual plants : Like in mass selection, the progenies of all the single plants selected in step 1 are raised in separate rows. Proper spacing is maintained between rows and between plants. On the basis of eye observation, defective and inferior progenies are rejected at this stage so that the remaining progenies which are promising in real sense are easily accomodated during the latter part of the programme. The selected progeneis are than grown under different environmental conditions to test their worth particularly in respect of resistance to diseases and insect pests and their adaptability and stability of performance over a range of environments.

This allows further rejection of undesirable progenies and selection of a few top promising lines. If felt necessary, the progenies may be artificially inoculated with the prevalent disease races and may be grown in off-season nurseries for clearcut screening and for reducing the period required for the programme to be completed. The top promising lines are harvested separately. This part of the breeding programme usually takes 3-5 years.

3. Growing the selected progenies in the replicated yield trials : Since in pure line selection, the best line is to be identified for constituting the new variety, the promising lines retained at the end of the second step of the programme are grown in replicated yield trials at more than one location with the best check (s) of the locality to make comparisons between the lines themselves. Statistical tests are usually applied to make these comparisons. All the important characters like mean yield, disease and insect pest resistance, quality, adaptability and stability are given due weightage. The best line having significant superiority over the check (s) and other lines is recommended for release as a new variety. This third step of the method usually takes 3-4 years.

The pure lines, in addition to be released as new varieties, are used for some other purposes also. For example, in case of self-fertilized crops, the parents used in most of the hybridization programmes are the pure lines and in this way desirable characters of different homozygous parents are combined to evolve superior varieties. Many a times, there are some pure lines which are not released as varieties because of one or the other defect, but the same are included in the hybridization programmes for combining their desirable characters with those of others.

The pure lines help in studying mutations, particularly in case of quantitative characters. Barring natural hybridization and mechanical mixtures, all genetic variability arising in a pure line may be attributed to mutations. Thus, by carefully excluding natural hybridization and mechanical mixtures, the rate of mutation in pure lines can be determined.

Pure lines or inbred lines also find an important place in many biological investigations. Inbred lines of some animals like rabbits, mice and guinea pigs form the basis of many studies of immunological, physiological, biochemical, nutritional and medicinal nature since such animals do not possess any genetic variation and whatever variation they show after treating with any vaccine, medicine, chemical, etc. is the clearcut effect of that particular treatment. In such studies, therefore, confusion about the mixed effects of treatment and the genetic variation can be avoided beyond reasonable doubts.

Functions of Pure Line Selection

Pure line selection performs a number of important functions like improvement of land varieties, existing pure line varieties, introduced material, a specific character and segregating material in the breeding of self-fertilized crops.

Improvement of land varieties : The local or desi varieties which are under cultivation since long periods possess considerable genetic variability. These varieties though are well adapted to specific environmental conditions are usually low yielding. The application of mass selection method to these varieties generally results in marginal improvement only. Pure line breeding has proved very successful in evolving new varieties from land varieties in many crops like wheat (NP 4, NP 6, etc.), barley (C 50, K 12, etc.), tobacco (NP 28, NP 63, etc.), linseed (NP 11, NP 12, etc.), okra (Pusa Sawani), cowpea (T 1) and mung (T 1).

Improvement of existing pure line varieties : Due to the occurrence of mutations and hybridization, the pure line varieties gradually become genetically variable. In the process, some of the important characteristics of the variety are lost, but some new types appear which may have some special featuers like earliness, dwarfism and resistance to a particular disease. The selection of such off-type plants in some cases has proved fruitful. For example, Shyama variety of rice was developed through pure line selection from a dwarf off-type plant of the variety Kalimoonch 64. Similarly, in U.S.A., some important pure line strains (Frazier, Kanota, etc.) were developed from the well known Fulghum variety of oats which itself was developed by J.A. Fulghum from an off-type plant in the field of Red Rustproof oats.

Improvement of introduced material : Kalyansona and Sonalika varieties of wheat are the most striking examples of this type. These varieties were developed from the introduced wheat material from CIMMYT (Mexico). The two varieties made significant contribution towards wheat revolution in India. Sonalika is still being grown in many parts of the country as a late sown variety. WH 542 a timely sown variety of wheat has also been developed from selection of CIMMYT material. Another example of the improvement in introduced material relates to the mung variety Shining Mung No. 1 which was also developed through pure line selection.

Improvement in respect of a specific character : The importance of a character may change from time to time. A character which is considered unimportant today may assume great importance in future. Some of the characters which were previously unimportant have now become important criteria for selection in a number of crops. For example, dwarfism in wheat which has revolutionized world agriculture was not attached much attention by the wheat breeders.

A clearcut example of improvement in such a character comes from sorghum. A new root rot disease called as sorghum disease or milo disease having catastrophic effect on the sorghum crop was discovered in Kansas, U.S.A. in 1926. A resistant strain to this disease was developed from plants selected from the variety Dwarf Yellow Milo. The new strain had almost all the characteristics of the mother variety except that it was completely resistant to the milo disease.

Improvement of segregating material : Pure line breeding is also applied to develop varieties from segregating materials obtained after crossing different parents. This aspect will be discussed later.

Advantages of Pure Line Selection

1. The variety developed through pure line breeding is homozygous and genetically homogenous. Its produce is uniform in all the characters and, therefore, the variety is favoured by the seed producing agencies, farmers and the consumers.

2. Because of the extreme uniformity in all characteristics, the variety makes its identification very easy in seed certification programmes.

3. Since the best line is evolved as new variety in this method, a lot of improvement over the original variety can be achieved.

4. All farming operations (harvesting, threshing, etc.), particularly in mechanized agriculture, are easy with a pure line variety than with a variety which is a mixture of different genotypes.

Disadvantages of Pure Line Selection

1. Pure line selection is relatively lengthy and costly method than mass selection. Therefore, breeder cannot devote much time to other breeding programmes.
2. The application of the method is confined to self-fertilized crops only.
3. Pure line varieties have usually narrow adaptability and low stability of performance under different environmental conditions.
4. A ceiling is imposed on the ultimate degree of improvement by the best line of the base material. No variety evolved from the base material can be superior to this best line. However, the extent of improvement over the original variety through pure line selection is more than it is through mass selection.

Achievements of Pure Line Selection

As regards the varieties developed by pure line selection in different crops, there is a very long list since for a long time this procedure was one of the most popular methods followed by the plant breeders and thus was most extensively and successfully used for developing varieties in self-fertilized crops. In U.S.A., each of the pure line varieties (Nebred, Kanred, Blackhull, etc.) developed from Turkey wheat showed improvement over the mother variety for one or more important characteristics (disease resistance, earliness, stiffness of straw, etc.). For other general characters like yield potential and adaptation, these lines were almost similar to the old variety, except Blackhull which could be easily identified from other lines and the mother variety on the basis of some distinct features. Development of pure line strains like Franklin, Frazier and Kanota from the Fulghum variety of oats and a root rot (caused by *Periconia circinati*) resisant strain from Dwarf Yellow Milo variety of sorghum are other important examples of the successful use of pure line selection.

In India, a very large number of pure line varieties have been developed in a number of crops species like wheat (C 13 or K 13, K 46, K 53, Pb 8, Pb 11, NP 4, NP 12, etc.), barley (K 12, C 50, C 251, etc.), rice (BR 1, Patni 6, T 1, T 3., T 29, S 155, Mtu 1, Mtu 7, etc.), mung (T 1, B 1, etc.) tobacco (Chatham, Surti, T 23, T 59, etc.) and cotton (MCU 1, Gadag, etc.).

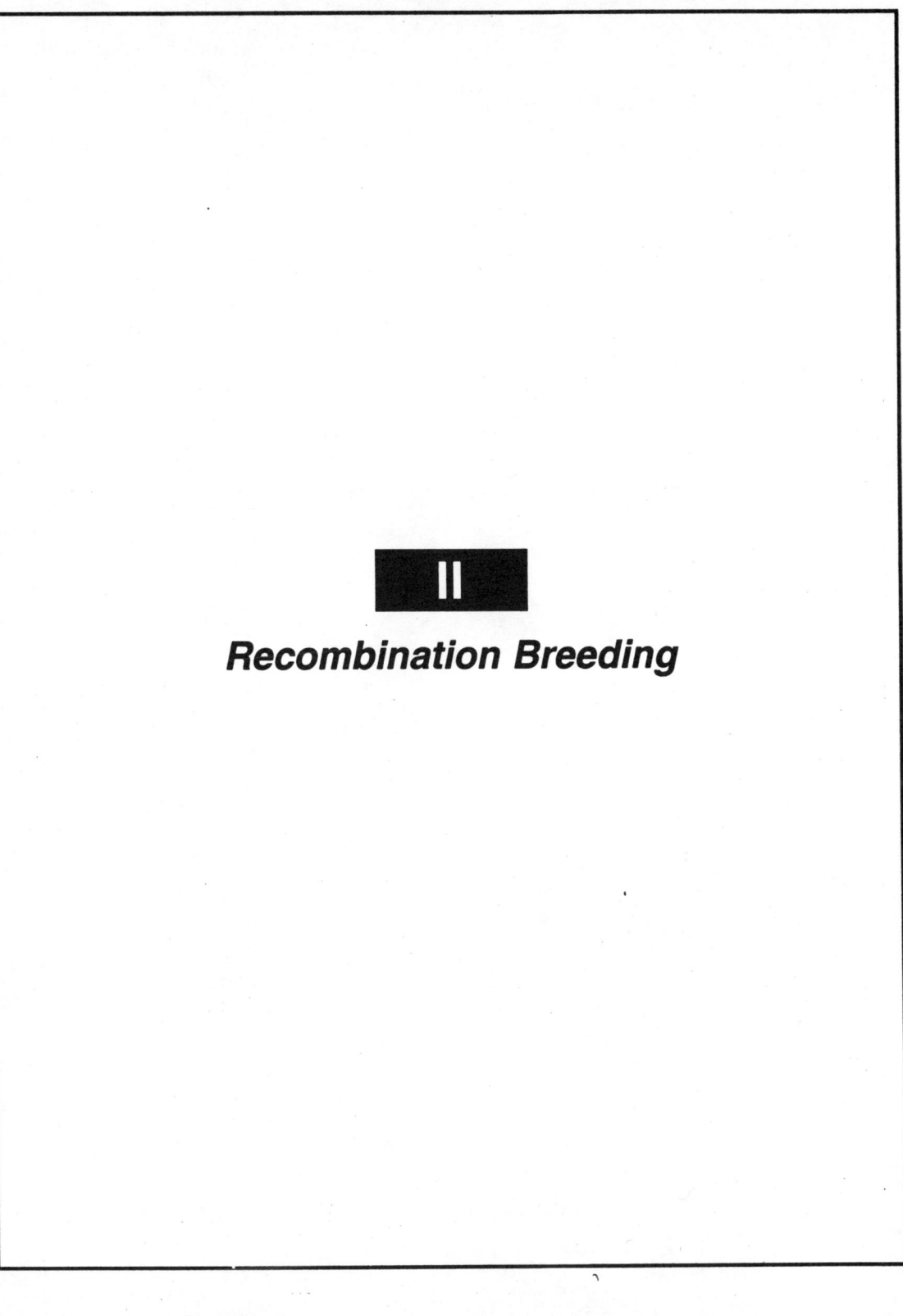

II

Recombination Breeding

CHAPTER 12

Pedigree Method and its Variants

Pure line selection and mass selection without hybridization have the serious drawback that the characters of two or more genotypes cannot be combined together. Therefore, hybridization becomes all the more important for the plant breeder. Most of the hybridization programmes relate to the crosses made between different varieties of the same species and handling of segregating generations in various ways. In self-fertilized crops, four methods are generally used to handle segregating generations. These methods are : (1) Pedigree method; (2) bulk method; (3) backcross method; and (4) single seed descent method. The basic objective of all these methods is to evolve a homozygous variety possessing desirable features of the parents involved in hybridization. These all methods come under recombination breeding for developing line varieties.

Although the methods mentioned above have their own merits and demerits depending upon the objective of the plant breeder, the kind of parents available for hybridization and the facilities (land, labour, etc.) available to the breeder, the pedigree selection method is most commonly used to handle the segregating generations. Pedigree method in self-fertilizing crops, is similar to pure line selection except that pedigree records (parent-progeny relationships) are not maintained in the latter and that the former is used to handle the segregating generations whereas the latter is applied to homozygous but genetically heterogeneous base materials. The main features of the pedigree method are :

1. Crossing of carefully chosen homozygous but genetically diverse parents (possessing complementary characteristics of breeder's interest like high yielding ability, wide adaptation, resistance to diseases and insect pests and high quality) to combine their desirable genes in a single genotype to be evolved as the new variety;

2. producing F_1 heterozygous for maximum number of loci so that F_2 generation has maximum possible heterogeneity;

3. selection of single desirable plants from space planted large F_2 population (depending upon the genetic width of the cross);

4. within and between families selections in the F_3 and subsequent generations;

5. more emphasis on between families selection than within family selection after F_5 as variability between families is more conspicuous and variability within family is quite less at this stage due to continued selfing;

6. final evaluation of a few promising lines in preliminary yield trials along with check varieties; and

7. maintenance of pedigree record (that is, record of all parent-offspring relationships).

It is this last feature which provided the basis for the name of the method. While recording pedigree (a series of notes about the relationship among the individuals or lines selected or description of the ancestors of each individual or family saved for propagation), notes in respect of each individual or family selected at every stage of the breeding programme are recorded for general vigour, resistance to diseases and insect pests, earliness, distinguishing features of families, or some other attributes depending upon the objective of the plant breeder. These notes should not be very lengthy and complicated since keeping record of each desirable plant or family is very much demanding of effort and time and is thus considered to be one of the demerits of this method. However, selection or rejection of any individual or family is decided on the basis of these notes. Recording of pedigree is especially useful when the breeder comes across some lines possessing almost equal potential. If the record of line relationships has been maintained accurately, the breeder would not feel any difficulty in choosing lines of diverse descent. In fact, the main advantage of maintaining the record of the lines of descent of all the selected plants or families is to discriminate between the lines that whether the equally performing families come from the same line of descent or from the different lines of descent. However, the information on line relationships does not ensure development of the same variety from the same cross in future. There is no point in keeping the records of the plants or families which have been rejected due to their obvious defects.

There are several different ways of recording pedigree. The breeder can choose any one according to his convenience and objective. However, the method of keeping record should be easily understandable and should give accurate information about the year in which the cross was made, the number of cross and the relationships among the plants or families selected. This information can be used in inheritance studies if needed.

While handling hybrid populations of self-fertilized crops, the breeder is expected to know the following :

1. The genetic structure of populations in different generations is different.

2. The variability present in F_2 generation indicates the maximum potential of a cross since no new gene combinations are expected to appear in F_3 and subsequent generations if the F_2 population grown is large enough.

3. Beginning at F_2 stage, the homozygosity will increase in each selfed generation with a corresponding decrease in heterozygosity. However, the homozygosity will increase rapidly in the first few generations of selfing and its rate of increase will be very slow in the later generations.

4. The presence of overdominance may create complications during selection since the plants seem to have a high potential at F_2 stage, may not do so in later generations.

5. Segregants exceptionally better or worse than the parents of the cross may appear in segregating generations.

Procedure of Pedigree Selection

In this method, starting from F_2 selection is practiced in each generation for all the desirable characters which the breeder wants to incorporate in the new variety. Therefore, a careful choice of parents and use of aids to selection (discussed in Chapter 6) will increase the degree of success of the programme. The whole procedure involves the following steps :

(1) Making crosses : Cross making takes usually least time than any other single operation of the breeding programme. However, the easiness with which the crosses can be made varies generally from crop to crop. In some crops (cotton, wheat, etc.) this operation is very easy, while in some others (chickpea, mungbean, urdbean, etc.) one has to face lot of difficulty in producing crossed seed because of the shedding of flowers, small size of flowers, etc.

In this operation, flower buds of suitable age on the parent to be used as female in the programme, are emasculated. When the stigmas of these flower buds become receptive, pollen from the male parent is put on the stigmas. For producing more quantity of crossed seed or attempting several crosses simultaneously, adjustments in the sowing dates can be made. This operation can be carried out in a green house or in the field.

(2) Raising F_1 generation : Usually a small quantity of hybrid seed (around 50 seeds) is needed to raise the F_1 generation. In usual practice, half of the F_1 seed produced is used to produce the desired F_2 population and the remaining half is kept reserved to grow a similar F_2 if the first crop fails due to any reason. Therefore, it should be ensured that the number of F_1 plants is enough to produce seed required for these two purposes and that the F_1 plants are the hybrids in real sense. Space planting of the hybrid seeds helps in producing more F_2 seed.

(3) Generations of individual plant selections : The first generation (segregating) in which single plant selections are done is the F_2. For exploiting maximum genetic variability provided by a cross and for convenient evaluation of plants, a large space planted F_2 population is grown. The size of the F_2 population will depend on the degree of wideness of the cross and the facilities available with the plant breeder, that is, how many F_3 families he will be able to handle ? Around 5000 plants in F_2 generation may be grown to have a good representation of different gene combinations and to produce a good number of F_3 families. The ratio of F_2 plants to F_3 families usually depends on the genetic diversity between the parents. If the genetic diversity between the parents is less, this ratio may be 10:1 and accordingly the number of the plants grown in F_2 will be much less. On the other hand, this ratio may be 100:1 or more in case of crosses between distantly related parents and accordingly a ruthless rejection of plants has to be done at F_2 stage to avoid unnecessary rush of materials. Since F_2 generation offers the first opportunity for selection to the breeder, fixing of some criteria of selection at this stage will help him in selecting outstanding plants. These criteria are :

(i) Rejection of all plants with easily and distinctly observable defects (such plants will carry undesirable major genes);

(ii) more emphasis should be given on the selection of plants that have high intensity of simply inherited characters;

(iii) identification and selection of transgressive segregants;

(iv) rejection of plants based on laboratory tests particularly for quality traits; and

(v) reduction of plants to an easily manageable number so that no difficulty is faced with regard to the space available, recording pedigree, etc. at the F_3 stage.

These criteria indicate that high is the narrow sense heritability of a character, more effective will be the selection. Therefore, selection for high vigour and yield which can present a confusing picture should be avoided. However, the success in selecting outstanding plants and rejecting all undesirable ones will depend on the practical ability of the plant breeder.

To minimize the adverse effect of inter-plant competition and to make selection of F_2 plants more effective and efficient, the honey comb method of planting arrangement suggested by Fasoulas (1973) and further improved by Bos (1983) may be used. The method allows selection of individual plants from all over the field and is thus least affected by soil heterogeneity.

In F_3 generation, each plant selected from F_2 will be represented in the form of a separate family. Since enough care is taken during selection in F_2, visible differences should appear between the F_3 families. For doing a critical evaluation of these F_3 families, enough number of plants (usually 10-30 or more) of each family should be space planted. Since most of the F_2 selected plants were heterozygous at several loci, they will segregate in F_3 and should produce F_3 families with visible within families differences. At this stage, therefore, emphasis should be given on both the within families and between families selections. By this time, however, the breeder may have an idea about the genetic worth of a cross. If the parents have nicked well, the cross will give rise to a large number of plants worth selecting. Contrarily, progenies of some other cross may perform so poorly that the breeder may have to discard the entire cross. In general, the number of plants selected at this stage should not exceed the number of the F_3 families raised. The family of a superior F_2 plant will show better average performance in F_3. To phrase it otherwise, the F_2 plants are evaluated on the basis of the average performance of F_3 families produced by them. To facilitate discrimination between F_3 families, a replicated honey comb design may be used.

Usually, there is not much difference in handling the F_4 and the F_3 generations. But, since the degree of homozygosity will be more in F_4 than in F_3 generation, relatively more emphasis is given on between families selections than on within families selections. In F_4 also (like in F_2 and F_3) the basis of selection is single plants. By this generation (F_4), the differences between families become more pronounced and clear and thus the number of families can be reduced to a considerable extent at this stage. This drastic reduction in the number of families is done on the basis of pedigree record as well as on the basis of visual evaluation to avoid the selection of two or more families of the same line of descent. Generally, 25-50 families are retained at the end of F_4 generation. The number of individual plants selected in F_4 should be much lower than the number of F_4 families.

In F_5 generation, although the basis of selection is usually the same as in F_4, the between families differences will become wider (since the different famileis become almost genetically fixed by this generation) and the within families differences much narrow. It is, therefore, advisable to grow F_5 families in large plots (preferably in multi-row plots than in single-row plots for better comparisons among families) with plant spacings similar to those under field conditions. At this stage, usually 70-80 plants or less are selected on the basis of visual appearance, pedigree record, laboratory tests and yield. If within families differences have become so narrow that breeder feels difficulty in discriminating some plants of the same family, two or more plants may be used to produce a family in the next generation. Some plant breeders are of the opinion of initiating preliminary yield trials of the F_5 families (lines) with 2-3 replications as an additional criterion of selection.

(4) Generations of line (family) selections : In F_6 generation, the single plants selected in F_5 are grown in large (multi-row) plots with plant spacings like those of a commercial crop for evaluation. By this stage, the families become almost homozygous and differences between them become so distinct that almost entire variation is accounted for by between family differences. As a result, the within families selection is no more effective. Thus, only between families selections are made on the basis of visual observation, pedigree record, yield and laboratory tests. Generally, 12-15 best families are selected in F_6 and the plot of each family is harvested in bulk rather than as single plants. If preliminary yield trials were not initiated in F_5, they are conducted in F_6.

In F_7, the operations are almost similar to those in F_6 except that the bulk seed of F_6 families are sown in replicated plots of standard size with check varieties in a simple experimental design preferably a randomized block design. The term 'family' which implies variability is appropriately replaced by the term 'line', or 'selection' at this stage (some use the term 'line' or 'selection' in place of family even at F_6 stage). The selections (lines) with moderate performance are drastically eliminated and a few (3-4) very promising lines are saved on the basis of yield and quality tests in addition to some other selection criteria like disease resistance, plant height, lodging resistance, maturity, etc.

(5) Final evaluation of promising lines : The final evaluation of lines saved in F_7 may also be discussed under step 4 (above), generations of line selections. However, since at the end of this evaluation only one best performing strain is selected after vigorous testing, it has been discussed separately here. The final evaluation takes a minimum of three years (F_8 to F_{10}) but may be extended up to five years (F_8 to F_{12}) in some cases. The same operations and tests which were done in F_7 generation, are continued at different locations with bit more precision. The tests involved in this evaluation are : (a) A critical evaluation of lines for obvious weaknesses that might have not appeared in the previous stages of the programme, (b) tests for quality, and (c) more precise tests for yield.

After these precise tests, a single most promising strain is finally selected, given a variety name and recommendation is sent to the variety release committee for its release. At the same time, the seed of the strain is increased by growing it in large plots.

Advantages of Pedigree Method

1. The method offers greater opportunity for the breeder to exercise his skill and judgement in the selection of plants as compared to other prominant methods (bulk population breeding and backcross method) used to improve the self-fertilized crops.

2. It allows drastic elimination of undesirable material in early generations and thus saves time and space for giving more attention to the promising materials.

3. Improvement for characters that are easily identifiable and controlled by simple genetic systems can be achieved very rapidly through this method.

4. The method allows evaluation of material under different locations and in different years.

5. It permits exploitation of transgressive segregants.

6. Since pedigree records of all the selected plants and families are maintained, the inheritance studies about characters can be carried out.

Disadvantages of the Method

1. It is the most expensive method in terms of land, labour, time, etc.

2. Individual plant evaluations take lot of time and therefore the breeder can handle a limited amount of material only.
3. There is early fixation of genotypes and thus selection for the characters like yield is not very effective in early generations. As a result, a large number of progenies are retained till later generations when selection becomes effective for such characters. If sufficient number of progenies are not retained in early generations, some promising genotypes may be lost during these generations.
4. The method does not provide any scope for natural selection to act upon the populations.

The pedigree selection has been and is being most extensively used for the improvement of self-fertilized crops particularly when a plant breeder wants to achieve most out of a limited amount of material in the shortest possible time. The list of varieties developed through this method is so big that it is not possible to give it here. The varieties K 65, K 68, WL 711, C 306, WH 147, WH 283, WH 291, WH 711, HD 2687, PBW 502 of wheat, Padma, Jaya, Krishna, Kauveri, Ratna and Karma of rice and Gaurav of gram are some examples from that long list.

Variants of Pedigree Method

The procedure of pedigree method described above provides general guidelines about the method. However, appropriate modifications can be made in the method as and when felt necessary to make it more efficient and more informative. It is now a general realization among the plant breeders that the choice of one method of breeding does not mean the rejection of the other, rather different methods should be considered supplementary to each other and thus a method combining merits of two or more methods will be more effective than using a single conventional method. Further, the plant breeder sometimes has to make modification in the breeding procedure according to the prevailing climatic conditions as weather also dictates its own terms. Some variants of the pedigree method are :

Progeny row selection : This method was suggested by Vinall and Cron (1921). The method was largely used for sorghum improvement (Quinby and Martin, 1954). In this method, the F_1 plants of a cross involving diverse parents are raised in rows and cross pollination is prevented by bagging the heads of the plants. The F_2 generation is also raised in rows and single plant selections are made for promising types. The heads of the selected plants are bagged and comparisons with parents and other commercial varieties are made. Progeny rows of F_2 selected heads are grown along with checks (parent varieties) in replicated blocks. Again, best plants are selected and heads are bagged. Simultaneously, a sufficient number of heads from extra-ordinarily performing rows may be bagged to make a plat test in the following season. If the rows become sufficiently uniform in F_3, each row can be treated as a unit. Seeds of best performing F_3 lines are grown in duplicate plats in F_4 and cross pollination with other sorghum varieties is checked by growing tall corn plants around the plat and final data are recorded for grain and forage yield. Seeds of other selections made in F_3 are raised in head-to-row as was done in F_3. All undesirable rows are discarded. In F_5, large plots of best performing plats are raised for increasing seed (Fig. 12.1).

Mass pedigree method : The method was suggested by Harrington (1937) to remove the deficiency of pedigree method when selection and recording of data become ineffective due to unfavourable environmental conditions. For example, in a drought year, selection for disease resistance and some other characteristics becomes ineffective and money and labour employed during such a season go waste. The best course under such circumstances is to use bulk population breeding till there are favourable environmental conditions for selection. After that pedigree method is resumed. Such use of bulk breeding can be done at any stage of the programme and for more than one season.

This method provides scope for natural selection and makes the programme less costly than pure pedigree method. In this way, the mass pedigree method combines merits of both the pedigree method and the bulk population breeding (Fig. 12.3).

Modified pedigree method : Breakwell and Hutton (1939) suggested a variant of pedigree selection and called it modified pedigree method. The method provides information regarding yield potential and product quality just after the collection of data and allows more intense and more precise selection at an early stage of the breeding programme. Here, F_2 generation is straightforwardly pedigreed. In F_3 generation, however, two programmes are run simultaneously. On one side, single plants are selected from the best performing lines for continued pedigree selection and, on the other side, bulking of the rest of the selected families is done for a yield trial in the next season. This process of continued pedigree selection and bulking of the rest of the selected families is repeated in F_4 and F_5 generations. The seed for each subsequent trial is taken from the progeny rows as well as from the previous trial. In F_6, most promising lines from the progeny rows are raised in small plots and seed of these plots is bulked for seed multiplication (Fig. 12.2).

Pedigree trial method : A method similar to modified pedigree method was proposed by Lupton and Whitehouse (1957) and was named as pedigree trial method. In pedigree trial method, the bulking of selected lines is started one step later, that is, at F_4 stage and the seed for subsequent trial is taken from progeny rows only. The process of continued pedigree selection and bulking of selected lines is repeated in the F_5 and F_6. In F_7, most promising lines from the progeny rows are raised in small plots and then the seed is bulked for increasing its amount.

Modified pedigree method and pedigree trial method have some advantages in common : (1) Both the methods provide quantitative estimates of yield and product quality at an early stage of the breeding programme, (2) the methods allow more intense and more precise selection at an early stage, and (3) both the methods are especially appropriate for computer development and thus make the experience and skill of the plant breeder less important so that a small group of workers can handle large amount of material with far greater precision. Both the methods, however, require more attention at each stage. Rathjen and Lamacraft (1972) realized the importance of such methods of breeding and developed a computer system for the improvement of cereals using pedigree trial method (Fig. 12.2).

Bulk-pedigree selection – Soybean breeders realized that though both pedigree and bulk breeding methods were popular in this self-fertilized crop, a method combining merits of both these methods was expected to prove better than using either method singly. Weiss (1949) proposed such a combination of pedigree and bulk methods for soybean breeding and called it bulk-pedigree selection. In this method, bulk method is employed upto F_3 or F_4 generation, then selections are made and these selections are classified according to maturity groups. After this, normal pedigree method is applied. The method is a simple modification of the mass pedigree method of Harrington (1937).

Culbertson (1954) emphasized the importance of pedigree method for developing wilt and rust resistant flax varieties despite the fact that all methods recommended for the improvement of self-fertilizing crops were being used for improving flax. In the pedigree method described by Culbertson, cross is made in the field and F_1 plants are grown in green house. A large number of F_2 plants (2000-

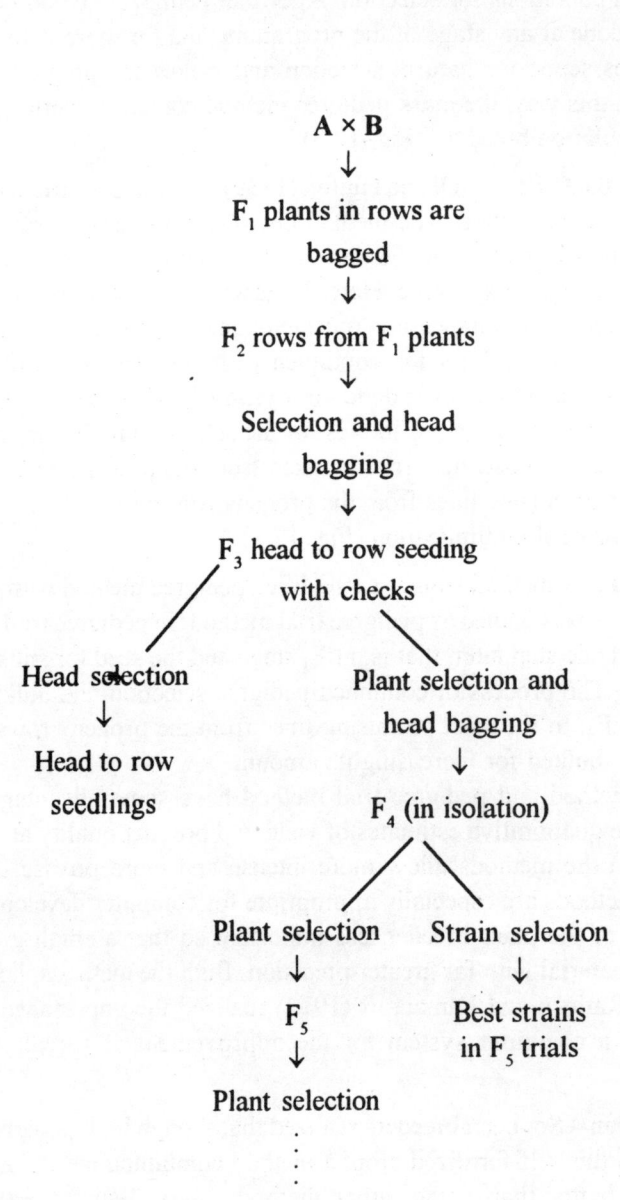

A × B

↓

F_1 plants in rows are
bagged

↓

F_2 rows from F_1 plants

↓

Selection and head
bagging

↓

F_3 head to row seeding
with checks

Head selection

↓

Head to row
seedlings

Plant selection and
head bagging

↓

F_4 (in isolation)

Plant selection   Strain selection

↓   ↓

F_5 Best strains
                        in F_5 trials

↓

Plant selection

⋮

Figure 12.1 Progeny row selection method

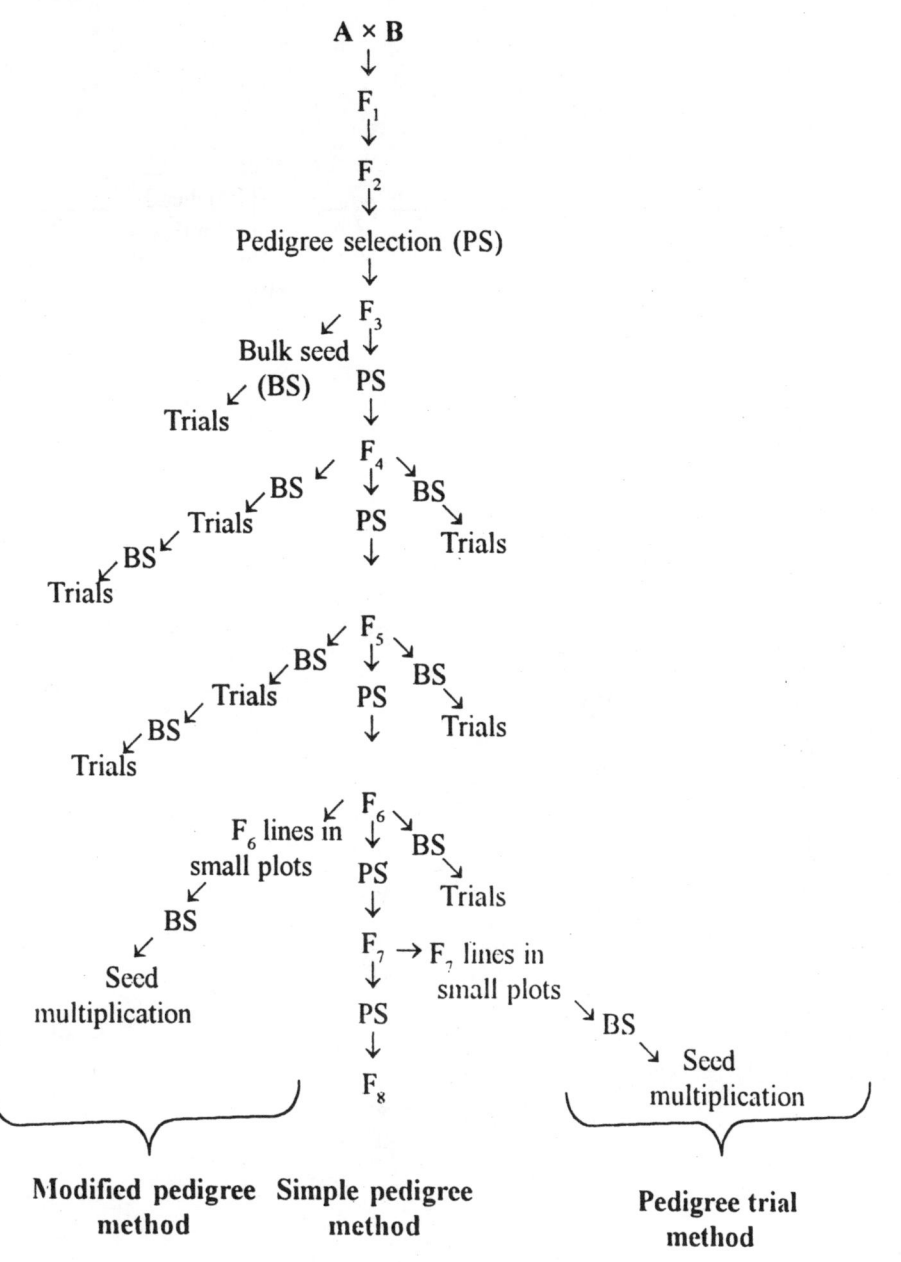

Figure 12.2 Simple pedigree method, modified pedigree method and pedigree trial method

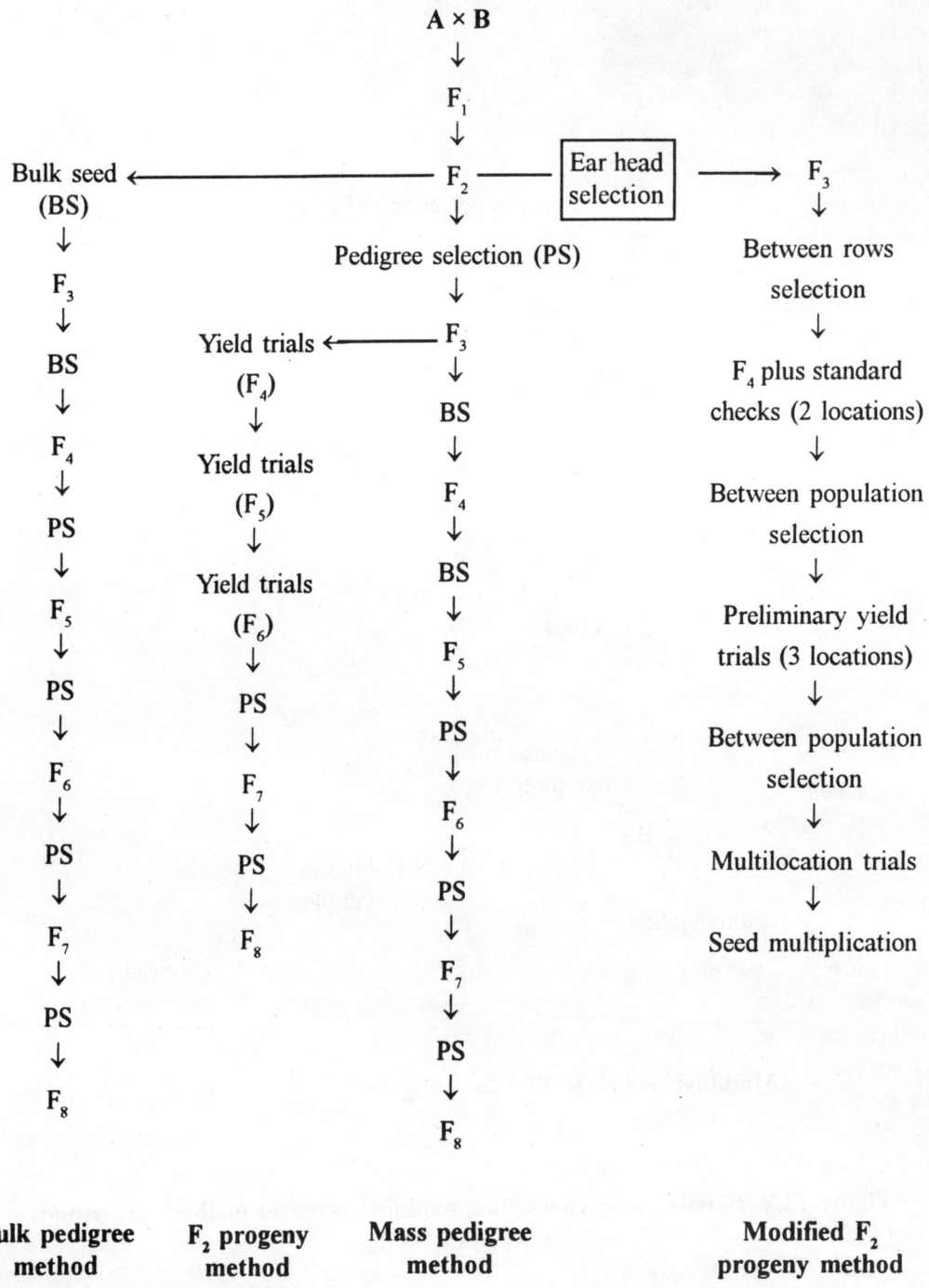

Figure 12.3 Some variants of pedigree method

10000) are space planted on wilt-infected soil to test wilt resistance. Plants showing resistance to wilt and prossessing desirable agronomic characteristics are selected from this population. F_3 seedlings are inoculated with rust spores and only those lines which possess resistance to rust are retained. In F_4 and F_5, selected lines are grown on wilt-sick soil and selections are made for wilt and rust resistance and other important characters. Desirable F_5 lines are harvested in bulk, tested for wilt and rust and are placed in preliminary yield trials.

F_2 progeny method : Another variant of mass pedigree method of Harrington (1937) was suggested by Lupton and Whitehouse (1957). The method was named as F_2 progeny method. The method involves growing of F_3 (pedigreed) families (progenies of selected F_2 plants) in yield trials without selection in F_4, F_5 and F_6 generation. After F_6, single plant selection (pedigree selection) is resumed. Like modified pedigree and pedigree trial methods, this method also allows an early assessment of yield potential and quality. The method saves space and labour at each stage particularly at F_4 to F_6 stage, of course, at the cost of precision in the identification of individual promising genotypes. The method has been successfully used by Rathjen and Pugsley (1978) in Australia (Fig.12.3).

A modification in F_2 progeny method has been proposed by Smith (1987). The variant is called as **modified F_2 progeny method** and is especially suited for wheat breeding in semi-arid regions. The method involves selection of ear heads from promising F_2 populations in a multiple of 100. The selections are based on general vigour, appropriate height, maturity and high fertility. The selection in the F_3 head rows (occasionally grown with standard check varieties) is confined to between rows only and is based mainly on important traits like general vigour, maturity and disease resistance. Then selected subpopulations along with standard check varieties are raised in F_4 at two locations (non-replicated) and identification and examination of promising subpopulations is done for bread making quality and grain weight. Promising subpopulations are then included in preliminary yield trials which are raised at three locations representing a good range of environmental conditions for drought stress. Selection is practiced in these trials for most promising subpopulations which are then evaluated in multilocation trials. The approach allows determination of yield potential and phenotypic stability at an early stage of the breeding programme. If needed, roguing and reselection are applied for reducing heterogeneity present in the finally selected subpopulations (Fig.12.3).

Accelerated pedigree selection : Valentine (1984) proposed an alternative approach to overcome the low efficiency of the otherwise very powerful pedigree breeding method for improving self-fertilizing crops and called it as accelerated pedigree selection. The method combines some of the best features of pedigree method and single seed descent breeding method and is especially appropriate for breeding winter cereals. Here, assessment of lines (derived by accelerated generation procedure) rather than individual plants is the basis of initial selection, that is, line selection is done in an early generation. The main advantage of early line selection is to overcome the risk of differential mortality or sterility. The method is superior to pedigree method and single seed descent method as it has shorter breeding cycle than both these methods. It is superior to pedigree method in two main aspects : (1) The length of time taken to reach the first generation raised as single or multiple rows in the field with normal times of sowing and harvesting is less in this method than in pedigree selection (whereas this length of time in pedigree method is three years after hybridization, the same is two years in accelerated pedigree selection), and (2) the assessment of several important plant characters (yield, maturity, adult plant resistance to diseases, lodging resistance, etc.), is more convenient and better when it is done on rows or plots basis than on close-spaced single plants basis. The method has three advantages over single seed descent breeding method : (1) Overcomes the drawback of small sample size (single seed) from each plant in single seed descent method and thus covers the risk of differential mortality or sterility and accompanying undesirable genetic shift carried by single seed descent method; (2) since the number of accelerated generations in single seed descent method is larger than in accelerated pedigree selection, the former method needs more resources per initial cross than the latter; and (3) the single seed descent method does not allow assessment of performance and adaptability as is done by the accelerated pedigree selection during several years of pedigree selection, the former method needs to assess lines over several years for high performance and wide adaptability after the termination of the programme. Accelerated pedigree selection is most suitable for winter cereals since the need for vernalization in these cereals makes the generation time longer. As regards the length of the breeding cycle in winter cereals, it is six years for pedigree selection and single seed descent method but only five years for accelerated pedigree selection and thus the latter method is one year shorter than the two former methods. However, embryo culture 2-3 weeks after anthesis coupled with 4-week cold treatment in winter wheat reduces the generation time by almost 40 days (Sharma and Gill, 1983).

The complete procedure of accelerated pedigree selection in winter cereals may be as follows:

Crosses are made in the month of June (normal season) and the embryos are grown on culture media or alternatively grains are harvested, dried and sown at the green stage in July. The F_1 crop is harvested in December and F_2 seeds are sown for random multiplication. In August, the F_2 crop is harvested and F_3 single rows are raised in October. Row selection rather than individual plant selection is made and the F_3 crop is harvested in August. In October, multiple F_4 rows are raised. Row selection is again practiced and the crop is harvested in August. In the fourth year, F_5 yield trials are raised. Yield assessment of these trials is done and the crop is harvested in August. In the fifth year, sowing of F_6 yield trials and purification plots is done in the month of October. The yield assessment and purification are done and the crop is harvested in August.

Valentine (1984) also suggested a modification in accelerated pedigree selection scheme to further shorten the length of breeding cycle. Such a modification is especially advantageous for exceptionally promising breeding materials or when there is fierce competition with other breeders. The modified scheme takes about four and a half years from cross making (March in zero year) to final harvesting of the material (August in fourth year) after yield assessment and purification. As regards the overall modified scheme, crosses are made in March, F_1 seeds are sown in April and the crop is harvested in August. The F_2 generation is raised in September (first year) for random multiplication and the crop is harvested in January. In February, sowing of F_3 plants or single rows is done for further random multiplication and the crop is harvested in August. Multiple F_4 rows are raised in October. The row selection is practiced and harvesting is done in August. In the third year, F_5 yield trials are sown. Yield assessment of the trials is done and the material is harvested in August. In October of the fourth year, sowing of F_6 yield trials and purification plots is done. The assessment for yield and purification is made and the final harvesting is done in August (Fig.12.4).

Simplified method of bulk and pedigree selection : Record keeping of each selected plant or family in pedigree method is very much demanding of effort and time and thus makes the method most expensive. One or two cycles of bulking after F_2 or F_3 and thereafter continuing usual pedigree selection will make the method simplified, provide relief to the breeder, minimize unreasonableness involved in the application of pedigree method and will combine advantages of both the pedigree selection and bulk population breeding. To incorporate some element of bulk population breeding in pedigree method, Sneep, Murty and Utz (1979) suggested bulking of F_3 selected rows in the F_4 of the pedigree programme and called it simplified method of bulk and pedigree selection. The bulking of material can be done at any appropriate stage of the breeding programme according to the breeder's desire. Simplified pedigree selection (bulking at F_3 stage) method has proved better than or equal to the traditional pedigree method in terms of mean performance, coefficient of variation and the number of high yielding lines selected in wheat crop (Pawar *et al.*, 1990).

Shuttle breeding system : A method similar to simplified method of bulk and pedigree selection is being used at CIMMYT to breed wheat for drought resistance for semi-arid regions. The method is called as shuttle breeding system and has been discussed by Osmanzai *et al.* (1987). The main objective of this method is to obtain information about the most important characteristics of CIMMYT germplasm (wide adaptation, high yield and high stability), that is, to combine input efficiency and input responsiveness using alternate sites of contrasting fertility and moisture conditions. Here, F_2 F_5 and F_6 generations are raised under optimal conditions of fertility and moisture and pedigree selection (individual plant selection as well as line selection) is followed based on good tillering, disease resistance, proper ear development, grain plumpness and high leaf retention capacity, while the F_3 and F_4 generations are grown under reduced moisture and low fertility conditions (stress environment) and modified bulk method is followed in both the generations. Most promising lines performing equally well under optimal as well as under stress environmental conditions are selected and evaluated in international trials for drought resistance and yield (Fig. 12.4).

Backcross pedigree method – In most of the cases, one of the parents used in hybridization programme has proven performance for majority of its characters in the areas of intended use, while the other parent is chosen to complement specific weaknesses of the first parent, that is, almost without exception, desirable characters are not equally distributed among the parents used in any hybridization programme. For this reason, restricted backcrossing in the beginning of the pedigree programme or in the early generations can serve a useful purpose in concentrating desirable characters in early generations without loosing much heterozygosity and thus increasing the efficiency of the method. One such method which combines advantages of both the pedigree and backcross methods has been suggested by Allard (1960). The method is called as backcross pedigree method.

In addition to the modifications in pedigree method described above, the efficiency of this method may be enhanced by following some practices like intermating among selections, early generation testing and within and between crosses evaluation (Fig. 12.5).

Random crossing among selections in early generations : Successive selfing in pedigree method (and bulk population breeding to be discussed later) after hybridization puts a limit on the appearance of new recombinants and enhances fixation of linkage blocks in early generations. Each generation of selfing imposes a ceiling on the ultimate value of the derived line. Crossing of selections at random in early generations will help in checking early fixation of genotypes and breaking linkages. By this, better recombinants can be obtained since it will provide further variability for making selections. If there are loose linkages between desirable and undesirable genes, they can be broken by following one or two cycles of random mating and desirable recombinants may be identified. Since the number of plants selected in early generations is large enough, random mating among all selections will not cause a great change in the mean and variance of the population, rather the two

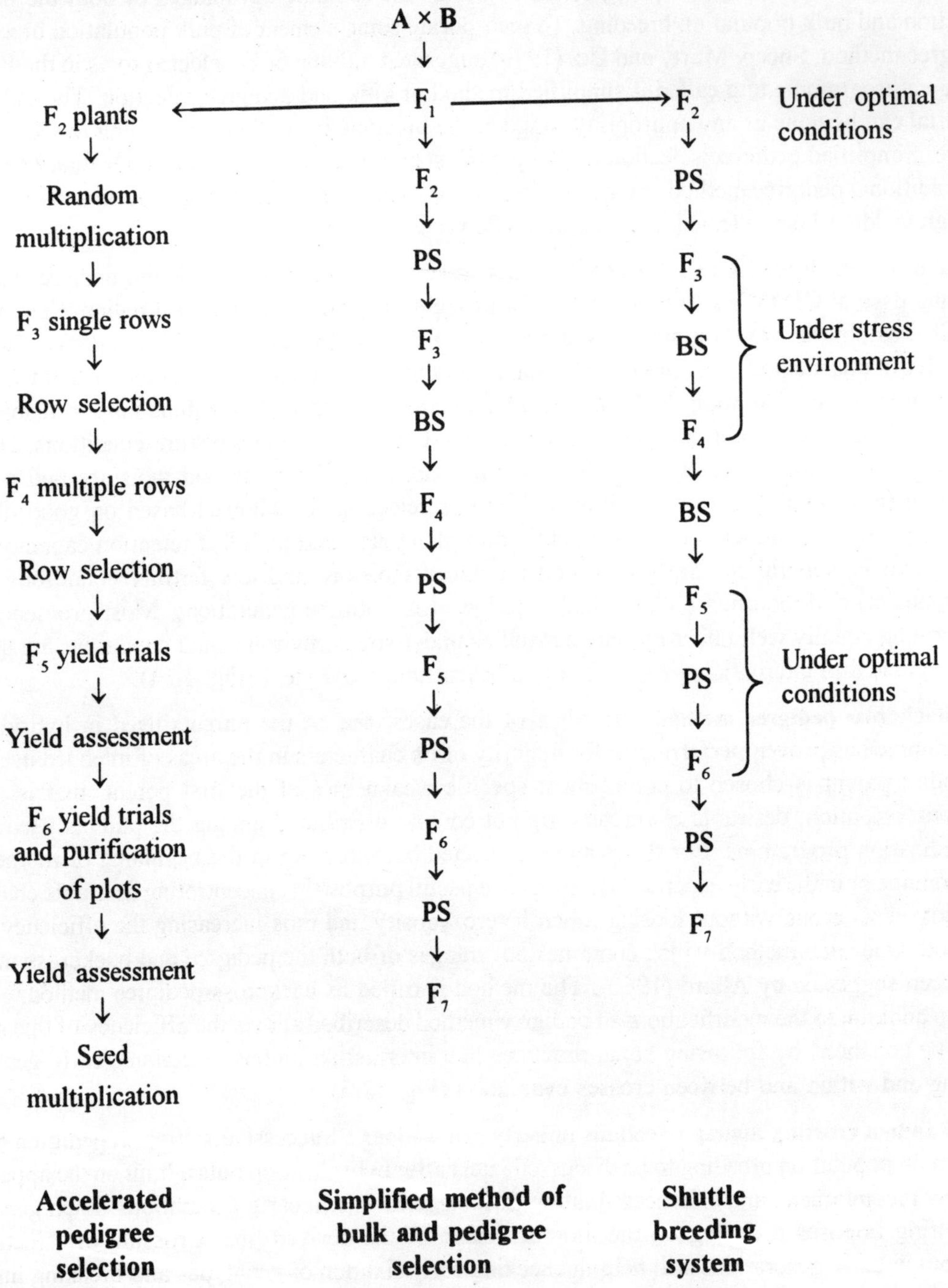

$$\mathbf{A \times B}$$

Accelerated pedigree selection	Simplified method of bulk and pedigree selection	Shuttle breeding system
F₂ plants	F₁	F₂ Under optimal conditions

$$
\begin{array}{ccc}
 & A \times B & \\
 & \downarrow & \\
F_2 \text{ plants} \longleftarrow & F_1 & \longrightarrow F_2 \quad \text{Under optimal} \\
\downarrow & \downarrow & \downarrow \qquad \text{conditions} \\
\text{Random} & F_2 & PS \\
\text{multiplication} & \downarrow & \downarrow \\
\downarrow & PS & F_3 \\
F_3 \text{ single rows} & \downarrow & \downarrow \quad \text{Under stress} \\
\downarrow & F_3 & BS \quad \text{environment} \\
\text{Row selection} & \downarrow & \downarrow \\
\downarrow & BS & F_4 \\
F_4 \text{ multiple rows} & \downarrow & \downarrow \\
\downarrow & F_4 & BS \\
\text{Row selection} & \downarrow & \downarrow \\
\downarrow & PS & F_5 \\
F_5 \text{ yield trials} & \downarrow & \downarrow \quad \text{Under optimal} \\
\downarrow & F_5 & PS \quad \text{conditions} \\
\text{Yield assessment} & \downarrow & \downarrow \\
\downarrow & PS & F_6 \\
F_6 \text{ yield trials} & \downarrow & \downarrow \\
\text{and purification} & F_6 & PS \\
\text{of plots} & \downarrow & \downarrow \\
\downarrow & PS & F_7 \\
\text{Yield assessment} & \downarrow & \\
\downarrow & F_7 & \\
\text{Seed} & & \\
\text{multiplication} & & \\
\end{array}
$$

Figure 12.4 Three variants of pedigree method

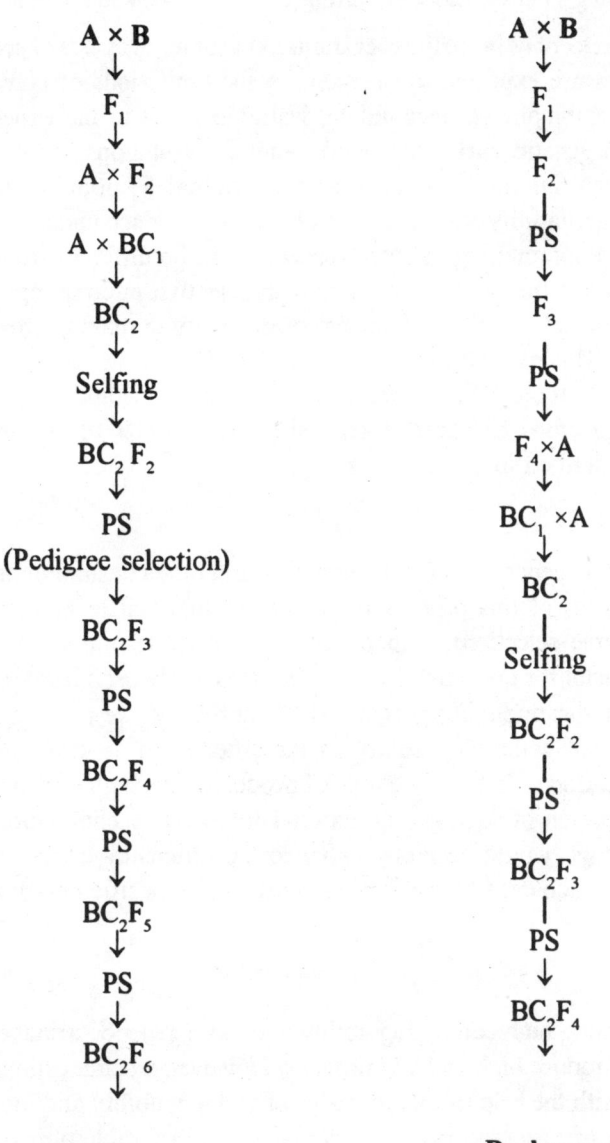

**Backcross
pedigree method I**

**Backcross
pedigree method II**

Figure 12.5 Backcross pedigree methods I and II

parameters will remain almost constant. However, random mating among large number of plants is a difficult task in many self-fertilizing crops (gram, mungbean, urdbean, cowpea, etc.). In self-fertilized crops where crossing is easy, random mating can be done without much difficulty.

Mating among superior selections in early generations : One or more cycles of selective intermating in between the selfing series are expected to overcome some limitations of conventional breeding procedures, faster gene recombinations, accumulate desirable genes in the experimental material, ensure utilization of hidden genetic variability, shift genetic correlations in the desired direction, provide better source of material for selection than the original F_2 population, and elevate the population mean, degree of heritability and the extent of genetic variance under selection. Therefore, the consequences of selective intermating in some respects, will be different from those of random mating among selections. Since the number of crosses in selective intermating will be much less than in random mating among selections, the intensity of difficulty in making crosses will be less in the former than in the latter. This method has been found effective for increasing the mean values of yield and its components in wheat crop (Pawar *et al.*, 1990, 2000). A method which involves selective intermating in self-fertilizing crops has been described by Jensen (1970) and is called as 'diallel selective mating system' (described in Chapter 17).

Early Generation Testing

Development of crop variety is generally a costly, complex and time consuming process. One of the factors determining the success of this process is the size of material to be handled by the plant breeder (especially in pedigree selection). Experience shows that despite a full care taken at the stage of the selection of parents for hybridization, several apparently promising parents do not nick well and as a result produce disappointing progeny showing that *per se* performance of parents is not always a good indicator of combining ability as the latter often depends on complex genetic systems. Therefore, identification of hybrids capable of producing the highest proportion of superior segregates and critical evaluation of segregating material followed by elimination of segregates of low potential at an early stage would certainly enhance the efficiency of the crop improvement programme. This goal can be achieved by estimating general and specific combining ability effects in an early generation.

Between Crosses Evaluation

Generally, the crosses exhibiting high combining ability and more genetic variance for yield in early generations are expected to produce high yielding progeny. However, a better estimate of the progress under selection can be had with the help of the estimates of yielding ability and those of the different components of variation in early generations. A predominance of a fixable type of genetic variance (additive genetic variance plus additive × additive epistasis), that is, high narrow sense variability in the early generations of a cross will indicate more effectiveness of selection in that cross. The effectiveness of selection will decrease with the increasing effect of environment. If the estimates of the components of variance are based on the experiments carried out at several locations in different seasons, these estimates along with those of genetic advance will provide a good criterion of selection for the identification of superior hybrids. Hybrids with high mean yields and high genetic advance should be selected and their progenies should be handled in the subsequent generations. The effectiveness of selection in some cases may be less at the F_2 stage but may increase in F_3 and F_4 generations. Therefore, information obtained from F_3 and F_4 trials is usually more reliable.

Within Cross Evaluation

Within cross selection is based on the characteristics of single plants. The performance of single plants is affected by the environment differently for different characters. For the simply inherited characters (presence or absence of awns, flower colour, grain colour, earliness, disease resistance, etc.), selection in early generations mostly proves effective since these characters are usually highly heritable. On the contrary, single plant yields in an early generation like F_2 have usually no bearing on the yields of their progenies in the subsequent generations, that is, selection in early generations for a complex character like yield is mostly ineffective. The studies on this aspect indicate that selection in F_3 and F_4 generations is effective for the characters with moderately high heritabilities. Space plantings in F_2, F_3 and F_4 and growing materials at different locations and in different seasons increase the reliability of the information. For yield, however, it is advisable to eliminate only the poorest plants or families (a truncated type of selection) in the early generations and reduce the lines drastically in later generations when they are almost homozygous.

CHAPTER 13

Bulk Method and its Variants

Bulk method is perhaps the oldest, simplest and the cheapest method. For the first time it was used in Sweden by Nilsson-Ehle (1908) who called it as 'population breeding' at that time. It is also known by some other names like bulk population method, bulk population breeding, mass method and evolutionary method of breeding. The method involves crossing of two carefully chosen parents, planting of F_2 generation in large plots (accomodating usually several thousand plants) with planting rates and cultural practices similar to those used in growing a commercial variety, harvesting the plot in bulk at maturity and using seeds for planting a similar plot in the following year or season. The number of bulked generations depends on the objective of the plant breeder and the type of the progeny produced by the cross. Thus, the method has involvement of both the nature and the man. The basic purpose behind the method is to gradually eliminate the weaker gene combinations (which fail to compete well with other combinations) and to identify and evolve the lines which show consistently good performance under natural conditions.

The method has a restricted application since it cannot be used in all kinds of crops with equal ease. Whereas it can be very easily applied to improve the seed crops like cereals, grain legumes and smaller millets, it is rather practically unfeasible to use this method in case of many vegetables (where a single plant occupies enough space) and fruit crops. Nilson-Ehle used this method for combining together two desirable characteristics of winter wheat, the winter hardiness and high yield, in a single variety. He crossed the variety Squarehead (possessing winter hardiness) with Stand-up (a high yielding variety) and planted F_2 generation in a large plot followed by bulk harvesting. He repeated the bulking process in several succeeding generations. Single plant selections were made when lines had become almost homozygous.

The important features of this method are :

1. Raising large populations (usually several thousand plants) in each generation increases the possibility of appearing many more gene combinations including those desired by the plant breeder, e.g., winter hardiness with high yield in wheat.

2. Since the pedigree records are not maintained and selection is not practiced in this method from F_2 to F_6, the plant breeder can handle a large amount of material as the time required to handle a cross is usually very little.

3. By growing material under specific environmental conditions the rate of shift can be increased towards a particular type desired by the plant breeder.

4. The lines selected after 4-5 generations of bulking are expected to breed true.

5. The rate of progress is slow but longer-lasting.

6. Except the duration of bulking the method is exactly similar to pedigree selection.

7. Since natural selection plays a major role in this method, the gene frequencies and the mean of the experimental material are expected to change over generations. The proportionate increase or decrease of a particular type during bulking period will depend upon its relative reproductive success.

The Procedure of Bulk Method

The method, in general, involves the following steps :

1. Making crosses : Like in pedigree selection, carefully chosen parents are crossed and enough crossed seed is produced because large populations are grown in F_2 and several subsequent generations.

2. Raising F_1 generation : The crossed seeds (around 50) are space planted to produce enough quantity of seed for planting a large F_2 population and for keeping half quantity of the seed reserved for raising a similar F_2 generation in case the crop fails due to some reason. The plants arising due to self pollination are eliminated. The material is harvested and the seed is bulked.

3. Generations of bulking : A large F_2 population (20000-25000 plants are usually enough) is planted like a commercial variety in respect of plant spacings, seed rate and cultural practices. At maturity, the material is harvested in bulk and F_3 generation of the same size is raised from the bulked seed. This process of bulking is usually continued upto F_6. However, there is no hard and fast rule regarding the duration of bulking. Depending upon the objective and the type of progenies produced by a cross, the breeder may stop bulking at F_5 stage or may continue it upto F_7 or F_8 stage. The number of bulked generations is generally more in wide crosses than in crosses between closely related parents.

Just like the F_2 generation of pedigree method, the last bulk generation is space planted and single plant selections (usually 1000-2000 plants) are made on the basis of visual evaluation, disease reaction and grain characteristics.

4. Generations of lines evaluation : The single plants selected from the last bulk generation are in fact homozygous or nearly homozygous lines. Hereafter, the method is exactly similar to the pedigree selection except that there is some change in the criteria of selection. Since no pedigree records are maintained in this method, they cannot become the basis of selection. As in pedigree method, the number of lines is drastically reduced at this stage of the programme and only a few promising strains are saved at the end.

5. Final evaluation of promising strains : Like in pedigree method, ultimately a single best performing strain is selected after vigorous testing and recommended for release as a variety.

The steps discussed above clearly indicate that in bulk method no artificial selection is practiced from F_2 to F_6 generation, whereas in pedigree method single plant selections are made in each of these generations.

Advantages of Bulk Method

1. It is the cheapest and most convenient method. As a result, homozygous lines can be obtained with least expense and effort by using this method.
2. The improvement attained through this method is longer-lasting as the method depends mainly on natural selection which starts acting right from F_2 generation. Suneson (1956) rightly called this method as 'an evolutionary method of plant breeding'.
3. Provides opportunity for handling large quantity of material.
4. Variety developed through this method need not to be tested at farmers' field.
5. The method provides more scope for the appearance and exploitation of transgressive segregants.
6. Provides information about the survival of genotypes and genes in populations.

Disadvantages of Bulk Method

1. The improvement attained through this method is slow.
2. Since pedigree records are not maintained, inheritance studies cannot be made.
3. The method does not provide opportunity to the breeder to use his full skill.

Comparison of Bulk Method with Pedigree Method

The comparison between these two methods used to improve self-fertilized crops can be made on two aspects : (a) theoretical aspect and (b) practical aspect. As regards the theoretical aspect (that is, what can be achieved by applying these two methods), the main differences between the two methods are as follows :

(i) The main force operating in pedigree method is artificial selection, while the bulk method is based mainly on natural selection.

(ii) The rate of improvement in pedigree method is higher than in bulk method.

(iii) The improvement achieved by bulk method is longer-lasting than that achieved by pedigree method.

(iv) Whereas pedigree method is most expensive, the bulk method is most simple, convenient and inexpensive method.

(v) The pedigree method permits the breeder to use his full skill and judgement but the bulk method does not provide any such opportunity.

(vi) Since the pedigree records are maintained in pedigree method, the method puts a limit on the quantity of material a breeder is able to handle. On the other hand, very large quantity of material can be easily handled by the breeder while using bulk method.

(vii) The bulk method provides more scope for the exploitation of transgressive segregants than does the pedigree method because the former method provides more opportunity for new gene combinations to appear.

(viii) Inheritance studies are possible in pedigree method, while no such studies can be carried out in bulk method.

(ix) In pedigree method, the segregating generations are space planted and single plant selections are made. In bulk method, on the other hand, the populations are grown just like commercial varieties and bulking is done in each generation.

(x) The bulk method provides information about the survival of genes and gene combinations in populations, whereas no such information is made available by the pedigree method.

The comparison between these two methods based on practical aspect (that is, what has already been achieved through the use of these two methods ?) does not present a very clear picture. There are reports against and in favour of both the methods. There are also reports which suggest that the two methods are equally efficient. Raeber and Weber (1953) and Torrie (1958) working with soybean found that the average performance of outstanding lines isolated from bulk populations was similar to those of obtained from pedigree populations. Copp (1957) advocated the superiority of pedigree method over the bulk method on the basis of experiments carried in wheat for a long period of time provided that disease does not act as a limiting factor. On the contrary, Florell noted a superiority of bulk method over pedigree method in barley on the basis of the performance of 45 promising lines isolated in 1934 from the bulk population of the cross Atlas × Vaughn. This cross was made by Florell in 1927 and grown in bulk in each year upto 1934. Four varieties, namely, Arivat, Beecher, Glacier and Gem, were released from these promising lines. Harlan (1956) has given list of a number of other barley varieties which had direct or indirect relation to composite cross populations.

Variants of Bulk Method

Some modifications of bulk method are described hereafter.

Composite crosses : The idea of composite cross method was initially given by Love (1927). But the method could get prominence only after Harlan and Martini (1929) and Harlan *et al.* (1940) had described it on the basis of barley data. In this method, a number of F_2 populations of different crosses are bulked together to make a **composite cross** or a **composite population** which is generally handled by a typical bulk population breeding.

Harlan and Martini prepared a composite cross by bulking all the F_2s produced by crossing 28 selected barley varieties in all possible combinations. No artificial selection was made upto F_7. From the bulk population grown in F_8 generation, 2921 desirable plants were selected on visual basis. The hybrid populations of all the 378 individual possible crosses ($28 \times 27/2$) were handled simultaneously by pedigree method and 2921 F_8 lines were selected from these individual crosses. A comparison between these two kinds of selections for yield indicated a higher efficiency of composite cross method. Harlan and Martini, however, attributed the higher average yield of composite selections mainly to the low survival value of low yielding two-rowed varieties whose number was relatively much less in the composite and thus their representation among the selections from the composite was very poor. On the other hand, among selections from the individual crosses, their representation was similar to that of other high yielding varieties because the two-rowed varieties were separately maintained in pedigree method. The poor representation of these low yielding varieties among the selections from the composite would naturally increase the average yield of these selections. However, a comparison between the composite and the sum of individual crosses indicated that the two methods showed equal efficiency as regards the yield, but the composite cross method may be considered superior since it requires lesser experimental efforts.

Further studies were carried out on this composite cross by Suneson (1956) by growing it at Davis (California) for 29 generations without selection. The yielding ability of the composite was compared periodically with Atlas a commercailly grown variety of barley. In the early segregating generations, the Atlas variety was conspicuously superior to the composite for yield and other agronomic characters. But as the generations advanced the composite started showing gradual improvement and by the F_{15} generation the bulk consisted almost entirely of coast types grown commercially in California and became almost equal to Atlas in yielding ability. By F_{20} generation, the composite surpassed the yield level of Atlas and showed 6% higher yield. The composite maintained this superiority consistently in later generations.

Suneson and Stevens (1953) isolated progenies from this composite bulk at different stages of bulking and found that individual lines selected in different generations of bulking were different in performance. For example, none out of 356 lines selected in F_{12} generation could outyield Atlas in replicated yield trials. On the other hand, out of 50 selections made in F_{20} generation (the stage at which the average yield of the composite was more than that of Atlas), two were found very promising in yielding ability and other agronomic characteristics. One of these two lines was so good that it showed superiority over Atlas in yielding ability (37 per cent) as well as for test weight (3.5 pounds higher). The superiority of this line over Atlas was observed in each of the seven years of their testing in yield trials. Similarly, out of 66 lines selected in F_{24} generation, the average yield of three top lines was 56 per cent higher than that of Atlas in replicated trials conducted for four years. These results indicate that the proportion of the outstanding lines in the composite increased with the advancement of generations of bulking. According to Suneson (1956), after 15 generations of bulking without artificial selection, the composite could be handled by any one of the three methods :

(a) Bulk population breeding with prospects of natural selection
(b) Pedigree method, that is, conventional selection and evaluation
(c) Intermating and selection, that is, making cyclic hybrid recombinations and selection (a kind of recurrent selection)

There are different ways of deriving a composite cross. Some are as follows :

(1) Method of Harlan *et al.* **(1940) :** In this method, a number of varieties are crossed in all possible combinations and the same amount of F_2 seed from each cross is bulked into one composite. This method, however, has one serious drawback in self-fertilizing crops that though each variety is crossed with every other variety, any hybrid plant of the composite cannot have germplasm from more than two varieties because there will be no cross pollination. The method, therefore, does not provide opportunity to combine germplasm from more than two varieties.

(2) Making multiple crosses (Harlan *et al.,* **1940) :** One can make 16 crosses from 32 selected varieties, involving each variety only once. The plants of 16 F_1s thus produced are crossed and 8 crosses are obtained. The F_1 plants of these 8 crosses are crossed to produce 4 crosses. Similarly, then two crosses are made and ultimately a single cross is made by crossing two parents, each parent possessing germplasm of 16 varieties. The number of varieties crossed in the beginning of the method may be reduced according to the convenience of the plant breeder (16 or 8 varieties may be taken).

(3) Suneson's (1956) method : In this method, crosses are made among promising lines of a composite to constitute a new composite. This procedure also suffers from a similar drawback as does the method of Harlan *et al.* Here also, no hybrid plant can have germpalsm from more than four varieties.

Allowing cross pollination through the use of male sterility, however, can greatly help the plant breeder in increasing the number of gene combinations in a composite hybrid population, particularly when bulking of composite is done for many generations. Such a use of male sterility was made by Suneson (1945) in barley.

Backcross bulk method : If the breeder is handling a complex cross, restricted backcrossing (one or two generations) in the beginning of the programme (before bulking the hybrid population) will help in accumulating the desirable genes of the parent possessing more favourable genes in the progeny. Thereafter, the normal bulk method is followed to handle the hybrid population. One such method which combines advantages of backcross and bulk methods, was suggested by Florell (1929) (Fig. 13.1).

According to some scientists, the **mass pedigree method** of Harrington (1937) which avoides wasted effort of the breeder in handling segregating generations by pedigree selection during the years unfavourable for effective selection for important characters (disease resistance, lodging resistance, drought resistance, etc.), is a variant of bulk method. This method has already been discussed in Chapter 12 as a variant of pedigree method. In fact, it is not always necessary that the plant breeder should rely exclusively on natural selection between F_2 and F_6 generations of bulk population breeding rather artificial selection can be practiced at any stage of the programme if needed. So, bulk population breeding can be done in between the pedigree method and, similarly, pedigree selection can be done in between the bulk method, that is, the mass pedigree method, depending upon the situation, can be considered as a variant of pedigree method as well as of the bulk method.

Single seed descent method : In this method, rather than bulking all seeds obtained from all the plants raised in F_2, a single random seed is taken from each F_2 plant and bulked. This bulked seed is used to grow the F_3 generation. The same process is followed in F_3 and succeeding generations. This method also is considered as a variant of bulk method as well as of pedigree method. The method has been described in detail in Chapter 15 (Fig. 13.2).

Modified mass method : In this method, bulk hybrid populations are space planted. Better plants are selected and bulked in each generation. Therefore, whereas the entire population is bulked in ordinary bulk method, only selected plants are bulked in each generation in this modified method. However, the modified mass method has been found suitable only for those characters which have high narrow sense heritability since basically it is nothing but a mass selection procedure in segregating generations. Thus, high the narrow sense heritability, more will be the effectiveness of selection (Fig. 13.2).

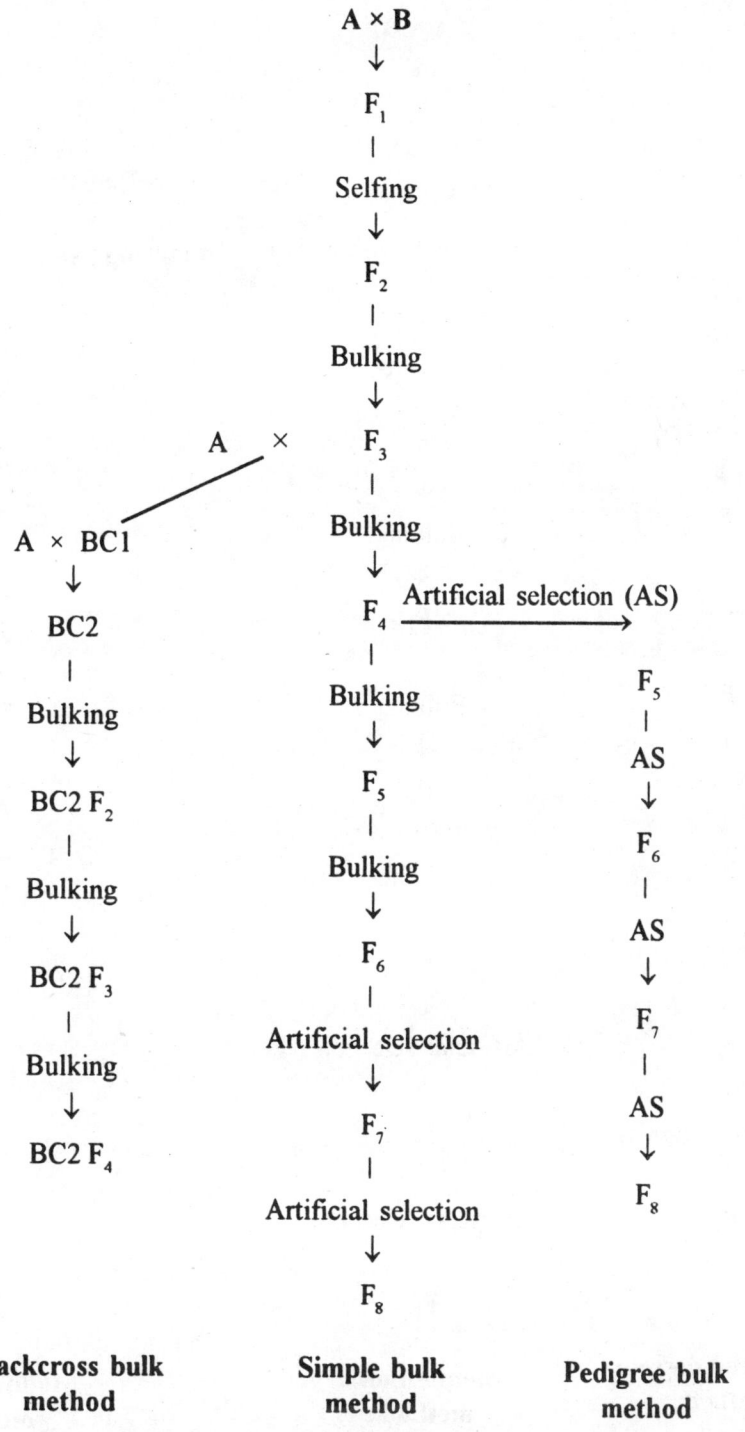

Figure. 13.1. Bulk method and its variants

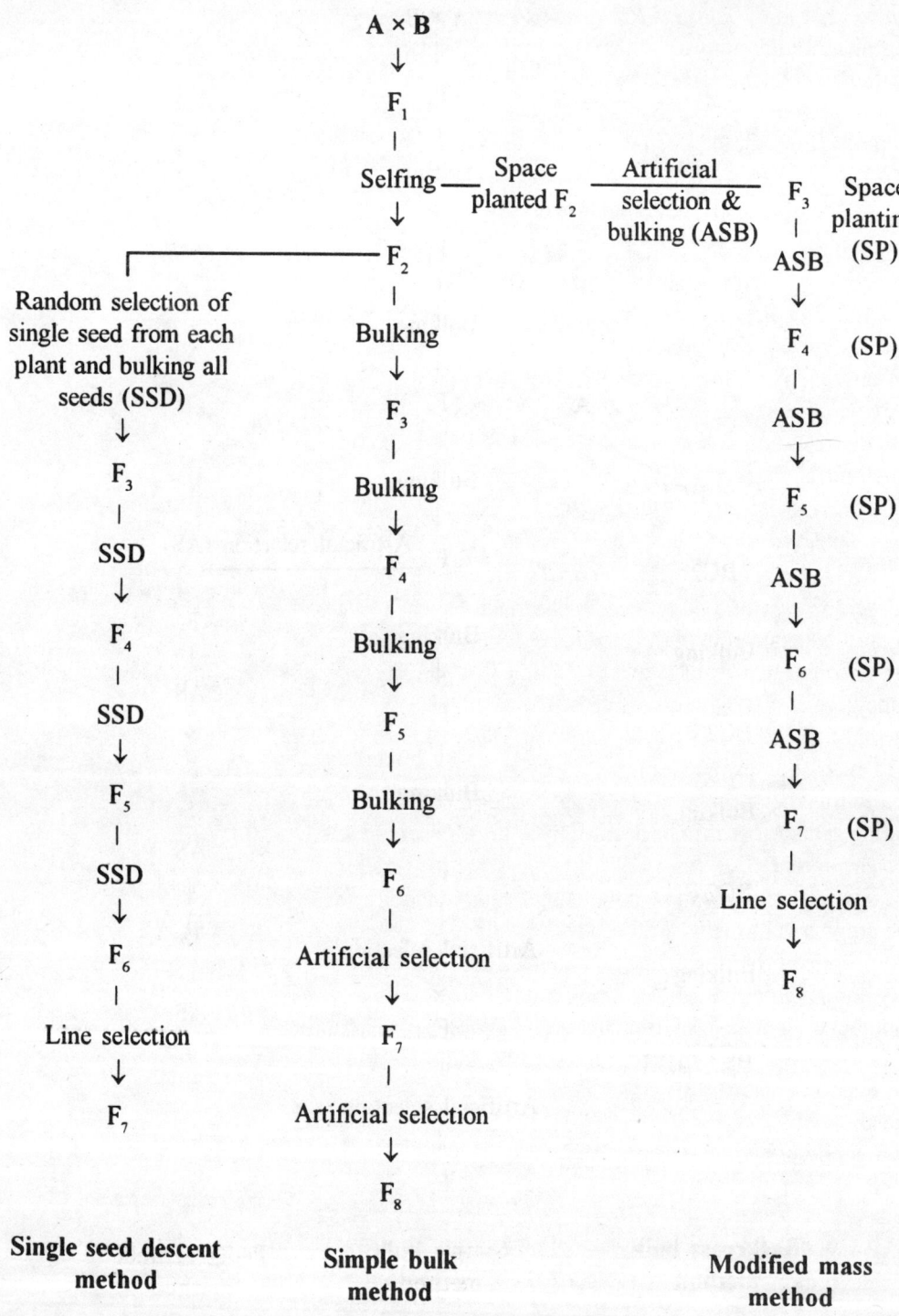

Figure. 13.2. Bulk method and its variants

Atkins (1953) applied this method to hybrid barley populations and recorded slightly higher average yield of lines obtained from selected populations. Further, though no selection was practiced for plant height and maturity the selected populations showed a tendency towards both the lateness and the tallness. In some respects, this method is similar to pedigree selection and in some respects similar to bulk method. Therefore, the method combines merits of both the methods and can be considered as modification of pedigree method as well as of bulk method.

Pedigree bulk method : A procedure similar to the mass pedigree method of Harrington (1937) has been proposed by Allard (1960). The method has some element of artificial selection in between bulk population breeding and thus may be called as pedigree bulk method. Whereas mass pedigree method was suggested to overcome one of the deficiencies of pedigree method, pedigree bulk method helps in more rapid elimination of unwanted genotypes and thus shifts the population toward agriculturally desirable type. Allard emphasized the importance of artificial selection in bulk populations particularly for the characters which are neutral in competition (disease resistance, etc.). Artificial selection also helps in selection of early maturing types and in roguing of susceptible plants to diseases and insect pests. In some respects, pedigree bulk method is also similar to modified mass method since in the latter method space planting and artificial selections are done before bulking in each generation (Fig. 13.1).

Effect of Natural Selection on Bulk Populations

Natural selection which is defined as the relative reproductive success of a type, has been and is playing an important role in the evolution of cultivated species. Any new type coming in nature has to face the consequences of natural selection and thus the perpetuation of the type would depend on its competitive ability with other types. However, nature may not favour the characteristics of plant species that are useful to man. It is also not always otherwise true and many of the characteristics useful to man (that is, contribution towards adaptation and productiveness) also make important contribution to the survival of the type. Since in the bulk population breeding the fate of plants in segregating generations is to be decided by natural selection, it is important to determine the correlation between the competitive ability of a type to survive and its agricultural worth.

The survival of a genotype competing against others in a sufficiently large bulk plot depends on : (a) the number (not the weight) of seeds produced by each genotype and (b) the capability of these seeds to produce offspring in the next generation, that is, how many of the seeds of each type reach maturity and produce progeny. However, the two factors mentioned above will in turn depend on a large number of genetic, environmental and genotype × environmental factors most of which either cannot be observed or cannot be measured by the plant breeder. These unknown factors cannot be avoided even in case of artificial selection.

The theoretical survival proportion of the poorer competitor in a particular generation when only two genotypes are competing, can be calculated as follows :

$$P_n = ic^{n-1}$$

Where,

P_n is the theoretically expected proportion of the poorer competitor in the nth generation of bulking;

i is the proportion of the poorer competitor in the beginning of the experiment (that is, initial proportion); and c   is the competitive ability or selective index or survival value of the poorer competitor.

The proportion of the better competitor will be 1- the proportion of the poorer competitor.

With a striking difference in the survival values of the two competitors, the proportion of the poorer competitor will decrease rapidly (with a corresponding increase in the proportion of the better competitor) in first several generations. With the advancement of generations, this rate of decrease in the proportion of the poorer competitor (or the rate of increase in the proportion of the better competitor) will become gradual and very low after many generations. With a small difference in the survival values of the two competitors, the rate of change in the proportion of the competitors in first generations is low and it is very low in the later generations.

If there are several competitors, the best and the poorest competitors will have the trends of their proportions similar to those of the better and the poorer competitors mentioned above. The proportion of the next best will increase in first several generations but will start declining thereafter. By that time, the poorest competitor will be at the verge of elimination. The next poorest competitor which did not show a very rapid decrease in its proportion in first several generations will now take the place of the poorest competitor. Ultimately, the next best will become the poorer competitor and in the end only the best competitor will survive.

The theoretical expectations discussed above are based on two assumptions : (a) The number of seeds produced by the good competitor throughout the programme is uniform; and (b) there is no effect of environment on competitors. However, these two assumptions are not usually fulfilled (particularly the second). The best competitor may produce relatively high number of seeds in the beginning of the programme but this number may be less in the later generations when its competiton with remaining competitors becomes tough. Further, there is no doubt that a number of environmental factors (temperature, soil moisture, light intensity, etc.) can affect the survival in populations.

Natural selection in varietal mixtures : Harlan and Martini (1938) conducted a comprehensive experiment on natural selection in 11 cultivated varieties of barley. Equal numbers of seeds of these varieties were mixed. The mixture was grown at 10 experiment stations for 4-12 years in different parts of the United States. A census of the number of plants of each variety was taken for each planting by growing out 500 spaced plants the next year. Therefore, enough bulk seed was saved each year to plant two plots in the following year, one for bulk harvesting and the other for taking census. The census clearly indicated that there was a wide range of variation for adaptation among the eleven varieties. Some of them exhibited wide adaptation, some specific adaptation and some others poor adaptation at all the 10 locations. Harlan and Martini noted that there was a good agreement between the theoretically expected curves of the different varieties according to their competitive abilities and the curves drawn on the basis of their actual proportions recorded in a number of generations. As expected, the best varieties generally showed curves tending upward, the better of the mediocre types a humped curve (tending upward for first several generations and then tending downward), the poorer of the mediocre types a straight horizontal line for few generations and then tending downward rapidly and the poorer varieties a curve tending downward very rapidly in the first several generations. Another important finding of their experiment was that, in general, the superior varieties also had a high competitive ability. This agreement between the agricultural worth of the varieties and their competitive ability was more striking in case of mediocre varieties.

Suneson and Weibe (1942) conducted an experiment to know the agreement between the competitive abilities and the average yields of four widely cultivated varieties of California barley, namely, Atlas, Vaughn, Club Mariout and Hero. The mixture of the four varieties (having equal numbers of seeds of each variety in the beginning of the experiment) was grown in bulk plot every year adjacent to the pure stands of the four varieties (for measuring yield). Although average yield differences among the four varieties were small, the sequence of the varieties on the basis of percentage of total yield (8 year average – 1933 to 1940) was Vaughn (27.0%) – Atlas (25.0%) – Hero (24.0%)– Club Mariout (24.0%). On the other hand, there were tremendous differences among the varieties for competitive ability. The proportions of the four varieties recorded by Suneson in 1948 (who continued the mixture of Suneson and Weibe for seven more years after 1941) were 88.0% (Atlas), 10.5% (Club Mariout), 0.7% (Hero) and 0.4% (Vaughn) (See Suneson, 1949). The results of this experiment, therefore, indicate that there was no very strong positive association between the agricultural worth of the varieties and their survival values. For example, the variety Vaughn which was highest yielding was virtually eliminated after 16 years of bulking (from 1933 to 1948).

Natural selection in bulk hybrid populations : The population structure of hybrid bulks is expected to be different from varietal mixtures. In early generations, the genotypes of the hybrid bulk will segregate and thus natural selection will have to act on different sets of genotypes in different generations. By F_6 or F_8 generation, the genotypes of the hybrid bulk are expected to become homozygous and at this state and in subsequent generations the population structure of the hybrid bulks will be similar to that of varietal mixtures with the difference that the number of homozygous genotypes in a hybrid bulk population will be generally much more than the number of the varieties in a varietal mixture. Three kinds of studies can be carried out to determine the effect of natural selection on bulk hybrid populations : (1) Determining changes in the frequencies of alleles at particular segregating loci on the basis of the frequencies of phenotypes in succeeding generations, (2) observing entire populations considering their yield performance and agronomic appearance and making comparison with standard varieties and/or among different generations, and (3) isolating individual lines from the bulk population periodically and using their performance as the degree of change expected in different generations of the bulk population. The points 2 and 3 have already been discussed earlier in this chapter (composite crosses). As regards point 1, Suneson and Stevens (1953) made a comprehensive study for measuring the competitive ability of alternative alleles at 5 loci (smooth versus rough awn, two-versus six-rowed, white versus black lemma, presence of lateral florets versus deficiens and awned versus hooded) in the barley composite bulk cross produced by Harlan and Martini. It was clear from the results of this experiment that there was a prompt reduction in the frequency of deficiencis, hooded, rough awn and black lemma alleles and after about 10 years of bulking, these four alleles were virtually eliminated. The proportion of two-versus six-rowed types, however, depended on the location. These results, therefore, indicated a strong positive association between agronomic superiority and high competitive ability and thus were in agreement with the findings of Harlan and Martini in varietal mixtures. However, one point should be made clear here that the survival of alternative alleles at a particular locus is controlled by the combined effect of that gene and the neighbouring loci and not by the effect of the gene alone. For example, the high survival value of smooth awn allele in the experiment described above was not only because of the effect of that particular gene but a linkage of this allele with genes for high yield was also responsible for its increased frequency.

Role of Artificial Selection in Bulk Population Breeding

Although the main force operating in bulk populations is natural selection but the use of artificial selection in these populations may sometimes greatly help the plant breeder in eliminating undesirable and inferior types. However, the importance of artificial selection in bulk populations varies according to the kind of programme (whether a short term bulk or a long term bulk) and the competitive ability of important character (s). Since the proportion of superior lines in bulk populations increases steadily over long periods of bulking, artificial selection in long term bulks does not seem to have much importance except for those characters which have poor or neutral competitive ability. In case of beans, for example, determinate types have low survival value than the indeterminate types, but the former types are often considered desirable because of synchronization in maturity and some other characteristics. Similarly, resistance to some diseases has a neutral survival. Artificial selection for such characters, therefore, will be helpful even in long term bulk programmes of more than 25 generations. In short-term bulks (8-10 generations), on the contrary, artificial selection is rather a must. Artificial selection thus provides assistance to natural selection in eliminating undesirable and inferior types more rapidly.

Artificial aids to selection may also prove helpful in achieving rapid improvement in bulk populations. Screening genotypes for earliness by bulk harvesting the crop at the time of its partial maturity, eliminating undersize seeds with the help of sieves of appropriate size and reguing susceptible genotypes by artificially creating disease epiphytotics are some examaples of such selection aids.

Further, the evaluation of hybrid populations in early generations can be done to choose varieties to be used as parents in future crossing programmes and to choose crosses expected to produce outstanding lines in later generations. However, early generation selection will be effective only for those characters which have high narrow sense heritability. Evaluation in early generations for characters with low heritability may present a confusing picture about the worth of a variety or a cross.

CHAPTER 14

Backcross Method and its Variants

In backcross method, a hybrid is crossed to either of its parents. Many a times, some specific defect (susceptibility to a particular disease, frost, drought or lodging, undesirable colour, shape or size of seed, etc.) occurs in a variety that possesses almost all other desirable attributes including high yield and wide adaptation. There are numerous examples where a variety (which is otherwise very promising) has gone out of cultivation due to one or a few such defects. The most striking example is of Kalyansona variety of wheat which has gone out of cultivation mainly because of its susceptibility to rusts. Similarly, Co1148, a wonder variety of sugarcane became susceptible to red-rot disease and has gone out of cultivation. The primary objective of using backcross method is to eliminate such defect(s) from a good variety (used as recurrent or recepient parent) by transferring one or a few characters (for which the promising variety is deficient) from some other variety (used as non-recurrent or donor parent) which may not possess other desirable characters and reconstituting the genotype of the recurrent parent through repeated backcrossing. The new variety thus produced will contain all good characters of the recurrent parent plus the character(s) transferred from the donor parent. To bring the transferred gene(s) in homozygous condition, selfing is done after the last backcross. Since, in this method, the breeder has a high degree of genetic control on his material, the results of the method can be predicted in advance and can be repeated if need be. Such prediction and repetition of results is not possible in pedigree selection or bulk method.

The backcross method has been used by livestock breeders for more than a century under the name **line breeding** to fix breed characteristics in a number of animal species. However, the application of this method in plant species was firstly proposed by Harlan and Pope in 1922 to improve small grain crops. They emphasized the importance of the method particularly under some specific situations when it could be better than other prominant plant breeding methods. Briggs in the same year (See Briggs 1930), used this method extensively to develop disease resistant varieties of wheat and barley. The method has a very sound scientific basis since the breeder can almost exactly predict in advance the morphological and agricultural characteristics of the variety to be developed and that the same variety could be developed again by using the same material and following the same steps. The method is being used quite frequently to improve the self-fertilized as well as cross-fertilized crops in three ways : (1) As an independent method, (2) in combination with other methods, or (3) as a supplement to other methods.

The Genetic Basis of Backcross Method

The genetic basis of backcross breeding is similar to that of selfing as regards the percentage of homozygosity and the percentage of homozygous individuals. However, backcrossing does not allow the progeny to split into 2^n (n being the number of heterozygous loci) homozygous genotypes at the end of the programme as is expected after selfing for a number of generations, rather all the genotypes after 5-6 backcrosses would resemble to the recurrent parent, that is, under backcrossing, the homozygotes are of desired type only.

Some undesirable genes from the non-recurrent parent along with desirable gene(s) may enter a hybrid at the time of crossing the two parents in the beginning of the programme. The elimination of undesirable gene(s) depends on the degree of linkage between the desirable and undesirable genes and the number of backcrosses and can be determined with the help of the following formula (Allard, 1960) :

The probability of elimination of
an undesirable allele $= 1 - (1-r)^{b+1}$

Where, r is the degree of recombination between the desirable and undesirable genes and b, the number of backcrosses made. If no selection has been practiced against the undesirable allele, the backcrossing is relatively more efficient method of eliminating undesirable genes than a selfing series since the former method provides more chances of desirable recombination than the latter. For example, with 20% recombination, the probabulity of elimination of an undesirable allele after five backcrosses is 74 per cent, whereas this probability with the same recombination fraction in selfing is 20 per cent only. Nevertheless, a decrease in the degree of recombination (that is, the linkage getting tighter) will be followed by a reduction in these probabilities.

But if selection is made against the linked undesirable allele and it is effective, the two methods (backcrossing and selfing) are almost equally efficient in obtaining desirable recombinations. Rather selfing series offers better opportunities for obtaining desirable recombinants than does the backcrossing because there is possibility of effective crossing-over in male as well as in female parent in selfing, whereas there will be no effective crossing-over in recurrent parent during back-crossing. The effectiveness of selection, however, depends on the degree of narrow sense heritability. Therefore, backcrossing is advantageous when no selection is practiced or where selection is not effective.

As regards the heritability of the character to be transferred to the recurrent parent, it should be high because selection for this character is to be practiced throughout the entire backcross programme.

Requirements of the Backcross Programme

For a successful use of backcross method, the following requirements must be met :

1. A superior variety with proven performance which can be used as the recurrent parent in the programme must exist.

2. The recurrent parent must be almost completely reconstituted by making reasonable number of backcrosses.

3. The character under transfer must have high intensity throughout the entire programme.

4. The character under transfer must be simply inherited with high heritability.

Choice of recurrent parent and its recovery : The availability of a suitable recurrent parent is one of the most basic requirements of a successful backcross breeding programme. Therefore, the recurrent parent should be an exceptionally good variety which has wide adaptation and whose worth is well proven particularly in respect of yield, but has developed some defect. However, it is not difficulot to find such varieties in case of well established crops each of which is usually dominated by a few such varieties. These varieties are not only the result of the technical efforts made by the plant breeder during their development and time to time selection by the farmers, but also they have faced consequences of selection under natural conditions. The farmers have become well acquainted with these varieties regarding their cultivation, yield potential, utilization of the produce in the best possible way, etc. These dominant varieties of the crops probably have the maximum claim to be used as recurrent parents in the backcross breeding programmes.

As regards the recovery of the genotype of the recurrent parent, it is mainly a function of the number of backcrosses required to develop a variety exactly similar to the recurrent parent except the character(s) transferred from the donor parent. Usually, six backcrosses are needed to almost completely reconstitute the genotype of the recurrent parent because after every recurrent backcrossing the average proportion of the germplasm contributed by the donor parent will be reduced to one-half with a corresponding increasing in the germplasm from the recurrent parent. The selection in early backcross generations for the characters of the recurrent parent will enhance the recovery of the recurrent parent and will thus reduce the number of backcrosses to be made. Further, it is not always necessary that the recurrent parent possesses complete genetic homogeneity, rather it may be a group of a few closely related homozygous lines. Thus, to exploit the genetic variability of the recurrent parent fully for assuring the development of a variety which has all essential features of the recurrent parent, a sufficiently large number of plants of the recurrent parent should be used in each backcrossing.

The character under transfer and its maintenance : For an easy and successful transfer of a character from the donor parent to the recurrent parent, the following two requirements should be fulfilled :

1. It must be possible to maintain a worthwhile intensity of the character under transfer throughout the backcross programme, particularly when the heritability of the character is low. Even if the character is governed by a very simple genetic system, some of its intensity may be lost during backcrossing. For example, the Martin variety of wheat which was used as a donor parent at the California Agricultural Experiment Station to transfer bunt resistance (monogenically controlled) is almost completely resistant even when artificially inoculated. But 3-4 per cent plants of a

number of varieties which were developed through backcrossing and carried resistant allele of this donor showed bunt infection after artificial inoculation. According to Briggs (1929) (see Briggs 1930) this loss in the intensity of bunt resistance was attributable to modifying genes. However, the studies carried out by Weibe and Briggs (1937) clearly showed that at these low infection rates the bunt organism could not maintain itself even when the conditions were favourable for the growth of the pathogen.

2. The character under transfer must have high heritability. Whether characters of the recurrent parent have high or low heritability is meaningless in a backcross breeding programme because they are automatically taken care of by repeated backcrossing. But since the donor parent is crossed only once in this method and the character under transfer is maintained through selection after every backcrossing, the heritability of this character must be high otherwise the selection will not be effective. Easy identification of the character under transfer in segregating generations on the basis of visual observation or by simple tests will greatly help the plant breeder in handling this character and thus carrying out backcross programme successfully. One more important point which the plant breeder should clearly know is that a character of high heritability controlled by several genes can be transferred more speedily and with greater precision through backcross breeding than a character which is controlled by one or a few genes but has low heritability. Therefore for such transfers, the degree of heritability is more important than the number of genes governing a character.

When two or more characters are to be transferred, separate backcross programmes should be run for the transfer of each character. After the full recovery of the characters, the progenies (lines) obtained from these separate programmes may be crossed together to combine all the characters in a single variety. Simultaneous transfer of two or more characters through the same backcross programme may create difficulty in handling the large quantity of material required in such a transfer and in recovering the characters simultaneously because one of the characters may not express itself under the conditions the programme is being run. Such a situation may delay the transfer of all the characters.

The dependence of the character under transfer upon a dominant allele makes the transfer more easy. In case the character is under the control of a recessive allele, selfing after every backcrossing is required to bring the desirable recessive allele in homozygous condition. Alternatively, selfing can be done after first backcross and then after every two backcrosses. The latter method will help in reducing the duration of the programme.

For the transfer of quantitative characters, it is advisable to do two successive selfings after every backcross for the proper identification of desirable plants since a quantitative character is usually governed by several genes and generally has low heritability. Under such a situation, the size of the population should also be large enough since the intensity of quantitative characters may not remain so high as is possible in case of qualitative characters. However, there are a number of reports where quantitative characters have been quite successfully transferred through backcrossing. Of course, the degree of success in such programmes depends on the experience and judgement of the plant breeder in identifying the desirable plants. Therefore, the application of backcross method is not only limited to the transfer of qualitative characters but it can be used to transfer the quantitative characters as well.

As has been mentioned earlier in this chapter, the linkage between the gene being transferred and some undesirable gene may complicate the backcross programme because selection in favour of the desirable gene will automatically ensure the presence of undesirable gene in the selected material. More tight the linkage between such genes, more trouble it will create. If there is some possibility of effective selection against the undesirable allele (that is, some identifiable desirable recombinants appear due to linkage breaks), a selfing series following a cross will provide better opportunities to eliminate the undesirable allele. If, on the other hand, selection is not effective against the undesirable gene due to its low heritability or due to its inability to express under conditions of selfing and backcrossing, the undesirable linkages will be broken more effectively by backcrossing than by selfing.

So far, the application of backcross method has been mainly confined to the transfer of disease resistant genes to the otherwise good varieties. It is probably because of the significant influence of diseases on crop production and the quality of the produce. But it does not mean that backcrossing is an ineffective method for transferring other characters like plant height, seed size and earliness, rather the method can be applied quite successfully to transfer any character that has moderate to high heritability.

It is clear from the points discussed above that while selecting a non-recurrent parent for backcross programme, the plant breeder has to be particular only about the intensity and heritability of the character (s) to be transferred and need not bother at all about other characters of the donor parent.

The Procedure of Backcross Breeding

The procedure of backcross method is very simple particularly when the character under transfer is governed by a dominant allele and the heritability of the character is high. The transfer of a recessive desirable allele complicates the procedure a little bit. The successive backcrossing programme with **dominant allele under transfer**, involves the following steps :

Crossing the parents : The recurrent parent is crossed with the donor parent (a variety possessing the character in worthwhile intensity for which the recurrent parent is deficient). It is advisable to use the recurrent parent as female since the new variety will have all genes of recurrent parent except that which has been transferred from the donor parent and thus the genotype will be in its old cytoplasmic environment. If the donor parent is used as female, all the genes except one which has been transferred will be in a new cytoplasmic environment. This presence of genes in new cytoplasm may not be as hormoneous as it was in the old environment.

The F$_1$ generation : The F$_1$ plants will contain 50 per cent germplasm (plus cytoplasm if the recurrent parent has been used as female) of recurrent parent and 50 per cent that of donor parent. These plants will be heterozygous at all the loci for which the two parents differ (including the locus controlling the deficiency of the recurrent parent) and thus each plant will carry the desirable allele to be transferred. Since the plant population will be genetically homogenous, there is no question of selection for any character at this stage.

Backcross generations : The F$_1$ plants are backcrossed to the recurrent parent to produce first backcross seeds. The first backcross plants raised from this seed will show variation for the character under transfer and the other characters. Fifty per cent of these plants are expected to carry the desirable allele under transfer and remaining 50 per cent will be homozygous for the undesirable recessive allele. At this stage, selection is practiced for the character under transfer and the selected plants are further backcrossed to the recurrent parent to produce second backcross seeds. This process of selecting desirable plants in respect of the character to be transferred from the backcross generations and backcrossing the selections to the recurrent parent is usually continued upto fifth backcross generation. However, selection for all desirable characteristics in first, second and third backcross generations would greatly help in early recovery of the recurrent parent. Selection after third backcross for the characters of the recurrent parent will be largely ineffective because segregation will be reduced to a very low extent at this stage.

The plants selected from the fifth backcross generation are backcrossed to the recurrent parent and sixth backcross generation is produced. By this stage, the plants will be almost completely homozygous for all the characters of the recurrent parent.

Generations of selfing : Plants possessing the character under transfer are selected and selfed. The seeds of these plants are harvested separately and individual progeny rows are grown. Intensive selection is made at this time for the character under transfer and the plant type of the recurrent parent and selected plants are harvested separately. Again, individual plant progeny rows of the selected plants are raised.

Lines homozygous for the character under transfer and similar to the plant type of recurrent parent are harvested in bulk, their seed is increased and released for commercial cultivation. The whole programme takes about 12 years or seasons.

Transfer of a recessive allele : If the character under transfer is governed by a recessive allele, it makes the backcross programme somewhat lengthy because in between the backcross generations selfing has to be done for the identification of plants possessing the desirable recessive allele. A recessive allele can be transferred in either of the following two ways :

(1) Selfing can be practiced after every backcross generation. Plants homozygous for the gene under transfer are selected from the selfed progeny and backcrossed to the recurrent parent. This method will take about 17 years or seasons.

(2) Selfing can be done after the first backcross and then after every two backcrosses, that is, first backcross-selfing-second and third backcrosses-selfing-fourth and fifth backcrosses-selfing-sixth backcross followed by two successive selfings. In this method, plants homozygous for the gene under transfer are selected from the selfed progeny obtained after the first backcross and are backcrossed to the recurrent parent. All plants in second backcross generation will be heterozygous for the gene under transfer and thus no selection is possible for this character. Selection may, however, be practiced for the characters of the recurrent parent and the selected plants are again backcrossed to the recurrent parent. Plants in third backcross generation are selfed without any testing for the character under transfer. Homozygous plants for the gene under transfer are identified and backcrossed to the recurrent parent. Like second and third backcrosses, the fourth and fifth backcrosses are made in succession. Plants of the fifth backcross generation are selfed, homozygous plants are identified in the following generation and they are backcrossed to the recurrent parent to produce sixth backcross generation. After sixth backcross, the method is exactly similar to the procedure explained for the transfer of dominant allele. This method takes about 15 years or seasons.

Whatever method is followed, selection should be practiced for all the desirable characteristics after selfing.

The variety developed through backcross method is not generally tested in replicated varietal yield trials because it is an old variety with improvement in some specific character.

As regards the number of plants necessary in backcross generations, about 50 plants from backcrossed seeds are enough if the character under transfer is governed by a dominant allele. The number of plants from selfed seeds will be, however, more (about 100 plants after first selfing and about 70 rows of 24 plants each after second successive selfing). If the character to be transferred is under the control of a recessive allele or a partially dominant allele, the number of plants required is still less because homozygous plants for the allele concerned may be recognized in the first selfed generation itself and the same are backcrossed to the recurrent parent. But when the character under transfer is governed by two or more genes and/or the gene under transfer is linked with some other undesirable gene, the size of the population at each step of the programme will accordingly increase.

The backcross breeding programme is least affected by environmental conditions because it can be carried out in all those environments where the character under transfer is able to express itself. Further, since a small amount of material is required at each step of the programme, the material can be grown in offseason nurseries, green houses, etc. without any difficulty and several generations can be grown per year. It will greatly reduce the duration of the programme.

Advantages of Backcross Method

1. There is no risk of loosing already achieved improvement.
2. The method is independent of the environment.
3. Since small quantity of material is required in this procedure, the method can be carried out in offseason nurseries, green houses, etc.
4. No extensive testing of the backcross derived variety is needed.
5. No agronomical research is needed for the new variety,
6. No publicity to convince the farmers for accepting a new variety is required.
7. The performance of the variety to be evolved is predictable well in advance.
8. The method is repeatable and the same variety may be again developed following the same steps.
9. The method is relatively rapid.
10. The method allows faster elimination of an undesirable allele (than pedigree method) if selection against that allele is not practiced.
11. The method is applicable to both the self- and cross-fertilized crops.
12. It is best method for introgressing germplasm from wild relatives into cultivated species.

Disadvantages of the Method

1. It is a conservative method. It does not permit any further improvement in the variety (recurrent parent) except the character which has been transferred from the donor parent. It is the biggest disadvantage of this method.
2. The method is not suitable for the transfer of characters which have low heritability.

3. A tight linkage between the gene under transfer and an undesirable allele will result in the addition of an unwanted gene in the recurrent parent.

4. In those species where emasculation and crossing are tidious and there is high degree of shedding of flower buds resulting in a very low percentage of seed setting, it becomes difficult to reconstitute the genotype of the recurrent parent completely.

5. All efforts made for carrying out the backcross programme will go in vain if some other new variety supersedes the recurrent parent in terms of yielding ability and other characteristics.

Comparison of Backcross Method with Pedigree Method

The comparison between these two methods can be made on the basis of the following points:

1. The main objective of backcross method is to remove one or a few defects of a dominant variety of the locality, whereas the pedigree method is aimed at to combine several characters of one variety with those of other variety.

2. In pedigree method, usually the cross is made only once (that is, in the beginning of the programme), while in backcross method the progenies are crossed again and again to the recurrent parent till the genotype of the latter is completely reconstituted.

3. The variety developed through backcross method is similar to the recurrent parent except the character which has been transferred from the donor parent, whereas the variety developed by pedigree method is superior to both the parents for several characters.

4. The variety developed through backcrossing need not be tested in the replicated varietal yield trials, while testing of new variety developed through pedigree selection is necessary.

5. Relatively large amount of material is required in pedigree method than in backcross method.

6. Starting from F_2, selection is practiced throughout the whole pedigree breeding programme for all the desirable characters, whereas in backcross method selection is practiced only for the character under transfer from first to sixth backcross.

7. . Whereas high heritability of various characters considered during pedigree method facilitates identification of superior gene combinations, the degree of heritability of the characters of recurrent parent is meaningless in backcross method.

8. No pedigree record of any plant or family is kept in backcross method, while it is necessary to maintain pedigree record of each selected plant and family in pedigree selection.

9. Whereas there is no effect of environment on backcross method, the effectiveness of pedigree selection changes according to the environment.

10. Neither the procedure nor duration of pedigree method is affected by the type of allele (dominant or recessive) governing a desirable character, but different methods and durations are required to transfer dominant and recessive alleles through backcrossing.

11. Backcrossing is a useful method for the production of substitution and addition lines, whereas no such lines are produced with the help of pedigree method.

Variants of Backcross Method

Various methods which are used to improve crop plants have their own merits and demerit. A merit of one method may be the demerit of the other and *vice versa*. Therefore, appropriate modifications can be made in any method to remove its demerits or make it more useful in some other aspect. Backcross method has one disadvantage that it does not permit the exploitation of transgressive segregation. Most of the modifications of backcross method suggested by different workers relate to the production of selfed generations at one or the other stage of the backcross programme so that new gene combinations may appear and the advantage of unusual segregants may be taken. Following are some variants of the backcross method :

Convergent improvement method : The method was suggested by Richey (1927) to improve established inbreds of maize. The convergent improvement name was given because the F_1 produced by crossing two inbreds is backcrossed independently to both its parents, that is, running two backcross programmes parallelly for improving both the inbred parents. For example, if I_1 and I_2 are the two inbred parents of an F_1 hybrid, the single cross is backcrossed to I_1 to improve it by transferring desirable genes from I_2 and maintaining them through careful selection, as well as to I_2 for improving this inbred by transferring desirable genes from I_1 and maintaining them through selection. Richey and Sprague (1931) found some degree of superiority in corn inbreds obtained after 3-4 backcrosses followed by 2-3 generations of selfing over the original inbreds used as parents. According to Murphy (1942), the convergent improvement can also be used as a method to improve the yielding ability of the single cross hybrid. However, the improvement in all these cases was not of considerable degree and therefore convergent improvement could not be identified as a good method of maize breeding.

Modified backcross method : This method was proposed by Mac Key in 1954 to prepare optimum parental material for initiating a multiple cross. The method involves crossing of a (may be two or more also) well adapted variety to several (say 8) unadapted strains and backcrossing the progenies thus obtained to the adapted variety once, twice, thrice or even four times for increasing the proportion of the germplasm of the adapted variety in the material to be used in further breeding programmes. If no backcrossing is done, the proportion of the germplasm of the adapted variety and that of all the unadapted strains together will be 50:50. This proportion of the germplasm after one backcross, two backcrosses and three backcrosses will be 75:25, 87.5:12.5 and 93.75 : 6.25, respectively. Lawrence (1974) followed this method to introgress germplasm from a wild species of oats (*Avena sterilis* L. from Mediterranean region) into the cultivated oats (*A. sativa*) and suggested that two-four backcrosses should be made with the cultivated species to bring the proportion of its germplasm to an appropriate level for using the material in further breeding programmes.

This method may also be called as **multiple non-recurrent parent backcrossing** where two or more non-recurrent parents are involved in the backcross programme.

An almost similar multiple non-recurrent parent backcrossing approach is followed in developing multiline varieties for disease resistance where a superior variety is crossed to a number of other varieties or strains possessing different kinds of genes for resistance. As many backcross programmes are run simultaneously as there is number of varieties crossed to the superior variety. In the end, all the component lines thus obtained are mixed together to incorporate different sources of resistance in the same variety (Fig. 14.1).

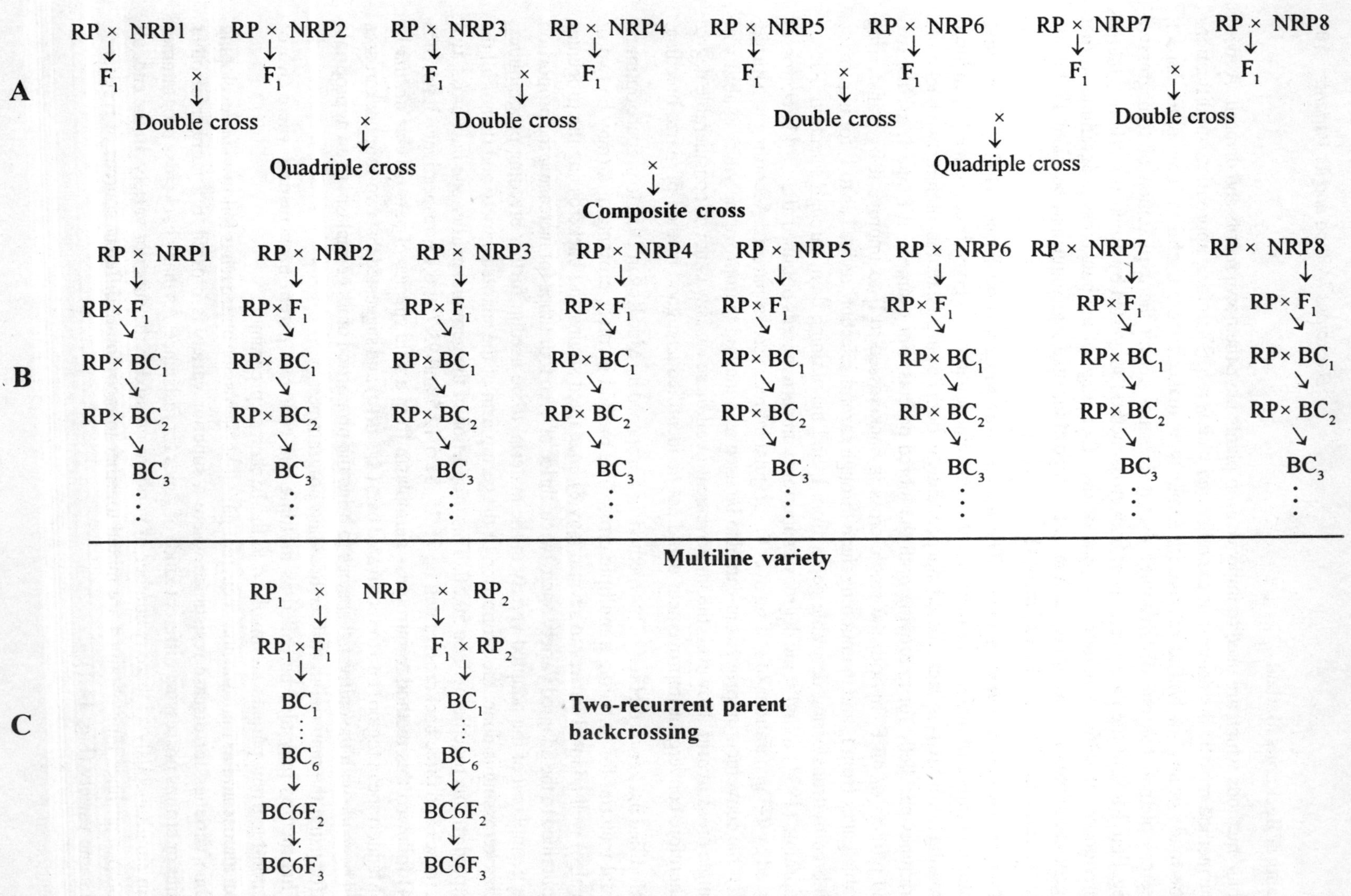

Fig. 14.1. Multiple non-recurrent parent and recurrent parent backcrossing. A, composite cross; B, multiline development; C, two-recurrent parent backcrossing

Multiple recurrent parent backcrossing : Sometimes most of the desirable characteristics are not found concentrated in a single variety but are scattered in two or more superior varieties. Under such a situation, different recurrent parents are used at different stages of the backcross programme so that all desirable characters of the recurrent parents as well as the character under transfer may be combined together. Such an approach has been used in case of tomato and sugarcane.

Intermittent backcrossing : This method has been described by Allard (1960). In this procedure, as usual, the F_1 of the two parents is backcrossed to the recurrent parent to produce first backcross generation. In the first backcross generation, selection is practiced for the character under transfer and the selected plants are selfed. Selection is made in first selfed generation (F_2) for all the characters including the character under transfer. Selected plants (from F_2) are again selfed and second selfed generation (F_3) is produced. Again, selection is made for all the characters and selected plants are backcrossed to the recurrent parent to produce second backcross generation. The plants selected in second backcross generation for the character under transfer are backcrossed to the recurrent parent to obtain third backcross generation. Now the same process of selection, selfing (producing F_2 and F_3 generations) and backcrossing which was followed after the first backcross is repeated and fourth, fifth and sixth backcross generations are produced. Similarly, first and second selfed generations (F_2 and F_3) are produced from the selected plants in the sixth backcross generation on the basis of the character under transfer. Intensive selection for all the desirable characters (that is, character under transfer and plant type of the recurrent parent) is made in both the selfed generations. Selected lines are harvested in bulk, their seed is increased and released for commercial cultivation. The whole method may be represented as – cross between the two parents, F_1, first backcross, first and second selfings, second and third backcrosses, first and second selfings, fourth, fifth and sixth backcrosses, first and second selfings. This method has two main advantages over the successive backcrossing :

1. Selection in the two selfed generations (F_2 and F_3) provides equal opportunity of improvement as is possible by making two additional backcrosses.
2. Selfing allows large population to grow for making effective selection for all the desirable characteristics.

    Similarly, other modifications can be made in the backcross method (Figs. 14.2 and 14.3).

Selection in early backcross generations : Usually, no selection for the characters of the recurrent parent is practiced in the successive backcross breeding programme except after the sixth backcross when selection is made for all the characters in selfed generations. However, selection for the character under transfer as well as for the characters of the recurrent parent in early backcross generations helps in accumulating the desirable genes of the recurrent parent at an earlier stage and thus reduces the number of backcrosses. This method was used by Briggs. According to him, selection for all the characters in first, second and third backcross generations is equivalent to two additional backcrosses. But, the selection for the plant type of the recurrent parent after third backcross generation is largely ineffective.

Application of Backcross Method to Cross-fertilized Crops

The application of backcross method to cross-fertilized crops is not basically different from its application in self-fertilized crops. However, since the recurrent parent in case of cross-fertilized

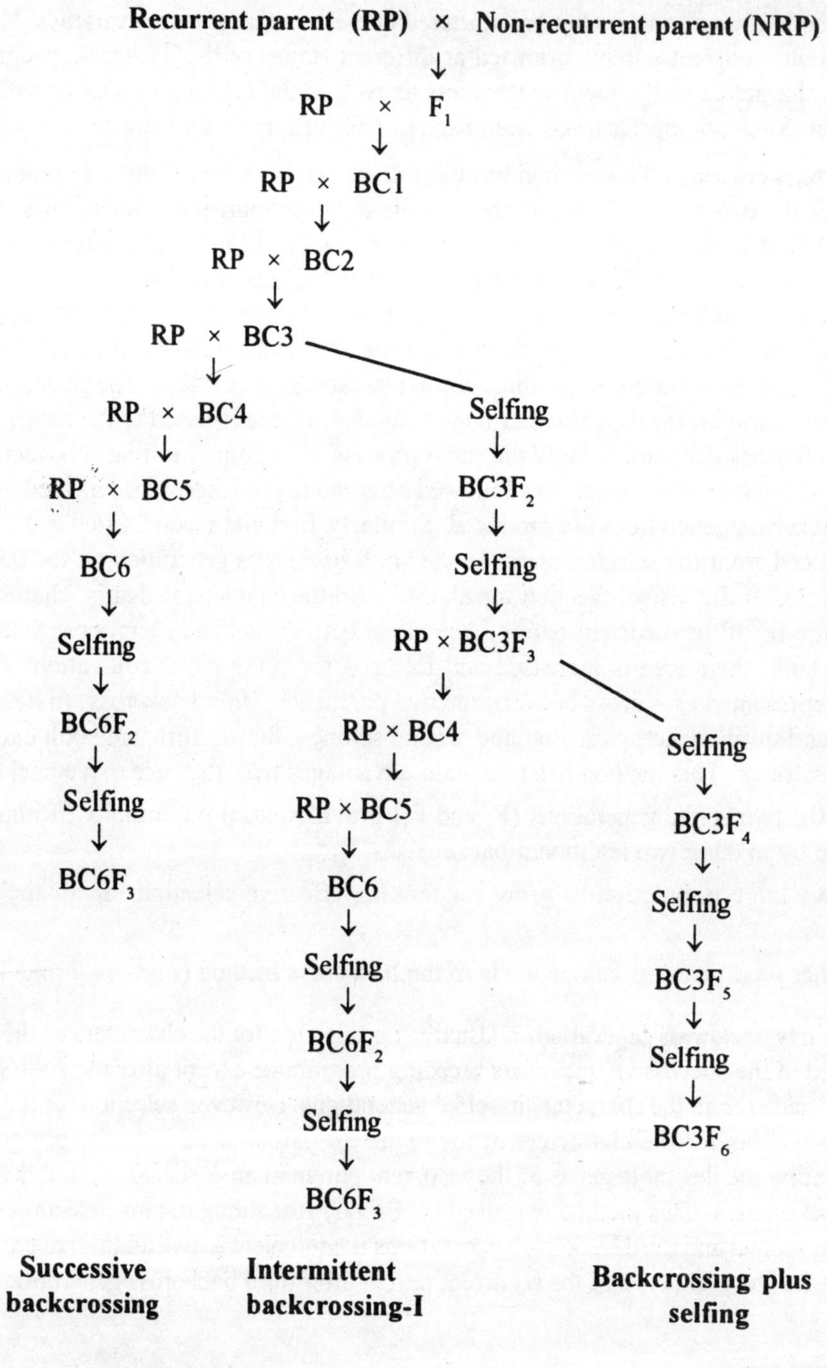

Recurrent parent **(RP)** × **Non-recurrent parent (NRP)**

↓

RP × F₁

↓

RP × BC1

↓

RP × BC2

↓

RP × BC3

RP × BC4   Selfing

↓   ↓

RP × BC5   BC3F₂

↓   ↓

BC6   Selfing

↓   ↓

Selfing   RP × BC3F₃

↓   ↓

BC6F₂   RP × BC4   Selfing

↓   ↓   ↓

Selfing   RP × BC5   BC3F₄

↓   ↓   ↓

BC6F₃   BC6   Selfing

↓   ↓

Selfing   BC3F₅

↓   ↓

BC6F₂   Selfing

↓   ↓

Selfing   BC3F₆

↓

BC6F₃

| **Successive backcrossing** | **Intermittent backcrossing-I** | **Backcrossing plus selfing** |

Fig. 14.2. Successive backcrossing and its two varients

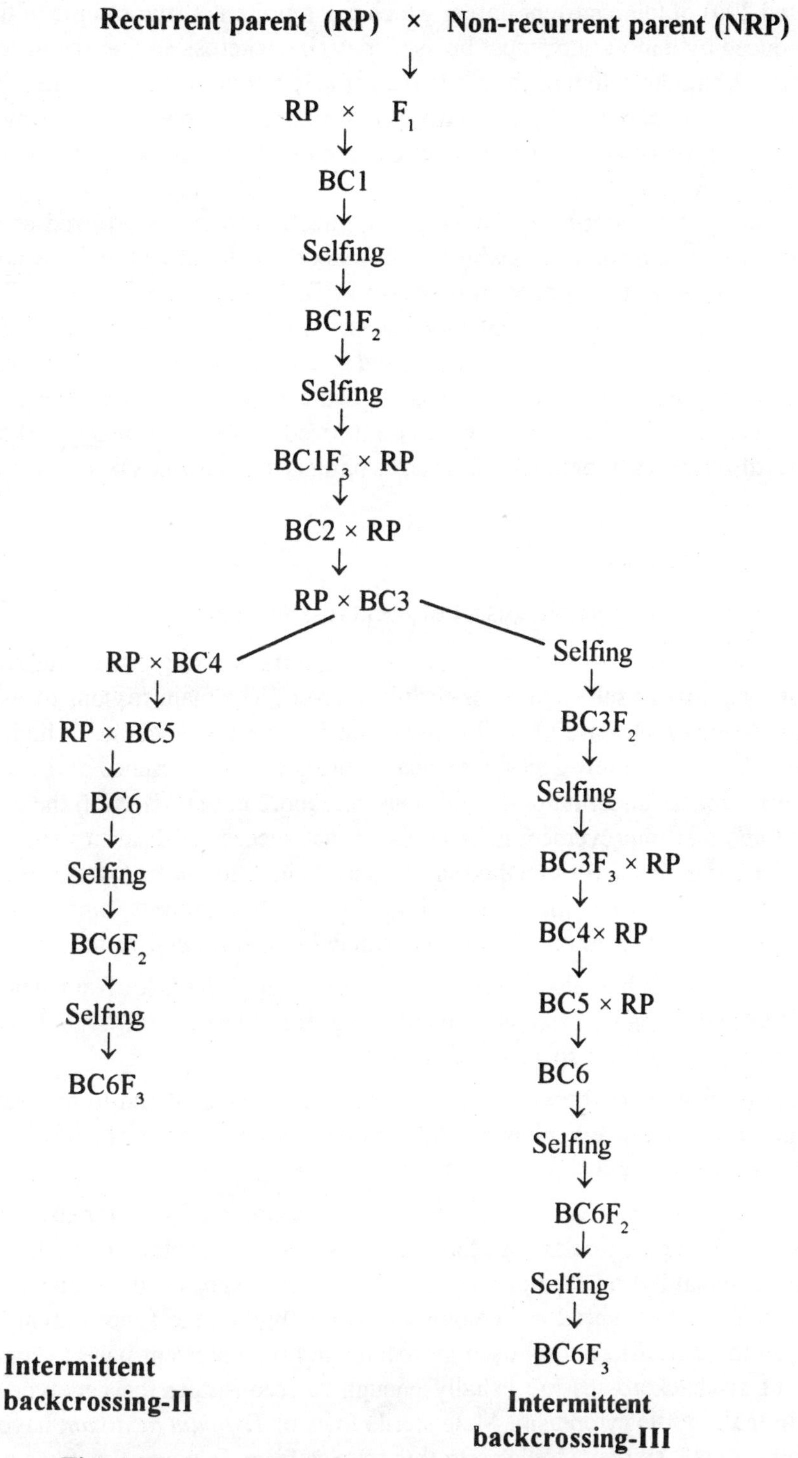

Recurrent parent (RP) × Non-recurrent parent (NRP)
↓

RP × F₁
↓
BC1
↓
Selfing
↓
BC1F₂
↓
Selfing
↓
BC1F₃ × RP
↓
BC2 × RP
↓
RP × BC3

RP × BC4
↓
RP × BC5
↓
BC6
↓
Selfing
↓
BC6F₂
↓
Selfing
↓
BC6F₃

Selfing
↓
BC3F₂
↓
Selfing
↓
BC3F₃ × RP
↓
BC4× RP
↓
BC5 × RP
↓
BC6
↓
Selfing
↓
BC6F₂
↓
Selfing
↓
BC6F₃

Intermittent
backcrossing-II

Intermittent
backcrossing-III

Fig. 14.3. Two kinds of intermittent backcrossing

species will be heterozygous, it will produce numerous kinds of gametes. Therefore, a large number of plants (around 200) of the recurrent parent which can represent a true sample of the total number of gametes produced by that variety, must be used in each backcross so that the genetic constitution of the new variety is similar to that of the old variety (the recurrent parent). Further, in case of cross-fertilized species it is not easy (rather it is extremely difficult or impossible in many cases) to grow backcross generations as selfed F_3 lines for more accurate identification of the plants that possess the gene under transfer.

One of the very good examples of backcross application to cross-fertilized crops comes from alfalfa. The variety Caliverde of alfalfa which is resistant to bacterial wilt, mildew and leaf spot was developed by transferring wilt resistance from the strain Turkestan to the California Common variety. The source of resistance to mildew and leaf spot diseases was a selected strain of California Common. The number of the plants of California Common used in each of the four backcrosses was about 200.

As has been discussed earlier, the backcross method can also be applied to improve inbred lines in cross-fertilized crops. Either a single inbred is improved by using ordinary backcross method as in case of self-fertilized crops or both inbreds used as parents in a cross can be improved by convergent improvement method.

Functions of Backcross Method

Following are the main functions (uses) of backcross method :

1. The greatest use of backcross method has been made in transferring disease resistance to superior varieties of a number of self- and cross-fertilized crops. The main reasons of its greatest use in the breeding for resistance are : (a) The disease resistance has moderate to high heritability, (b) it is governed by one or a few genes with clear expressin in the presence of the disease, (c) it has usually neutral fitness under natural conditions, and more importantly, (d) the disease resistance is such an important improvement in crop plants that success or disaster usually depends on it. However, the use of backcross method is not only limited to the transfer of disease resistance, but any character which has moderate to high heritability (earliness, plant height, etc.) can be transferred through this method. This transfer may be intervarietal or interspecific.

2. The backcross method has also been used to introgress germplasm from wild plants to crop plants. The introgression has been interspecific (e.g. from *Averna sterilis* to *A. sativa*) as well as intergeneric (e.g. *Tripsacum* species to *Zea mays*).

3. The transfer of characters through backcrossing is not only confined to qualitative characters, but the quantitative characters (plant height, seed size, earliness, etc.) which have high heritability can also be transferred by using this method.

4. Backcrossing has proved a very successful method in transferring cytoplasm from one variety or species to another particularly in those species where advantage of male sterility is being taken in commercial hybrid seed production. In such crop species, the transfer of a male sterile cytoplasm to a superior inbred is a common practice. In this, the donor parent (from which the cytoplasm is to be transferred) is used as female and the recurrent parent (superior inbred) as male. About six backcrosses are usually enough to reconstitute the genotype of the superior inbred with male sterile cytoplasm. Male sterile lines of *Triticum aestivum* have been produced by transferring male sterile cytoplasm to this species from *Triticum timopheevii*.

5. Another important function of backcross method is its use in the production of isogenic lines (genotypically identical lines except for one locus). The isogenic lines may be used for doing basic genetical studies (the effect of individual gene, etc.) and for producing multilines, etc. to incorporate disease resistance from different sources in the same variety.

6. Backcross method is not always used as an independent method of plant breeding, but in several cases it is used either in combination with other methods (like pedigree and bulk breeding) for exploiting transfgressive segregants or for some other purpose or to prepare suitable material to initiate some other breeding programme like multiple cross (modified backcross method of Mac Key).

Achievements of Backcross Breeding

The achievements of backcross breeding are numerous. The method has been used to perform a number of functions like transfer of disease resistance, introgression of germplasm from one species to another, transfer of cytoplasm, transfer of quantitative characters with moderate to high heritability and the production of isogenic lines. As regards the transfer of disease resistance, one of the best examples is the development of wheat multilines in the background of Kalyan Sona, some time back a wonder wheat variety which became highly susceptible to leaf rust. The disease resistant genes from different sources (Bluebird, Frecor, Robin, Tobari, etc.) were transferred to Kalyansona variety through backcrossing and three multiline cultivars, MIKS II, KSML 3 and UPKML 7406, were released for commercial cultivation. Further, the wheat varieties such as Sonak, HW 2004, HW 2034, HW 2044 and HW 2045 have also been developed and released.

Similarly, Tift 23A, a male sterile line of bajra introduced from Georgia was used most extensively in the hybrid seed production in the beginning. But this line became highly susceptible to downy mildew. A number of male sterile lines of bajra have been developed by transferring the cytoplasm of Tift 23A to the lines resistant to downy mildew.

CHAPTER 15

Single Seed Descent Method and its Variants

The single seed descent (SSD) method is, in fact, a modification of bulk method. It was firstly suggested by Goulden (1941) for overcoming the problem of sampling in bulk population breeding and the slowness of pedigree selection. The method consists of selecting a single seed randomly from each plant of F_2 and succeeding generations and bulk harvesting all the seeds for growing next generation till the plants become virtually homozygous. Since in SSD method single seeds are harvested randomly from each plant of each generation to take care of the total range of variability throughout the propagation period, this method has also been called as **random method** or **complete bulk**. The method has attracted attention of many plant breeders and has been elaborated and applied by Grafius (1965) in oats and by Brim (1966) in soybean. Whereas Grafius called SSD method a short-cut in plant breeding, Brim called it a modified pedigree method.

The SSD method allows more rapid advancement of segregating material than other methods. The main objective of suggesting this method was to minimize the effect of natural selection on the genetic structure of original population so that an F_2 genetic spectrum in a homozygous form could be obtained in shortest possible time. Snape and Riggs (1975), on the basis of the distribution of means and variances of a population advanced from F_2 to F_6 through SSD method, suggested that transgressive segregants can be fixed and genetic advance can be obtained in such a population irrespective of the genetic architecture of the trait concerned. Although the range of phenotypes in F_2 distribution and in the distribution of F_6 lines is similar irrespective of the genetic architecture of a characater, these phenotypes fall below the most extreme homozygote possible except when there are duplicate interactions. Under such a situation, the SSD method would be equivalent to pedigree method with selection in each generation as regards the genetic gain obtained. On the other hand, if heterosis is due to the dispersed dominant genes and complementary epistasis, the F_6 population will fall below the expectation of F_2 distribution. In such a situation, the pedigree method would be better than SSD method.

Since single seed from each plant is harvested and no selection is practiced for the general vigour of the plant or any other character from F_2 to F_6 generation in this method, the distance between plants and between rows can be reduced to a great extent and thus a large plant population can be accomodated in small area. This possibility of growing segregating generations with very high plant densities also provides opportunity to raise segregating materials in small off-season nurseries and in green houses for advancing generations very rapidly and to handle a sufficiently large number of crosses at a time.

The characteristic features of the SSD method are as follows :

1. The broadest possible representation of F_2 genotypes can be maintained during segregating generations. The near homozygous F_6 lines have genetic spectrum almost similar to that of F_2, regardless of the genetic architecture of the trait under consideration.

2. Selfing in each generation splits the population into homozygous lines and thus brings the hidden genetic variability on surface to a maximum possible level since seed from each plant of F_2 and succeeding generations was harvested.

3. From F_6 onward, the method is like bulk population breeding or may be modified depending on the objective of the plant breeder.

The Procedure of SSD Method

The two selected parents are crossed. The F_1 generation is space planted and enough quantity of seed is produced to raise a large F_2 population at a high plant density to accomodate as many plants as possible.

Generations of single seed harvesting : Single seed from each plant of F_2 population is harvested and bulked. The F_3 generation is grown from this bulked seed again at a high plant density. The same process of harvesting single seed from each plant and bulking of all seeds thus harvested is repeated till F_6. The F_6 generation is space planted.

Generations of single plant selection : Quite a large number of plants (usually several hundred) are selected from the F_6 generation on the basis of general vigour and other important characteristics. In F_7 generation, individual plant progenies are raised and drastic reduction is made in the number of progenies at this stage. A few promising lines are selected and bulk harvested. If some individual plants with outstanding agricultural worth are available, such plants may be handled by pedigree selection. The selected progenies are grown in preliminary yield trials. The rest of the programme is just like bulk population breeding.

Comparison with Other Methods

The SSD method which has become popular in recent years has been compared with other plant breeding methods like pedigree method, bulk method and early generation testing by a number of workers. Tee (1971) found that the SSD method showed equal or more efficiency when compared with bulk method for the maintenance of variability in wheat. Similarly, Kumar (1973) while comparing the efficiency of SSD and pedigree methods noted that the two methods were equally efficient as regards the variability for yield. However, for retaining the variability for other characters the SSD method proved better.

Luedders *et al.* (1973) compared SSD method with pedigree method, bulk method and early generation testing. The yield potential of the selected lines developed through different methods was almost the same. However, the SSD method and early generation testing proved better than the other two methods in terms of the number of lines which constituted the top 25 per cent of lines based on two year mean.

Boerma and Cooper (1975) compared the effectiveness and efficiency of three methods, namely, SSD method, pedigree selection and early generation testing, in four segregating populations of soybean and found that there were no consistent differences among the three methods for yield. Almost similar results were obtained by Knott and Kumar (1975) in two wheat crosses. They compared SSD method with early generation yield testing and found that the top 20 per cent of SSD lines were either equal or bit superior to F_5 yield test lines obtained from top 20 per cent of F_3 families.

Tee and Qualset in 1975 noted substantial differences between SSD method and bulk population breeding in terms whether the competitive effects for a particular character were important or not. If competitive effects were important, SSD method was superior and vice versa.

Yap *et al.* (1977) compared SSD method with visual mass selection and visual and index (combining pod length and number) pedigree selection. Although there were no significant differences between different methods for yield, the SSD method was undoubtedly superior to other methods in maintaining variability when the heritability was low. On the other hand, mass selection method was better at moderate heritabilities (25 to 50 per cent) but poor at low (10 per cent) or high (75 per cent) heritabilities.

There are some reports where SSD method has been found inferior to other methods. For example, Peirce (1977) recorded modest differences between pedigree selection, SSD method and a combination of these two methods in tomato. The SSD method was found inferior to the two other methods. Using three selection methods in two wheat crosses selected from a 9×9 diallel set on the basis of high heterosis and least inbreeding depression for grain yield in F_2 generation, Pawar *et al.* (1985a) obtained higher mean values in selected populations under pedigree selection than SSD and bulk populations for grain yield and its related traits in both crosses. The SSD method proved to be superior to bulk method when large populations and limited resources are involved. Using the same three selection practices, similar results were obtained by Pawar *et al.* (1985b, 1986) and Singh *et al.* (1987) in wheat. Also, bulk method proved to be inferior to both pedigree and SSD methods on the basis of range of distribution of genotypes and phenotypic coefficient of variation (Pawar *et al.* 1987). The latter two procedures were equally effective for maintaining range but SSD method has an edge over pedigree selection for maintaining variability. The SSD method was reviewed by Wright and Thomas (1976) particularly on three aspescts : (1) The number of lines worthy of inclusion in advanced trials, (2) the labour involved in the production of lines, and (3) the degree of resistance to diseases the lines possessed. They concluded that as regards the first aspect, the SSD method was as good as pedigree selection. But the method involves comparatively more labour in the production of lines and the lines produced through this method showed relatively more susceptibility to diseases because in this method no selection is practiced and since disease resistance usually has a neutral survival under natural conditions. According to Wright and Thomas, these are the main reasons why SSD method could not become a major breeding technique ? However, Sneep (1977) pointed out one more drawback of this method that a number of genotypes are lost in the method through genetic drift.

Advantages of SSD Method

1. SSD is the best method for rapidly advancing generations.

2. The method requires less labour and space.

3. Since the material in each segregating generation is grown at very high plant densities (because the plants are not to be tested for their general vigour and other important characteristics during this time), the method allows best use of off-season nurseries, green houses, etc.

4. The method allows to obtain the F_2 genetic spectrum in the homozygous state and thus more genetic variability is maintained from F_2 to F_6 generations.

5. The effect of natural selection in changing the genotypic array in the original population is minimized.

6. The method is least affected by environment because one seed is harvested from each minor patch of land represented by a single plant. Thus, the method is independent of the degree of heritability.

7. The method overcomes the problem of sampling in bulk population breeding.

8. Since almost entire range of variability is preserved during the segregating generations, the method provides better opportunities for the exploitation of transgressive segregants than all other methods.

9. The method is superior to bulk method when competitive effects are important.

Disadvantages of SSD Method

1. Harvesting a single seed from a plant particularly in early segregating generations is a too small sample to represent within plant variability.

2. Since no selection is usually practiced for any character from F_2 to F_6 and since disease resistance shows usually neutral fitness, SSD lines in general show greater susceptibility to diseases than the lines produced through pedigree method.

3. Many genotypes are lost through genetic dirift.

Variants of SSD Method

To overcome some of the weaknesses of SSD method (too small sample, loss of genotypes due to genetic drift and failure to exploit maximum potential of a cross), some modifications have been suggested in this method. Yonezawa and Yamagata (1981) proposed **one-plant-one-line method** at F_4 or F_5 stage of SSD method for rice breeding. According to them, the modified method is expected to prove better than one-seed-one-plant method in this crop. Similarly, harvesting **three to four seeds** instead of one from each plant of F_2 and succeeding generations or **one-spikelet-one-plant method** in some cereals like wheat and triticale and **one-pod-one-plant method** in some pulse crops like mungbean, urdbean and pigeonpea in each segregating generation will certainly help in exploiting within plant/family variability and minimize the risk of loss of genotypes due to germination failure and other factors.

Multiple seed descent (MSD) method : Perhaps the best variant of SSD method is the MSD method of Ndondi (1987). The investigator took a random sample of 100 plants with good agronomic characteristics from a space planted F_2 population. From each of these plants, 12 seeds were planted separately in the green house to raise F_3 generation. A single seed taken from each F_3 plant was seeded in green house for obtaining F_4 generation followed by growing seeds from F_4 plants in the field in 3 m head rows for seed increase. At maturity, 360 lines (that is, 36 MSD families each with a minimum of 10 sibs) were retained for further testing and the method was appropriately called as multiple seed descent (MSD) method.

Ndondi compared three selection procedures, namely, MSD method, SSD method and bulk population breeding, for yield and protein improvement in wheat and found that MSD method was better than the other two procedures in obtaining high yielding lines *per se*, combining high yield potential and high protein content, and in producing lines with greatest phenotypic variance among their means. As regards the production of lines with high protein content, MSD and SSD methods were equally effective. The bulk population breeding proved better than SSD method in producing high yielding lines *per se*. But the reverse was true for the phenotypic variance among the line means, that is, phenotypic variance among the line means was greater in SSD lines than in bulk population lines.

The MSD method has three main advantages over SSD method :

1. This method maximizes genetic variability of a cross as the method permits retaining a large number of lines until they attain reasonable homozygosity.

2. The method provides opportunity for exploiting maximum genetic potential of a cross as it allows between as well as within families selection on the advanced generation lines.

3. It minimizes the risk of loss in genetic variability due to germination failure, seed survival and the ability of each plant to produce at least one seed as the method is not restricted to harvestng single seed in each segregating generation.

The greatest disadvantage of MSD method is that it requires more space than the SSD method for evaluating the advanced generation lines. However, considering savings in labour and space over the generation advancement period, MSD method is a viable method for improving yield potential and protein content.

CHAPTER 16

Multiline Approach

A **multiline** can be defined as "mixture of near-isogenic lines". In fact, the term **multilineal** is more appropriate than multiline, but the latter term is more commonly used to make it comparable with its counterpart, **pureline**. The objective of producing a multiline variety is to develop a plant population which is uniform for agronomical characters (plant height, time of flowering and maturity, grain colour, yield potential, etc.) but heterogeneous for genes for disease resistance. Such a genotypic constitution of a varity of a self-fertilized crop helps in preventing an epidemic spread of air-borne diseases (e.g. rust in wheat). Further, the development of a multiline variety by incorporating resistance genes from different sources into the same genetic background through backcrossing is much easy than to incorporate multiple disease resistance into one genetic background. The development of multiline varieties provides a long term control of diseases in self-fertilized crops since such varieties are expected to stabilize the pathogen's race structure as the development of races with multiple virulence genes in this approach will be minimized.

The History of Multiline Approach

The multiline concept is basically a concept of diversification. Diversification may be interspecific (that is, mixture of different species) or intraspecific (that is, mixture of different varieties of the same species). The diversification in multiline varieties is still a narrow type of intraspecific diversification, rather it is an intra-varietal diversification. Here, the different varieties are replaced by near-isogenic component lines, that is, diversification in respect of genes controlling a disease (heterogeneous for genes that condition reaction to a disease organism) but uniformity in respect of all other agronomic traits.

The concept of diversification, in fact, is very old. Even before demonstration of plant species, different plant species existed together in the same natural habitat. This kind of situation can easily be seen still now in those natural habitats which have not been disturbed (or destroyed) by man. With the beginning of interference of man with nature, the diversification started reducing. Of course, in the old days, the farmers mostly used to grow different crops together as mixed crops or inter crops (wheat and gram, cotton and urdbean, etc.). Similarly, the land varieties of self-fertilized crops were nothing but the mixtures of pure lines. Even now, some farmers in central India are growing desi cotton (*Gossypium arboreum*) and American cotton (*G. hirsutum*) together. The main advantage of growing mixture of these two kinds of cotton is that there is less damage in desi cotton due to *Fusarium* wilt and in American cotton due to red leaf and leaf roll. However, with the advancement of agriculture, there was a drastic decline in interspecific diversification and when the land varieties were almost completely replaced by the pure line varieties, the intraspecific diversification was also considerably reduced. The diversification was further reduced due to the introduction of single genes into varieties for vertical resistance and thus coexistence of host and pathogen was completely disturbed.

Rosen (1949), for the first time, realized the importance of growing mixed population of plants as a cultivar for combining resistance to crown rust in oats. However, minimum importance was attached to agronomic uniformity. On similar grounds, the Harland variety of barley was developed by Suneson (1968) by mixing lines drawn at an advanced stage of a composite cross bulk. Although the approach suggested by Rosen made the basis for the cultivation of multiline varieties, the concept of multiline in its true sense was firstly proposed by Jensen in 1952 in oats to combine resisance to the crown rust disease in a single cultivar from different sources. According to Jensen, both the agronomic uniformity and the heterogeneiy for resistance to disease were equally important. Borlaug (1959) suggested a similar approach for combining stem rust resistance in wheat.

Requirements of Multiline Approach

To develop a good multiline variety, the following three requirements must be satisfied :

1. Availability of a good recurrent parent : The recurrent parent must be of proven yielding ability and should possess wide adaptability since the multiline variety will be developed in the genetic background of this parent.

2. Availability of resistance donors with well known genetic make up : Resistance donors whose genetic make-up in terms of different resistance genes (that is, resistance genes from different sources) is well known, must be available. For example, in wheat, donor varieties or strains carrying Lr (brown rust), Sr (stem rust) and Yr (yellow rust) resistance genes have been identified and used in the multiline cultivar development programmes of this cereal.

3. Availability of information on the race flora of the pathogen : For a successful multiline development programme it is essential to have complete information about the number and frequency of races of the pathogen.

Characteristics of a Good Multiline

1. A good multiline must have a high genotypic diversity in respect of the vertical genes for resistance.

2. The yield level of each component line must be higher than that of the recurrent parent under epiphytotic conditions.

3. Except heterogeneity for resistance genes, the multiline must have fairly high phenotypic uniformity for all the agronomic traits (height, flowering and maturity time, colour, shape and size of seed, etc.).

4. Each component line of the multiline cultivar must be strong to ensure that the reduction of the exodemic infection is maximum.

5. The component lines must also show resistance aginst other diseases of the crop for which the multiline has not been bred.

The Basis of Disease Resistance in Multilines

In the formation of a multiline, genes with vertical resistance (high-type race specific resistance) to the pathogen are incorporated from various sources into the same genetic background. Thus, a multiline cultivar with such resistance is expected to show complete susceptibility to certain races. But it has been observed that when a number of genes with vertical resistance are placed in a multiline variety, the plant population shows a degree of resistance to all the races of the pathogen. Such a resistance has been named as **population resistance** or **synthetic horizontal resistance**. This kind of resistance causes reduction in the rate of disease build-up in the field and, as a result, the occurrence of an epiphytotic is delayed.

The development of disease epiphytotic in the field can be explained on the basis of the following equation of the compound interest.

$$X_t = X_o e^{rt}$$

Where,

X_t = is the amount of disease at a particular time,

X_o = the initial amount of disease or capital,

e = the constant (=2.718)

r = the rate of development, and,

t = the time

In case of vertical resistance, since it is race specific, only the initial amount of the disease (X_o) will be reduced, but the rate of development of the disease (r) which is usually high remains unchanged. Therefore, the vertical resistance is of value only under a situation when it is able to keep X_o very small for all the prevalent races of the pathogen, that is, it gives resistance to all the prevalent races of the pathogen. But such a situation is not generally found and this is the reason why vertical resistance generally fails to control great epidemics. Of course, local outbreak and mild epidemics can be easily controlled by this type of resistance.

On the other hand, a cultivar possessing horizontal resistance is supposed to show resistance to all the races of the pathogen and, therefore, it causes reduction in the rate of development of the disease but does not reduce the initial amount of disease. As a result, there is a reduction in the rate of the epiphytotic development and thus crop loss is also greatly reduced.

In multiline cultivars, however, since several different vertical genes are placed in the same genetic background, the characteristics of both the vertical resistance and the horizontal resistance are combined together, that is, both the initial amount of the disease (a characteristic of vertical resistance) and the rate of its development (a characteristic of horizontal resistance) are reduced. This was clearly demonstrated by Browning and Frey (1969) in case of oat stem rust. They found that when the number of media per unit were counted in the mixture of isogenic lines and in the isogenic lines grown singly, their number was much less in the former case than in the latter. This reduced number of uredia in muliline was because the resistant plants served as **spore traps**, that is, resistant components acted as barriers to the spread of pathogen by checking the contribution of potential parental spores in he future cycles. In the pure line cultivar, the spore production was so high that there was occurrence of epiphytotic resulting in the premature death of plant tissues, whereas spore production in the multiline cultivar was considerably less and the rust development was terminated by the host maturity.

Gill *et al.* (1979) noted similar situation for yellow and brown rusts in wheat.

Breeding Multiline Varieties

For creating a multiline variety, seeds of several component lines that have different genes for resistance to the disease but are uniform for agriculturally important characters are mechanically mixed. The development of a multiline generally involves the following steps :

1. Development of component lines : The component (iso-genic) lines are developed in the background of a well proven commercial variety by using it as the recurrent parent. The component lines may be developed through : (1) backcrossing, (2) multiple crossing, (3) mutation, or (4) heterozygote selfing.

The development of isogenic lines through backcross method involves backcrossing of F_1 (of the recurrent parent and donor parent) and the segregating generation progenies to the recurrent parent and practicing selection for disease resistance in each generation. Some plant breeders prefer limited backcrossing (2-3 doses of the recurrent parent) rather than repeated backcrossing so that there is scope for the improvement in all the important characteristics including disease resistance.

In multiple crossing programme, different donor parents are crossed with the same recurrent parent, that is, [(A×B) (A×C)] [(A×D) (A×E)] where A is the recurrent parent and B, C, D and E are donors. The number of donor parents would depend upon the dispersion of resistance genes in the germplasm. It is a more rapid method and, like limited backcrossing, provides scope for improvement in other characteristics also in addition to disease resistance.

In mutation method, the parental pure line is treated with a mutagen and homozygous mutant lines obtained in M_3 and later generations are identified.

In case of heterozygous selfing, heterozygotes for a particular locus are selected after the initial cross. These heterozygotes are selfed for a number of generations to derive pairs of homozygotes with alternate alleles at the locus concerned. In this way, a number of line pairs homozygous for two alternate alleles may be produced in different genetic backgrounds.

Among these four methods, the backcross method is simple and less demanding. However, depending upon he objective of the plant breeder, the method may be modified accordingly.

2. Screening of component lines : The screening of component lines for their reaction to the individual races of the pathogen is done and the resistance genes carried by the different lines are identified under controlled conditions. However, the lines are also assessed under field conditions for their reaction to various diseases of the crop.

The component lines are also evaluated for their performance in respect of other important characters like plant height, earliness, seed characters and yield. It must be ensured that these lines have at least equal or more yield potential as compared to that of the recurrent parent in a disease free environment so that under epiphytotic conditions their yield is decidedly better than the recurrent parent.

3. Composition of the multiline : Out of the large number of isogenic lines produced, superior lines in respect of the level of resistance and other agronomic characteristics are selected to form a multiline variety. As regards the number of component lines to be included in a multiline variety, though there are controversial reports (15-20 according to Borlaug and 8-10 according to Browning and Frey), 8-16 component lines are usually mixed to form a multiline variety.

The proportion of the different components in a multiline usually depends on : (1) The prevalence of race flora of the pathogen and the genes carried by the components, (2) the relative agronomic performance of the component lines in a disease free environment, and (3) adult plant reaction to specific races under field conditions. One simple method is to mix all components in equal proportions.

Various mixtures (combinations) of components may be grown and the combination giving best performance may be selected to form the multiline.

It should be emphasized that a multiline variety, in general, gives better yield, has wide adaptability and shows comparatively low incidence of disease than the individual component lines.

Proportion of Resistant Population

A number of views have been given about the proportion of resistant population in a multiline variety. According to Borlaug (1959), the percentage of resistant host population should be 87.5 to 94, while Suneson (1960) put this limit upto 75 per cent, Jensen and Kent (1963) were still more liberal and said that 60 per cent host resistance is enough. Leonard (1969) advocated the view that it is more important to stabilize the rust population rather than stabilizing the level of resistance and thus a particular level of susceptibility should be maintained in a multiline. Under such a situation (when 35-40 per cent of the plants are susceptible to all the races), simple races will be maintained in the pathogen poulation and, therefore, more complex races will have a normal rate of multiplication since they will be virulent on all the components of the multiline. According to Leonard, the multiline varieties should not have a superiority over the pure line varieties as regards the disease control unless the simple races of the pathogen are able to suppress the reproduction of the complex races. He further advocated that a multiline with different genetic backgrounds will be more effective than a multiline with the same genetic background of the recurrent parent.

Clean and Dirty Crop Approaches

There is another controversy about the degree of resistance in the component lines which are to be mixed to form a multiline variety, that is, whether the components must be completely resisant to all races of the pathogen or they should have some degree of susceptibility. Borlaug has advocated that all the components should be completely resistant to all prevalent races of the pathgen and called it **clean crop approach**. According to this approach, even those component lines which are high yielding but have some degree of susceptibility do not find place in a multiline. Frey *et al.* (1975), on the other hand, have advocated that each component line should have some degree of susceptibility to disease and they called it **dirty crop approach**. This approach ensures that none of the components will be completely resistant to all the races of the pathogen and thus is in agreement with the views of Suneson, Jensen & Kent and Leonard discussed above.

Whatever approach one is following, it is essential to ensure that the component lines used in a multiline variety carry diverse genes for resistance.

Durability of Multilines

The durability of the resistance of a multiline cultivar depends on the rate of evolution of complex races of the pathogen. This rate of evolution, according to Parlevliet (1979), is affected by the following factors :

1. The number of component lines in a mixture : More is the number of components in a mixture, lesser is the erosion of resistance genes. However, there is always a practical limitation regarding the handling of the number of components in a multiline variety and the number of resistance genes available.

2. Stabilizing selection : This kind of selection prevents the development of complex races in a multiline cultivar (Van der Plank, 1968).

3. Epidemiological situation : With a high disease pressure (when the difference in disease reaction level between fully virulent and and fully avirulent races is greater), the rate of erosion of the multiline resistance will be more.

4. Replacement of component lines : The replacement of multiline component (s) whenever necessary because of susceptibility to new races of the pathogen, tends to increase the durability of the multiline.

Advantages of Multilines

The multiline strategy, though of a conservative nature, is undoubtedly an improved method through which resistance genes can be very well managed. This approach has following advantages :

1. The race structure of the pathogen population is stabilized.
2. Tends to slow down the epidemic.
3. Extends the life of useful resistance genes.
4. Allows exploitation of genes showing tight linkage in repulsion phase.
5. It is possible to exploit the multiple alleles.
6. Helps in reducing yield losses.
7. Allows exploitation of genetic diversity.
8. It is a dynamic system as it is possible to replace component (s) whenever felt necessary.

Disadvantages of Multilines

1. It is an agronomically conservative approach.
2. The method is expensive.
3. Provides breeding ground for new complex races.

Achievements

Miramar 63 was the first multiline released as a commercial variety. It was released in Columbia in 1963 against wheat stripe rust in the background of the variety Frecor. Browning *et al.* (1969) released a number of multiline cultivars of oats (E 68, E 69, E 70, M 68, M 69 and M 70) against crown rust. Three multiline cultivars of wheat, namely, KSML 3, MLKS 11 and UP KML 7406, were released in India for commercial cultivation.

CHAPTER 17

Diallel Selective Mating System

Conventional breeding methods used for improving self-fertilized crops generally suffer from some serious drawbacks (small gene pool size, early fixation of genotypes and no provision of intermating in segregating generations) as successive selfing after hybridization hinders the appearance of new recombinants, enhances accumulation of linkage blocks in early generations and thus does not provide adequate opportunity for selection. Further, since single plant selections are made in each generation to propagate the line, a new ceiling is imposed on the degree of further improvement of the line in every generation. Favourable genes once lost are lost for ever. Some less intense form of selfing (that is, one or more cycles of selective intermating in between the selfing series) is expected to ensure exploitation of hidden genetic variability by fostering gene recombination, increase frequency of desirable genes in the experimental material by reducing the chance of random loss of favourable genes, shift genetic correlations according to the will of the breeder and thus might help in the selection of theoretically best combining genotypes.

Realizing the general weaknesses (narrow genetic base with a risk of unnoticed elimination of desirable parental genes, accumulation of linkage blocks and lack of intermating which blocks recombination options) of the conventional breeding methods employed in self-fertilizing crops and the importance of selective intermating in segregating generations, Jensen (1970) outlined a complex but dynamic and unique breeding approach to improve self-fertilizing crop species where the production of crossed seed is difficult. The approach which was named as diallel selective mating system (DSMS) is a supplement to conventional breeding methods used in self-fertilizing crops and was proposed specifically to broaden the genetic base of populations. DSMS is a combination of several breeding procedures (conventional bulk population breeding, mass selection in each population series and recurrent selection procedures) and allows simultaneous insertion of multiple genotypes into a few central populations, infusion of germplasm from outside the original populations and extraction of new varieties at any stage of the breeding programme, additional recombination by intercrossing elite genotypes and screening of breeding material at different stages of development. Jensen's aim in proposing DSMS was to identify, discuss and overcome the drawbacks of conventional breeding system while retaining its advantages.

Procedure of DSMS

Jensen (1970) gave a general outline of DSMS for cereal breeding which involves two phases : (1) planning phase, and (2) implementation phase.

Planning phase : The initial step in this phase is to decide the project objectives to be accomplished. In addition to the general objectives of conventional breeding system used in autogamous crops, there are three main objectives of applying DSMS : (a) Widening of the genetic base of the breeding project, (b) breaking of linkage blocks, and (c) the transfer of desirable genes from the parents to progeny populations.

The second step in the planning phase is to identify desirable genes (their nature, number and spread among genotypes) in the germplasm to achieve the aforementioned objectives.

The choice of parents with proven quality (genotypes possessing desirable genes) and the number of parents to be incorporated (depending upon the spread of genes among genotypes) in the project form the third step of the planning phase. According to Jensen, inclusion of seven diverse parents at the initial stage, production of 21 F_1 hybrids by crossing these parents in a diallel fashion (n(n-1)/2 combinations) and then attempting complete diallel cross (CDC) or partial diallel cross (PDC) among the F_1 hybrids to produce double cross hybrids, is an appropriate amount of germplasm to achieve all kinds of objectives. However, the number of parents to be incorporated at the initial stage and the number of double cross hybrids to be produced may change from material to material depending upon the availability of desirable genes in the genotypes, type of crop and the facilities available to the investigator.

Implementation phase : DSMS is a variant of composite breeding where five composite populations are produced at the end of breeding programme. Jensen (1970) divided the implementation phase of DSMS into four stages : (1) Producing single cross hybrids (basic parent series or P_1); (2) producing double cross hybrids (F_1 diallel series or P_2); (3) applying mass and recurrent selection procedures (selective mating series or P_3, P_4 supplementary hybrids seeds for P_3 and P_5); and (4) standard line selection from P_1-P_5 populations.

Producing single cross hybrids : As the number of crosses in full diallel mating system (n^2 combinations) increases tremendously with an increase in the number of parents, the investigator cannot take liberty regarding the number of parents to be included in the project. If one has to accomodate larger number of parents, either some modified form of diallel should be used or the advantage of male sterility should be taken. Further, if the parents included in the project belong to a single group, unrestricted crossing is done among all the parents. But if the parents come from two or more groups (Indian wheats and Mexican wheats, spring wheats and winter wheats, etc.) than this stage is divided into two parts : (a) Unrestricted within group crossing, and (b) restricted between group crossing.

While making within group crosses, one may cross the parents in all possible combinations and produce n (n-1)/2 F_1 hybrids (that is, involving each parent in n-1 cross combinations) if the number of parents is less. According to Jensen, this number is seven or less so that 21 or less F_1 hybrids and 210 or less $F_1 \times F_1$ crosses are produced. Since the main aim of producing single cross hybrids is to set up the F_1 diallel series, the feasibility of making and managing $F_1 \times F_1$ crosses should be kept in mind. If the desired input of parents is more than seven, the number of $F_1 \times F_1$ cross combinations generally becomes unmanageable. For example, if the number of parents is 10, there will be 45 F_1 hybrids and 990 all possible $F_1 \times F_1$ crosses. However, if some device like male sterility is available, a large number of F_1 and $F_1 \times F_1$ hybrids can be produced without any difficulty. But if such device is not available and the investigator has to accomodate larger number of parents, partial diallel cross or its modified form should be followed depending upon the combinations of choice or convenience. The partial diallel can accomodate a larger number of parents than diallel for producing the same number of crosses. For example, if the number of paents is 10 and the sample size(s) to make random number of crosses is 5, the number of F_1 hybrids will be ns/2=25 instead of 45 in diallel. Further, if s=3, the number of F_1s will be 15 only. Partial diallel can accomodate 8-22 parents (Jensen, 1970). If this number is more than 22, some form of modified partial diallel should be used but it should be ensured that each parent is used at least once.

In restricted between group crossing programme, all desired between group crosses are made if the number of parents is 21 or less. But if the number of parents is 22 or more, selective two-parent crosses should be made so that the number of crosses does not exceed 21 and each parent is used at least once in the crossing programme.

The purpose of producing basic parent series is not only to set up the F_1 diallel series but also its (basic parent series) subsequent generations are handled by compositing populations through mass selection or by processing individual crosses through conventional breeding to develop line variety. Since the seeds produced on a plant are technically part of the next generation, Jensen divided each generation by using plant and seed symbols.

Producing double cross hybrids : Generally, large number of crosses are made at this stage. Here, composite diallel population (P_2) is produced by crossing the F_1 parents in all possible combinations [n(n-1)/2] if the number of F_1s is 21 or less. If the number of F_1 parents exceeds 21, some pollination control device (if available), partial diallel or modified partial diallel is used to produce $F_1 \times F_1$ crosses. The subsequent generations of P_2 are handled generally through mass selection to develop line variety.

Applying mass and recurrent selection procedures : One important aspect of the application of diallel selective mating approach in self-fertilizing crops is to involve selective mating in the breeding of these crops. There are two main objectives of involving selective mating : (1) to transfer and shift desired gene frequencies to the coming generation, and (2) to dissipate isolating mechanisms and within population linkage blocks. In DSMS, selective mating (with minimum interim selfing) is done at intrapopulation as well as at interpopulation level. At intrapopulation level, it is done at two stages, namely between $P_2 F_1$ and $P_2 F_2$ and in F_1 and/or F_2 or P_3 to produce P_3 and P_4 series, respectively. At interpopulation level, selective mating is done between $P_1 F_3$ and $P_2 F_2$ to produce supplementary hybrid seeds for P_3. In addition, $P_3 F_1$ or $P_3 F_2$ is back crossed to adapted F_1 hybrids to produce P_5.

Standard line selection from P_1-P_5 populations : It is the final stage of this breeding approach and includes identification and selection of desirable line and composite varieties from P_1-P_5 populations. From basic parent series, line variety may be developed either through mass selection from all F_1 hybrids or by processing individual F_1s using conventional breeding system. Similarly, best lines are identified and selected in the end from P_2 population. Composite varieties are developed from populations produced through selective mating. Parent lines are also produced for their use in subsequent breeding programmes.

The aforementioned details of the implementation phase of DSMS clearly indicate that the breeding approach rests on a sound scientific basis. However, looking at the complexity of this procedure, the breeder can make appropriate modification in the method depending upon the facilities available (land, labour, money, etc.). For example, the breeder may produce population only through one type of selective mating and/or may completely eliminate production of P_5 and may call the procedure as simplified DSMS.

Advantages of DSMS

1. Like other composite breeding procedures, DSMS is an efficient user of germplasm and effort.
2. The procedure strengthens the conventional breeding methods employed in self-fertilizing crops with new force.
3. The approach permits simultaneous input of multiple parents into a few central populations and thus widens the genetic base of the breeding project.
4. Elite genotypes can be introgressed into the breeding populations at any stage of the programme.
5. New experimental varieties can be extracted at any time.
6. The germplasm can be screened at different stage of development.
7. The procedure permits intermating within as well as between populations.
8. The approach facilitates breaking of linkage blocks and freeing of genetic variability by enhancing genetic recombination and helps in creating new gene pools.
9. Several varieties, each with different objective, are developed at the end of the breeding programme.

Disadvantages of DSMS

The approach is circumscribed by the following limitations :

1. A large number of crosses are required in the procedure and making large number of crosses in self-fertilizing species is not an easy task.
2. Parent identity is lost.
3. A number of plant breeders have reservation regarding general suitability, efficiency and success of this procedure in self-fertilizing crops.
4. The approach requires more time before plant selection.
5. Cross- or allogenotypic effects are more severe and negative than self- or autogenotypic competition effects.
6. An uncertainty prevails among plant breeders regarding the powerfulness and usefulness of DSMS.

These limitations obstruct the use of DSMS and its general accepance by plant breeders. As a result, the use of this approach in crop improvement programmes so far is virtually nil. However, to overcome the drawback of large number of crosses required in DSMS and to enhance the use of this approach, Jensen (1970) suggested two ways : (a) The use of simply inherited male sterility, and (b) to grow the breeding material under specialized environmental conditions. Whereas the availability of simply inherited male sterility in the material eases the problem of making crosses and thus removes the restriction on maximum number of parents to be included in the basic parent series and in the F_1 diallel series, the growing of breeding material under specialized environmental conditions maximizes genotypic expression of the characters to be selected.

Jensen (1978) discussed the integration of DSMS in relation to overall project breeding system, amplified certain features of DSMS (objective base planning for the system, forming DSMS composites, early generation handling methods and procedures of selective mating) and advocated, after considering each of the common weaknesses of the composite breeding given in the literature (more time required before plant selection, severity and negative of allogenotypic competition effects, ambivalence about composite breeding that it is not a powerful and useful tool and reservation regarding the efficiency, suitability and success of composite breeding), that these weaknesses cannot be proved. He opined that initial selection of pure line in composite breeding can be done exactly at the same time stage as it is done in the traditional pedigree method. According to him, some worst case situations support the view that loss of genotypes during competition is a welcome objective in breeding programmes to overcome the problem of numbers. Jensen says that the ambivalence about composite breeding that it is not a powerful and useful breeding tool is due to the inability of making clearcut distinction between short term and long term breeding programmes. Regarding suitability, efficiency and success of composite breeding, he gave examples of a number of cultivars of different autogamous crops which have been developed through this procedure and reiterated the point that composite breeding is a very efficient and less time consuming procedure as it takes exactly the same time period in producing cultivars as is taken by other short-range methods.

B

Hybrid Breeding

CHAPTER 18

Development of Hybrid Varieties

Hybrid breeding means development of hybrid varieties. Hybrid is an individual differing in its genotypic constitution at one or more loci and makes better use of heterosis than any other breeding method. A hybrid variety may be a single cross hybid (an F_1 produced by crossing two genetically different homozygous inbred parents), a three-way cross hybrid (a variety produced by crossing a desirable F_1 with a third inbred) or a double cross hybrid (a desirable F_1 crossed with another unrelated desirable F_1) and can be obtained by crossing genetically dissimilar inbred lines, F_1s, clones, etc. Thus, a single cross hybrid differs in its allelic constitution at all the loci for which the two parents involved in that cross differ. The number of single cross hybrids with n parents will be $n(n-1)/2$.

Nevertheless, hybrid breeding should not be confused with heterosis breeding. Whereas hybrid breeding is the development of hybrid varieties, heterosis breeding means exploitation of heterosis in heterozygous (usually unfixable heterosis) as well as in homozygous (fixable heterosis) condition. Heterosis breeding, in fact, is a very wide term and hybrid breeding is a part of hetrosis breeding. Therefore, heterosis breeding and hybrid breeding should not be used as synonyms.

As has been emphasized in Chapter 4, it is meaningless to classify plant breeding methods or to choose method for improving a crop on the basis of reproductive system alone since any breeding method may be used in any crop and heterosis can be exploited across the mating systems. In fact, the choice of the optimum type of variety (whether to evolve a line variety or a hybrid variety) mainly depends on the genetic basis of heterosis (whether the main cause of heterosis is dominance or overdominance). Unfortunately, barring some exceptions, the current knowledge of the contribution of the mechanisms involved in heterosis is poor. Of course, the research work done on *Nicotiana rustica* for plant height (Jayasekara and Jinks, 1976; Pooni and Jinks, 1981; Jinks, 1983; Jinks and Pooni, 1986) and some studies carried out on maize (Sprague, 1983) indicate that dominance of favourable dispersed genes rather than over-dominance is the main cause of heterosis. Once the mechanism of heterosis is known in a crop, it becomes easy to decide about the optimum variety to be evolved and to use and develop most appropriate breeding method in that crop. For example, if overdominance is the main cause of heterosis, it is always better to develop a single cross hybrid.

History of Hybrid Breeding

The history of hybrid breeding is, by and large, the history of hybrid maize breeding. A lot of work has been done on maize on this aspect mainly because the position and the morphological structure of male and female inflorescences of this crop made selfing as well as crossing very easy and thus this plant attracted many scientists to carry out hybrid breeding research in the late nineteenth century and early twentieth century. However, Koelreuter (1763) perhaps was the first to observe hybrid vigour in plants and realize its importance in tobacco. Darwin's (1877) conclusion about the beneficial effect of cross fertilization and deleterious effect of self fertilization encouraged many scientists (Beal and others) to produce hybrids of different crops. The production of varietal hybrids of maize by Beal in 1880 by detasseling technique proved to be a remarkable method for producing single cross hybrids. Nevertheless, the most systematic efforts in this direction were made by G.H. Shull (1909). He proposed pure line method for developing hybrid varieties of maize and is rightly considered to be the originator of hybrid breeding. His proposal brought a complete revolution in maize breeding and provided way for developing hybrids in many other fields, horticultural and vegetable crops. But this scheme of producing hybrid varieties got a setback due to three reasons : (1) Good combining inbreds were not available and thus hybrids produced using poor performing inbreds did not show much superiority over the best open-pollinated varieties; (2) hybrid seed was costly because the inbred line from which hybrid seed was collected was low yielding and the line used as male parent did not produce large quantity of pollen; and (3) hybrid seeds were usually ill-shaped and small and showed poor germination.

The contributions made by E.M. East, D.F. Jones and H.K. Hayes towards maize breeding were also equally important. The proposal of Jones (1918) for producing double cross hybrid variety by crossing two unrelated single cross hybrids, is considered to be a milestone in the history of maize breeding. This procedure of hybrid seed production helped the maize breeders in overcoming all the three drawbacks mentioned above. In this procedure, the hybrid seed used for commercial cultivation is produced on a high yielding F_1 hybrid used as female parent instead on a poor yielding inbred parent. Also, abundant pollen is produced by the other single cross hybrid used as the male parent. Further, the size and health of the seed produced on an F_1 hybrid are normal and thus do not create any difficulty in germination, etc. Despite this all, there was a considerable time lapse between the first commercial production of a double-cross variety (in 1921) and growing of hybrid maize varieties in large areas (in 1940 and after). Nevertheless, nearly complete adoption of hybrid maize varieties by the growers in the corn belt of U.S.A. resulted in virtual disappearance of open-pollinated varieites in that area.

However, a lot of improvement has now been made in the performance of maize inbreds using population improvement methods. The new inbreds are several times higher yielder than the old inbreds and have made the production of single cross hybrids comparatively much easy.

Procedure of Producing Hybrid Varieties

Production of hybrid varieties in cross-fertilizing crops generally involves five steps :

1. Selection of potential plants from heterozygous and genetically heterogeneous base material
2. Selfing/sib mating of selected plants to develop inbred lines
3. Lines evaluation of inbred lines
4. Crossing best inbreds for producing a desirable hybrid variety
5. Production of hybrid seed in large quantity for distribution to growers for commercial cultivation.

The crossing pattern in step 4 would, however, differ depending upon the type of hybrid variety to be produced (whether a single cross, three-way cross, double cross or a triple cross hybrid). Single cross hybrid, three-way cross hybrid, double cross hybrid and triple cross hybrid have involvement of two, three, four and six unrelated inbreds, respectively. If the identity and the genetical purity of the inbred parents are maintained, all the four kinds of hybrids are reproducible in succeeding years.

Selection of potential plants from the source material : The effectiveness of selecting desirable plants from the base material is determined by the proportion of potential genotypes occurring in that source material and the type of selection exercised. The proportion of potential genotypes in the source material, in turn, depends upon the frequency of favourable genes in the material. If the frequency of favourable genes in the material is high, the selection will be highly effective. But too low frequency of desirable genes in the material may lead to little or no selection response. Therefore, a high frequency of favourable genes in the source material will greatly help the plant breeder in selecting superior genotypes and developing outstanding inbreds. However, if the frequency of favourable genes is low, the use of some sort of cumulative selection which could gradually increase the frequency of desirable genes would check random fixation of genes due to inbreeding and thus would be a more reasonable solution to the problem.

Development of inbred lines : The procedure of the development of inbred lines in cross-fertilizing crops is similar to the development of pure line varieties in self-fertizing crops. The selected plants from the base material are selfed or sibmated (if selfing is difficult) and the progenies (generally 20-30 plants) of these plants are raised in the following season. Within as well as between progenies selections are made to choose best plants. Ultimately, the progenies will become homozygous (after 5-6 generations of selfing or sib-mating) with a marked decrease in their vigour and other characteristics and marked increase in the uniformity of the plants within line. In the end, all lines showing deficiencies for important characters are rejected and only those which meet the requirement of the breeder are retained. This method of developing inbreds may be modified by terminating selfing after five or six generations and continuing lines by sib-mating. Further, to accomodate a large number of inbreds, the size of the progeny of each inbred ear of maize may be reduced to 3-4 plants only (Jones and Singleton, 1934). The method maximizes the possibility of selection between lines and minimizes the chances of selection within lines (which is relatively less important after 5-6 selfings). If first cycle inbreds (inbred lines developed directly from the base material) fail to meet the level of improvement required by the breeder, second cycle inbreds (inbreds developed from the progenies of first cycle hybrids, that is, hybrids produced by crossing first cycle inbreds) or third cycle inbreds (inbreds developed from the progenies of second cycle hybrids, that is, hybrids produced by crossing second cycle inbreds) may be developed.

Evaluation of inbred lines : After the development of inbreds the next step is the evaluation of these inbreds and to identify the most promising ones for the commercial production of hybrids. In fact, the evaluation of inbreds is relatively more difficult than their development. If we look at the total number of inbred lines developed and the number of hybrids produced, we find that the number of outstanding inbreds is too small a fraction of the total number of inbreds tested. Only a few inbreds are being used repeatedly in the commercial production of large number of hybrids. It means that only a few of the thousands of inbreds developed are able to produce promising hybrids. The evaluation of inbreds can be done on the basis of their *per se* performance (visual selection), general combining ability and specific combining ability (their performance in hybrid combinations).

Evaluation based on visual or phenotypic selection : A controversy exists among the scientists with regard to the effectiveness of visual selection during inbreeding. According to some, a general relationship (significant phenotypic correlations) exists between the yield potential of inbreds and the hybrids they produce, whereas others have a considerably divergent opinion and say (on the basis of evidence they found in maize) that visual selection for yield is ineffective during the inbreeding process. In fact, the effectiveness of visual selection depends upon the genetic system governing a character. This selection is effective for characters with high narrow sense heritability and there is a higher positive correlation for such characters between the performance of inbreds and their hybrid progeny. But for the complex characters like yield which generally have low heritability, visual selection in most of the cases is ineffective.

The first operation in the development of hybrid varieties (that is, selection of desirable foundation plants from the source material) is nothing but a single generation of mass selection. But the experience of maize breeders indicate that this selection cannot be considered as the sole factor responsible for the production of potential hybrids. Further, it has been found in maize that visual selection during the development of inbreds also does not play a very significant role in achieving genetic gain. It means that identification and selection of outstanding inbreds at the third stage of hybrid development (evaluation of inbred lines) have a greater role to play in the production of potential hybrid varieties.

Nevertheless, whether visual selection is effective or ineffective during the process of inbreeding, it serves a useful purpose in the development of inbreds as the lines with major defects (susceptibility to some prominent disease, very low yield potential, etc.) can be easily eliminated through this selection. Visual selection also helps in reducing the number of inbred lines to a manageable limit.

Evaluation based on general combining ability : A huge number of reports are available in the literature where *per se* performance of inbreds does not show significant correlation with their general combining ability. Since the agronomic worth of any inbred is determined by its ability to produce superior progeny in combination with other inbred lines or a heterozygous tester, the *per se* performance does not always provide a correct value of an inbred. It is, therefore, imperative to measure the value of inbreds in terms of their general combining ability. One method of testing general combining ability of inbreds is the use of inbred × variety top crosses as suggested by Davis in 1927. Jenkins and Brunson (1932) compared general combining ability of inbreds (general performance of an inbred in top crosses) with their average combining ability (average performance of an inbred in a series of single crosses) and reported comprehensive data related to the method suggested by Davis. Now-a-days, general combining ability and average combining ability are being used as synonyms. But some scientists have reservation about the reliability of this method for testing the general combining ability of inbreds. However, the method can be made more reliable by growing the top crosses over a range of environments consisting of several seasons and several locations. Even if the method is used in its original form, it is a safe procedure for discarding a large portion of the lines showing poor performance.

Another method by which general combining ability of inbreds can be tested is line × tester mating design where the number of testers is more than one. Here the lines and testers are crossed in all possible combinations.

A more reliable method of testing the general combining ability of inbreds is the diallel mating design where the same set of inbred lines is crossed in all possible combinations. If the number of inbreds is p, there will be total p^2 combinations (p selfs, p (p-1)/2 a set of single crosses and p(p-1)/2 their reciprocals). But even if only one set of single crosses is made, the number of crosses tremendously increases with an increase in the number of inbreds to be tested. For example, with 8 inbreds there will be 28 crosses and with 16 inbreds there will be 120 crosses. If the number of inbreds is large (as is generally the case), it is not practically feasible to use this method since the method puts a limit on the number of inbreds to be tested. It is, therefore, advisable that in the begining top cross method is used for discarding a large portion of inbreds and the remaining portion of the lines may be tested using diallel method.

Evaluation based on specific combining ability : High × high general combining ability combinations of inbreds do not necessarily produce high yielding hybrids, that is, the performance of any hybrid combination cannot always be predicted on the basis of the general combining ability of the inbreds involved in that combination. It means specific combining ability which is the deviation in performance of any hybrid from that predicted on the basis of general combining ability of its parents, can be an important cause of heterosis.

Specific combining ability can be estimated from the phenotypic performance of single crosses, three-way crosses, double crosses, etc. using line × tester, diallel, triallel, quadriallel, etc. mating designs. But the number of crosses in triallel and quadriallel mating designs [p(p-1) (p-2)/2 and p(p-1) (p-2) (p-3)/8, respectively, where p is the number of inbreds] increases tremendously with the increase in the number of inbred lines. With 10 inbreds, for example, the number of single crosses, three way crosses and double crosses will be 45, 360 and 630, respectively. However, the performance of three way and double crosses can be predicted from the mean performance of single crosses without producing an unmanageable number of these crosses, that is, the performance of 360 three-way crosses and 630 double crosses can be predicted from the mean performance of 45 single crosses only. This approach, therefore, changed an almost impossible task of the production and testing of a huge number of three-way or double crosses into a simple one. To illustrate it let us consider 10 single crosses produced by crossing 5 inbreds (A, B, C, D and E) in all possible combinations. The crosses will be –

A×B	A×C	A×D	A×E	B×C
B×D	B×E	C×D	C×E	D×E

There will be 30 three-way crosses (A×B) × C, (A×B)×D, (A×B)×E, (A×C) ×B, (A×C) ×D, (A×C)×E, ... (D×E) ×A, (D×E) ×B and (D×E) ×C. The yield potential of each of these crosses can be predicted on the basis of the performance of its non-parental single crosses. For example, non-parental single crosses of the three-way cross (A×B) ×E will be A×E and B×E. Similarly, the yield potential of (D×E) ×C cross can be predicted from the yields of C×D and C×E single crosses. Further, if A×C and A×E are the best yielder single crosses, A(C×E) will be the best yielder three-way cross as C and E inbreds are in different parental crosses.

The 15 double crosses produced by these five inbreds will be (A×B) (C×D) (A×B) (C×E), (A×B) (D×E), (A×C) (B×D), (A×C) (B×E), (A×C) (D×E), (A×D) (B×C), (A×D) (B×E), (A×D) (C×E), (A×E) (B×C), (A×E) (B×D), (A×E) (C×D), (B×C) (D×E), (B×D) (C×E) and (B×E) (C×D). Jenkins (1934) suggested four methods for predicting the performance of double cross hybrids :

1. On the basis of the mean performance of six possible single cross hybrid combinations among any set of four inbred lines : For (A×C) (B×D) double cross, for example, the predicted yield will be (A×C) + (A×B) + (A×D) + (B×C) + (C×D) + (B×D) :

2.   On the basis of the mean performance of four non-parental single cross combinations : The predicted performance of the double cross (A×C) (B×D), for example, will be

$$\frac{(A×B) + (A×D) + (B×C) + (C×D)}{4}$$

3.   On the basis of the average performance of four inbreds involved in a double cross over a series of single cross combinations : Accordingly, the predicted performance of the double cross (A×C) (B×D) will be :

$$\frac{(A×Q) + (A×R) + (A×S) + A×T) + (B×Q) + \ldots\ldots + (D×T)}{n}$$

Where Q, R, S and T is a set of other inbreds and n is the total number of single cross combinations

4.   On the basis of the average top cross performance of the four inbreds involved in a double cross : The predicted performance of the double cross (A×C) (B×D), for example, will be :

$$\frac{(A×T) + (C×T) + (B×T) + (D×T)}{4}$$

Where T is the heterozygous tester used in top crosses.

Although all the four methods described above have been found to give good agreement between the predicted and the observed performance of any double cross, the method 2 gives most accurate estimate of the performance. Using this method, Anderson (1938) predicted yields of 15 maize double crosses from the yields of 10 single crosses produced by crossing five inbreds in all possible combinations. These predicted yields showed a remarkably close agreement with the actual yields of these double crosses. Several other investigators came to a similar conclusion and consequently this method became a standard breeding procedure for predicting yields of double crosses. The accuracy and reliability in the prediction of performance of double crosses can be further increased by minimizing the effect of genotype × environment interaction, that is, by conducting tests in several seasons and at several locations (Sprague and Federer, 1951).

Another factor which has an important bearing on the productivity of double cross hybrids is the order of pairing in these crosses (Allard, 1960). Eckhardt and Bryan (1940) had studied this aspect of genetic diversity in detail. Crossing between genetically diverse inbreds is expected to produce high yielding single crosses. Similarly, crossing between genetically diverse single crosses generally results in the production of high yielding double crosses. For example, if inbreds A and C are derived from the same source and D and E from other source, the productivity of the single crosses A×D, A×E, C×D and C×E will be higher than that of A×C and D×E crosses. On the contrary, the double cross (A×C) (D×E) will be highest yielder among the three possible double crosses poduced by this set of four inbreds. Therefore, to identify the highest yielding double cross, the best single crosses should be treated as non-parental single crosses.

A controversy exists among the plant breeders about the time of testing inbred lines. According to Jenkins (1935) and several others (Sprague, 1939, 1946; Lonnquist, 1950; Wellhausen, 1952; Wellhausen and Wortman, 1954), early testing of inbreds for general combining ability by top cross method at S_0 (original selection) or S_1 (first generation selfed) stage can help the breeder in detecting lines expected to produce good combining inbreds and thus eliminating unpromising lines. However, some other investigators (Singleton and Nelson, 1945; Richey, 1945, 1947; Payne and Hayes, 1949) have shown doubt about early testing of inbreds. In their opinion, there is risk of loosing many worthwhile lines due to early elimination of breeding material.

Improvement of Inbred Lines

The inbred lines to be used in the production of hybrid varieties should give the most reliable performance in a wide array of environmental conditions. For this, they should fulfil the following three requirements :
(1) Their general productivity (based on *per se* performance) should be high.
(2) They should posses resistance to major diseases and insect pests of the crop.
(3) Their combining ability should be high.

If any inbred is deficient in respect of any of these characteristics, it should be improved. Several methods have been suggested for improving inbred lines.

Pedigree method :– In many cases, the first cycle inbreds have a poor productivity. Their productivity can be increased by developing second cycle inbreds using pedigree selection. The procedure involves crossing of two genetically diverse inbreds and selecting superior recombinations in the segregating generations.

Backcross method : If any inbred line is deficient for some specific character (susceptibility to a particular disease, etc.) which is governed by a simple genetic system, that defect can be easily removed by transferring the desirable gene to the inbred from some other source using backcross method. Further, the backcross method is especially suited for incorporating restorer genes into the genotypes of the matching inbred lines and substituting the genotypes of inbreds in male sterile line. The method has been used in a number of cases to remove bottleneck genes. For example, this method has been successfully used to improve the male sterile line Tift 23A of pearlmillet which is highly susceptible to downy mildew. One of the improved versions of this line, namely ms 5141A has been extensitvely used by the Indian pearlmillet breeders in their breeding programmes.

Convergent improvement : It is a modified type of backcrossing. The method was suggested by Richey (1927) for improving the inbred parents simultaneously by running two parallel backcross programmes. Here, the F_1 produced by crossing two inbred lines complementing each other in some desirable attributes, is backcrossed to both the inbreds. In each of the two backcross programmes, the desirable genes from the donor inbred are maintained by selection. The inbred parent which is recipient in one backcross programme, is donor in the other programme. Although there are some reports in favour of this method for improving inbreds (Richey and Sprague, 1931) and the F_1 hybrid (Murphy, 1942), the method did not get support from many maize breeders and thus has not been used frequently in maize breeding.

Mutation breeding : An inbred line deficient in some specific character can be improved by developing desirable mutants. This approach has been used in pearlmillet for developing downy mildew resistant male sterile, restorer and maintainer lines.

Biotechnological approach : Biotechnology is a compendium of most sophisticated and innovative laboratory techniques and permits transfer of genes across sexual barriers. An inbred line can be transformed by transferring a desirable gene using an appropriate biotechnological technique. For example, a maize inbred lacking in the quality of storage protein can be improved by this method.

Gamete selection : It is a modified form of early testing and was suggested by Stadler (1944) for improving inbred lines. The method is based on the population genetics principle that the frequency of desirable gametes (p) in an open-pollinated variety is much higher than that of the desirable zygots (p^2). It means that gametic sampling is a better way of sampling open-pollinated varieties than routine inbreeding methods. In gamete selection method, a good inbred line is crossed with an open-pollinated variety. Each F_1 plant of this top cross is crossed to an appropriate heterozygous tester as well as selfed. The F_1 plants showing better performance in test crosses (that is, have received superior gamete from the open-pollinated variety) are continued through inbreeding. The method got a mixed response from the plant breeders regarding its usefulness in plant breeding. Whereas Pinnell *et al.* (1952) and Lonnquist and Mc Gill (1954) advocated the usefullness of this method for further selection of material from open-pollinated varieties, Gieshrecht (1964) found it an inefficient method of improving inbred lines.

Recurrent selection : This system of breeding maintains the genetic variability in the breeding material and also increases the frequency of desirable genes and gene combinations. Therefore, the method can be used for both the identification of outstanding inbreds from the source material and improvement of good inbreds. Depending upon the objective of the breeder, the inbred lines can be improved in respect of their general combining ability, specific combining ability or both by using an appropriate type of recurrent selection.

Producing Single Cross Hybrids

The product of a cross between two genetically dissimilar inbred lines or homozygous varieties is termed as a single cross hybrid. For producing a single cross hybrid, the plant breeder needs an outstanding homozygous female parent and a similar male parent. The two parents should complement each other for desirable attributes. In dioecious plant species, since the whole plant is either male or female, F_1 hybrid can be produced by growing the female and male parents in isolation and allowing open pollination. In monoecous plant species also, it is not generally difficult to make a plant as female parent and as a male parent. But in case of hermaphrodite species especially where flowers are small, some special device (male sterility, self-incompatibility, etc.) is to be used to make a plant as female. Depending upon several factors (morphological structure of male and female inflorescences and their positions on the plant, type of flowers-unisexual or bisexual, degree of easiness in selfing and crossing, availability of pollination control system, number of seeds produced per fruit or per pollinated flower, etc.), different methods have been used to produce single cross hybrids.

Hand emasculation and pollination : In plant species where both emasculation and pollination are easy and/or the number of seeds produced per pollinated flower is high, the production of single cross hybrids by hand is economical. This method is being used in case of cotton for commercial hybrid seed production. In this plant species, the flowers are large and the staminal column can be easily removed on the day before flowering. The emasculated bud is then covered and ripe anthers are brushed over the stigma of the bud on the next day. Hand emasculation and crossing can be done with the same degree of easiness in okra. The method of hand emasculation and pollination is equally successful in vegetable crops like brinjal, chillies, tomato and cucumber where number of seeds per fruit produced by a single emasculated and pollinated bud is very high. The okra plant has both the advantages of easiness in emasculation and crossing and high number of seeds per fruit.

Detasseling : The morphological structure of male and female inflorescences and their positions on the maize plant provide a unique opportunity for the plant breeder to make the whole plant female just by removing the male inflorescence (tassel). In maize, single cross hybrid seed is produced by growing the inbred parents in an isolated plot in alternate rows (generally 4:2 ratio of female and male), detasseling the inbred to be used as female parent and allowing free pollination. The seed is collected from the detasseled rows. Removal of male flowers is also easy in cucurbits but this group of plant species does not show conspicuous heterosis.

Using male sterility : Availability of usable male sterility in any crop species is a boon for the plant breeder. The cytoplasmic-genetic male sterility (a male sterility-fertility restorer system) is being used for the commercial production of hybrid seed in several crops (pearl millet, sorghum, onion and many others) by developing male sterile (A), maintainer (B) and restorer (R) lines. Efforts are also being made for the use of photoperiod and temperature dependent genetic male sterility (a two-line system) in some crops.

Using self-incompatibility : Because of some inherent difficulties, self-incompatibility has not been used as frequently in plant breeding as male sterility. For the production of single cross hybrid seed, the two parent lines (either both lines self-incompatible but cross-compatible or one line self-incompatible and the other self-compatible) are grown in alternate rows and F_1 seed is collected from the self-incompatible rows. This method has been used in some crops of *Brassica* genus.

Using chemical hybridizing agents (gametocides) : Chemically induced male sterility can be used as an alternative method of hybrid seed production in crops where usable male sterility or some other efficient device of pollination control is not available. A number of chemicals (maleic hydrazide, ethrel, RH 531, RH532, LY195259, SD84811, etc.) have been suggested for this purpose. LY 195259 and SD84811 of these chemicals induce complete or near complete male sterility and possess almost all the characteristics of a good chemical hybridizing agent. These chemicals have been successfully used for hybrid seed production in wheat, a self-fertilizing crop species. The application of gametocides can give equally good results in other self-fertilizing crops.

Using nuclear male sterility : Driscoll (1972) proposed an ingenious method for hybrid wheat seed production using chromosomal male sterility and called it XYZ system. The method involves three lines : X, a disomic addition line with on alien chromosome of rye (5R pair); Y, a monosomic addition line with one alien chromosome of rye (5R); and Z which contains the normal 21 pairs of wheat chromosomes, is homozygous for a recessive male sterile gene. The alien chromosome has gene for fertility and does not pair with the wheat chromosomes. The Y line, therefore, produces two types of pollen and the 21-chromosome type is largely favoured in fertilization. As a result, selfing of Y line produces 25 per cent Y and 75 per cent Z plants. Driscoll (1985) suggested modification in his method by not involving the X line and using the progeny of a selfed Y line instead of Y line itself. Chromosomal system of hybrid wheat though has some advantages (uninfluenced by environmental conditions, deficiency in fertility restoration affects the system at less serious stage and the line used as male contains no nuclear fertility restoring genes), is more complicated than cytoplasmic male sterility system.

Using BTT system : Ramage (1965) proposed an alternative method for the continuous availability of female (male sterile) parent for commercial hybrid seed production in barley. The method is known as BTT (balanced tertiary trisomic) system and can be used in other self-fertilizing crops with equal success. The characteristics of this system are :

(1) The presence of an extra translocated chromosome

(2) the presence of a recessive ms allele for male sterility on the two normal chromosomes of the diploid complement and that of Ms allele on the extra interchange chromosome very close to the break-point,

(3) the presence of one or more marker genes with dominant allele on the extra chromosome,

(4) the transmission of extra chromosome should be through the female parent only so that the male parent produces only one type of functional gametes,

(5) a suitable male sterility gene with normal expression in all environments should be available, and

(6) the BTT should have normal growth and development so that it is competitive and is able to produce abundant pollen.

The BTT when selfed produces about 70% homozygous recessive diploids, about 30% BTT and rare (upto 1%) homozygous recessive primary trisomics. For commercial hybrid seed production, the female rows are sown using diploid male sterile seed and the male rows are sown with the seed of a superior inbred showing high combining ability with the female parent. Hembar was the first hybrid produced by BTT method and was released for commercial cultivation in Arizona state of USA in 1969. Nevertheless, the method suffers from some serious draewbacks like difficulty in separating diploid sterile seeds from trisomic seeds and eliminating all the BTT plants from the female block and the rare transmission of extra chromosome containing fertile allele through pollen. Ramage and associates suggested a speculative scheme to mitigate the above metnioned difficulties. The scheme involves genetic prevention of the pollen transmission of dominant fertile allele and obtaining homozygous male sterile diploid stocks.

Advantages and disadvantages of producing single cross hybrids : The main advantages of producing single cross hybrid varieties are :

(1) All the plants of a single cross hybrid are genetically identical and uniform in appearance and thus such a hybrid has uniformity in produce.

(2) Since single cross hybrid is heterozygous for all the loci for which the two inbred parents differed, it generally has high yield potential as compared to other types of hybrids involving more than two inbred parents.

(3) With the same experimental efforts more parents can be accomodated than in three-way and double cross programmes.

(4) With the same set of parents the between crosses variance is larger than in case of three-way and double crosses.

The disadvantages associated with the production of single cross hybrids are :

(1) Since only two inbred parents are involved in the production of single cross hybrids, they have a narrow genetic base and thus may have a relatively poor adaptability than three-way and double cross hybrid varieties.

(2) The seed production in case of single cross hybrids in cross-fertilizing crops is costly. But with the availability of high yielding and good combining inbreds in many of these crops, the cost of production of hybrid seed is remarkably reduced.

(3) The farmers have to purchase hybrid seed every year/season.

(4) The risk of selecting the best single cross is more since the detection of the cross is based on phenotype and not on genotype.

Producing Three-way and Double Cross Hybrids

As has been mentioned earlier, a three-way cross has involvement of three genetically distinct inbreds and a double cross of four such inbred lines. In three-way cross, the female parent and in a double cross both the parents are single crosses. The double crosses have highest degree of heterogeneity among the three kinds of hybrids (single, three-way and double crosses). The three-way crosses fall in between the double crosses and single crosses. Single crosses produced by crossing completely homozygous inbreds are completely homogeneous. Therefore, between crosses variance is largest among single crosses and smallest among double crosses. The choice of best order of parents is more important in three-way crosses than in double crosses since the contribution of third inbred is greater than the first and second inbreds used to produce the single cross, while the contribution of all the four inbred parents is equal in a double cross. The production, evaluation and improvement of inbreds for producing three-way and double cross hybrids are the same as in case of the production of single cross hybrids. A double cross hybrid can be produced by employing two inbreds with restorer gene or by employing one inbred with restorer gene if male sterility is to be used for hybrid seed production. In the former case, the whole hybrid progeny will be fertile since the pollinator F_1 hybrid has restorer gene in homozygous condition, whereas in the latter case half of the hybrid progeny will be fertile and half will be sterile since the pollinator F_1 is heterozygous for the restorer gene.

Advantages and disadvantages of producing three-way and double cross hybrids : The main advantages of producing three-way and double cross hybrids are :

1. Because of the involvement of more inbred parents, the three-way and double cross hybrids have broader genetic base than single cross hybrids.

2. Unlike single cross hybrids which are usually sensitive to change in environment, the three-way and double crosses show greater phenotypic stability. The three-way crosses fall in between the double crosses and single crosses in this respect.

3. Since the seed parent in a three-way cross and both the parents in a double cross are single crosses and are vigorous, adequate quantity of hybrid seed can be produced without any difficulty and with less expenses.

4. In a double cross since the pollinator parent is also a single cross, abundant pollen is produced and thus male to female rows ratio can be increased (2:6 or 1:4).

5. The risk of selecting best cross is reduced since the prediction of a three-way or a double cross is based on the performance of more than one single cross.

The disadvantages of producing three-way and double crosses are :

1. The number of crosses increases tremendously with an increase in the number of parents. Even with a moderate number of parents, the number of crosses becomes unmanageable.

2. There is less uniformity in the produce.

3. The yield potential of these crosses is less than that of single crosses.

4. Between crosses variance is less than it is among single crosses.

Producing Triple Cross Hybrids

Thompson (1964) suggested development of triple cross hybrid in kale (*Brassica oleracea* var. *acephala*) using sporophytic type of self-incompatibility. The method involves use of six different inbreds as parents each containing a different self-incompatibility allele in homozygous condition. The method of producing triple cross hybrid seed is shown in Figure 18.1. Two single crosses are made after transplanting selfed inbreds 1 and 2 (1×2) and 4 and 5 (4×5) into isolation plots. The seeds are then drilled in 4:1 ratio of (1×2):3 and of (4×5) :6 in large plots (1-2 acres) to produce two three-way crosses. In the end, the seeds of these two three-way crosses are mixed and are given to seed agencies to produce the final cross. For the production of pure triple cross hybrid seed, it is necessary that the S_3 allele is active with both the alleles S_1 and S_2 in pollen as well as in style and S_6 also shows similar activity with alleles S_4 and S_5. The inactivity or recessiveness of S_3 to S_1 or S_2 or that of S_6 to S_4 or S_5 would result in cross compatibility of 1×3 and 2×3 plants from the three-way cross when raised with the other three-way cross for producing triple test cross hybrid seed. However, the condition of the presence of different S alleles in six different inbreds is not essential.

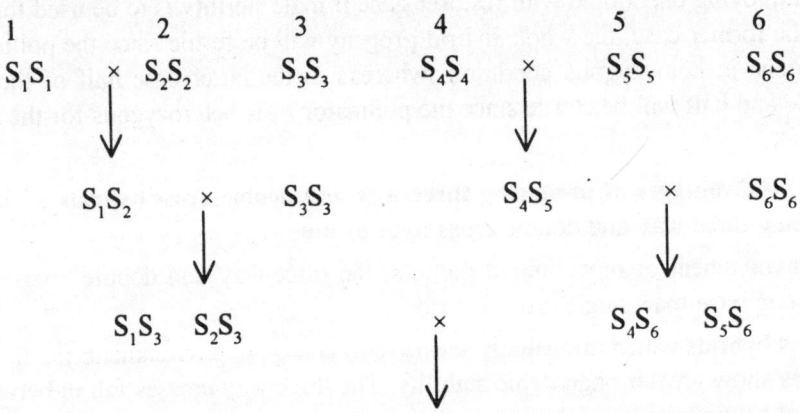

Triple cross hybid seed

Fig. 18.1. Production of triple cross hybrid

Advantages and disadvantages of producing triple cross hybrids : The advantages of producing a triple cross hybrid are :

1. The triple cross hybrids have a better combination of characters.
2. The choice of inbred lines is wider.
3. The survival of these hybrids in winter season is higher.
4. The amount of bud pollination is reduced as compared to that in a double cross.

The disadvantages of producing these hybrids are :

1. The method is restricted to sporophytic self incompatibility only.
2. The forage yield of these crosses is generally lower than that of double cross hybrids.
3. Dominance relationships between two alleles in pollen and style may complicate the multiplication of this hybrid.
4. The rearrangement of inbreds required in a triple cross creates difficulty particularly when the choice of inbreds used in the hybrid is not wide.

Difficulties in Producing Hybrid Varieties in Self-fertilizing Crops

Cross-fertilizing crops are either dioecious, monoecious or have some pollination control device and thus do not pose any serious problem in producing hybrid varieties. On the other hand, self-pollinators have several inherent difficulties (presence of natural homozygous balance, intergenomic heterosis in some polyploids possessing homeologous chromosomes, preponderance of additive genetic variation and widespread absence of overdominance and unfixable epistasis, dispersion of genes as a major cause of heterosis, non-availability of good pollination control system, drawbacks associated with the use of chemically induced male sterility, floral morphology leading to limited cross-pollination, non-synchronization in pollen shedding and stigma receptivity, economic considerations like high cost of F_1 hybrid seed, etc.) in exploiting heterosis at a commercial scale. Nevertheless, in selfers where single emasculated bud produces quite a large number of seeds (tomato, brinjal, chillies, okra, etc.), the breeder does not face much difficulty in producing single cross hybrid seed. But in crops where hand emasculation and pollination both are difficult and costly, the production of hybrid varieties is a remote possibility. Further, the improvement in self-pollinators has been slow as compared to that in cross-pollinators since the former group lacks a built-in mechanism for self improvement as continued selfing leads to early fixation of linkage blocks and to hindrance of free exchange of desirable genes between plants due to restricted or no recombination. Therefore, the breeding methodology presently being used in self-pollinators needs a radical change. This viewpoint has gained further strength after the successful development of hybrid rice at field scale in China.

To make F_1 hybrids of a self-pollinator a commercially viable proposition, three pre-requisites must be fulfilled by the crop. **First,** since the magnitude of heterotic yield effect is crucial for the commercial success of a hybrid programme, a high level of commercial heterosis must be present in the crop germplasm. **Secondly,** an effective pollination control system must be available. **Thirdly,** there should be a balance between the level of commercial hetrosis and the cost-effectiveness of the pollination control system. Once these pre-requisites are met, the development of hybrid varieties of self-fertilized crops can become a routine procedure. Nevertheless, many of the problems related to the development of hybrid varieties of self-pollinators are crop specific.

Efforts are being made to develop hybrid varieties in self-fertilizing cereal crops like wheat, barley and rice. However, the prospects of producing such varieties in rice are better than in wheat and barley since

(1) rice is a diploid species (barley is also diploid but wheat is hexaploid),

(2) pollination control system (male sterility and fertility restoration) is available,

(3) the seed rate in low,

(4) the crop can be transplanted,

(5) the crop has ratooning ability, and

(6) the crop can be vegetatively propagated through stubbles.

The last two characteristics of rice help in fixing heterosis in this crop.

The maximum success in producing hybrid varieties of rice has been achieved in China where hybrid rice has now become a ground reality. Breeding of hybrid rice in China started in 1970 with the availability of a male sterile wild plant of *Oryza sativa* f. *spontanea* on Hainan Island of the country. The plant had aborted pollen and became an important source of cyto-nuclear male sterility. This source was named as WA (wild abortive) and is still proving to be the most stable source of male sterility in rice. Two dominant nuclear genes (with additive effects) present in the ecographic race *indica* provided restoration of fertility. The three-line model, A line (cms line), B line (maintainer line) and R line (restorer line), is being used to produce hybrid rice seed. After about five years (in mid 1970s) of the start of hybrid rice programme in China, hybrid rice seed was distributed to the growers and by mid 1980s, a large area (more than 50% of the total rice area) was under hybrid rice. The hybrid varieties showed 15-20% yield superiority over the best conventional cultivars. In China, hybrid rice is now being utilized extensively by developing efficient seed production techniques and employing labour-intensive practices (to overcome the problem of low seed set). The lack of synchronization in flowering time of male sterile lines and pollinator lines can be overcome by planting the latter earlier than the former or by developing early-maturing pollinator lines. However, in countries where labour cost is high, employment of labour intensive practices would not be economical.

A good piece of research work on hybrid rice is being done at IRRI (International Rice Research Institute, Philippines) which was founded in 1960 as an international research and training centre aimed to help the people of developing countries by bringing improvement in rice. In addition to the development of short-statured (semidwarf), photoperiod-insensitive, high yielding and disease and insect resistant rice cultivars using Taichung Native 1 (Taiwan's semidwarf type) plant type, rice germpalsm is being screened for the characteristics promoting allogamy in the crop (large anthers with abundant pollen, large protruding stigma, long glume opening period, and more gap between spikelet opening and anther dehiscence). Some wild relatives of *Oryza sativa* (e.g., *O. perennis* and *O. sativa* f. *spontanea*) possess these characteristics. Some Indian collections from Assam also have been found to have such characters and can be helpful to the rice breeders of this region.

The Indian Council of Agricultural Research (ICAR) launched a massive programme in 1989 to develop hybrid varieties of not only cross-fertilizing crops (sunflower and pearl millet) but also of often cross-fertilizing (cotton and pigeonpea) and self-fertilizing (rice) crops. As a result, four rice hybrids were released for general cultivation in Andhra Pradesh (APRH 1 and APRH 2), Karnataka (KRH 1) and Tamilnadu (TNRH 1). Nevertheless, because of the poor growth of the parental lines and their susceptibility to insect pests and diseases, non-impressive yield vigour of the hybrids and substandard rice quality, the Chinese cms lines could not be used very successfully in India. This indicates that there is need to diversify the present male sterility source, enhance the magnitude of heterosis for important characters by making crosses between the three ecographic races (*Indica, Japonica* and *Javanica*), induce male sterility by other sources (like chemicals), take advantage of ratooning and vegetative propagation and to use two-line system (photoperiod-sensitive genic male sterility-PGMS or thermosensitive genic male sterility-TGMS) for developing better hybrids.

In contrast to the role played by cytoplasmic genetic male sterility for hybrid seed production in rice, genetic male sterility appears to provide better opportunity to barley breeders in their hybrid breeding programme. Cultivated barley, like rice, is diploid and thus has better prospects of the development of hybrid varieties than wheat.

As described earlier in this chapter, Ramage (1965) devised an ingenious method for hybrid seed production in barley and advocated (Ramage, 1983) the utility of the method in all other self-fertilizing crops. The procedure is called as balanced tertiary trisomic (BTT) method and ensures continuous availability of male sterile line.

However, in view of the drawbacks (difficulty in the separation of diploid and trisomic seeds on the basis of weight, problem in the elimination of BTT plants from the female block and transmission of extra chromosome through pollen in few cases), Ramage proposed a speculative scheme based on diploid plants rather than on trisomics. Here, pollen transmission of dominant fertile Ms allele from diploid heterozygous plants (Msms) is genetically prevented due to the linkage of a micro-gametic lethal mutation with Ms allele and a seedling lethal factor linked with male sterility helps in producing homozygous male sterile diploid parent of hybrids.

Bread wheat is a highly self-fertilized (96-100% self-pollination) diploidized hexaploid species with three genomes (A, B and D) having homeologous chromosomes. Therefore, there can be intergenomic heterozygosity at the common loci (homologous regions) of the chromosomes of the different genomes even when there is complete homozygosity within each genome, that is, due to intergenomic heterozygosity with intra-genomic homozygosity, heterosis is fixed in polyploid wheats and this condition makes these wheats permanent hybrids. Any degree of selfing cannot reduce the degree of this permanent heterosis in wheat. This is one of the main reasons that why the level of heterosis in F_1 hybrids of wheat is not high. Other reasons of low heterosis in wheat are :

(1) Predominance of additive variation for yield since dominant genes mainly responsible for increased yield in wheat behave additively;

(2) widespread absence of overdominance and unfixable epistasis in wheat populations; and

(3) dispersion of dominant favourable genes in the parents as a major cause of heterosis for yield in wheat.

Nevertheless, since there are numerous uncommon loci in the three genomes of wheat and a large portion of genetic variability present in wheat and in its relatives has not yet been exploited, it is not justified to rule out the possibility of developing potential F_1 hybrids of wheat. For example, use of Veery varieties of CIMMYT and that of the vast amount of unutilized variability present in D genome in hybrid wheat programme may be helpful in producing desirable F_1 hybrids.

Like other crops, the first step in hybrid wheat technology is the availability of a usable pollination control system. Three types of male sterility, namely, cytoplasmic-genetic (Wilson and Ross, 1962; Livers, 1964 and others), nuclear or genetic (Driscoll, 1972, 1985) and chemically induced (Chopra *et al.,* 1960; Law and Stoskopf, 1973; Hughes *et al.,* 1974; Rowell and Miller, 1974; Tschbold *et al.,* 1988; and others), described earlier in this chapter, are being used as systems of pollination control in wheat breeding programmes to develop F_1 hybrids for various purposes. Although the three kinds of pollination control systems have their own merits and demerits, the first type (cytoplasmic-genetic) is the best. The model for producing hybrid wheat using cytoplasmic-genetic male sterility is similar to the model used for hybrid seed production in pearl millet and sorghum where A line (male sterile), B line (maintainer line) and R line (restorer line) are developed. Nearly all hybrid wheat breeding research continues to be based on the male sterility from *Triticum timopheevii* which is least affected by environmental change, effectively inhibits pollen production, is usually genetically stable and generally does not provide adverse side effects. Hybrid wheat is the F_1 progeny of a cross between A line (carrying male sterile cytoplasm from *T. timopheevii*) and R line (carrying Rf_1 and Rf_2 restorer genes and normal cytoplasm from *T. aestivum* or male sterile cytoplasm from *T. timopheevii*). Nevertheless, fertility restoration in wheat is a complex phenomenon and cannot be described on the basis of a simple two-gene model. It appears that in addition to Rf_1 and Rf_2, a third major gene and a number of modifiers (minor genes) are required for good fertility restoration in stress environments.

Although cytoplasmically induced male sterility in wheat was identified long back (Kihara, 1951; Fukasawa, 1953; Wilson and Ross, 1962) but the importance of this pollination control system and that of hybrid wheat technology could not be fully realised as these findings coincided with the use of dwarfing genes in the development of semi-dwarf, photoinsensitive and fertilizer responsive varieties of wheat which showed tremendous yield superiority over the traditionally grown tall varieties. The extra-ordinary performance of these dwarf varieties diverted the attention of wheat breeders to the exploitation of dwarfing genes resulting in a setback to hybrid wheat technology. However, since the advantages of the dwarfing gene approach are now levelling off, the hybrid wheat technology should get more attention of the wheat breeders.

Singh and Pawar (1998), while reviewing recent advances in breeding of self-pollinated crops, discussed the development of hybrid varieties in rice, wheat and barley and suggested remedies for producing hybrid varieties in these crops. The remedies are :

(1) Increasing yield potential,

(2) gaining better understanding of physiologic mechanisms causing heterosis for higher yield potential,

(3) expanding genetic base by exploiting untapped genetic resources,

(4) searching usable male strility and diversification of cms lines,

(5) restructuring floral morphology for enhancing cross pollination,

(6) enhancing input use efficiency,

(7) fixing heteosis permanently through apomixis or temporarily through ratooning and vegetative propagation, and,

(8) using 2-line method of breeding.

C

Population Breeding

CHAPTER 19

Recurrent Selection

In the early decades of the twentieth century, the breeders of cross-fertilizing crops, especially the maize breeders, got so much involved in the development of hybrid varieties that they did not pay much attention to alternative or better methods of improving these crops. The main reason of paying too much attention towards hybrid varieties was the tremendous difference in the yield potential and other characteristics between the first cycle F_1 hybrids and their inbred parents isolated directly from open pollinated varieties (that is, first cycle inbreds). However, the production of second cycle and third cycle inbreds (isolated from segregating generations of first cycle and second cycle hybrids, respectively) in maize and F_1 hybrids produced from these inbreds led to some interesting conclusions:

1. The second cycle inbreds were conspicuously superior to first cycle inbreds and so were the third cycle inbreds to second cycle ones; and

2. the differences between first, second and third cycle hybrids were mostly non-significant.

That is, there was considerable improvement in the yield potential of inbred lines isolated from the segregating generations of hybrids, but the hybrids produced by crossing these improved inbreds did not show significant superiority over the hybrids produced by crossing inbreds isolated directly from open-pollinated varieties. These facts led to question mark about the so-called undisputed efficiency of the hybrid breeding system. Several maize breeders felt that hybrid breeding in maize was not the most efficient system of improving this important cereal crop.

Nevertheless, the development of outstanding inbreds depends upon the frequency of desirable genes in the base material and the effectiveness of selection during inbreeding. In general, three kinds of drawbacks were associated with the traditional system of hybrid breeding :

1. Low proportion of desirable genes in the base material (that is, open-pollinated varieties);

2. rapid fixation of genotypes causing hinderances to recombination between favourable genes; and

3. limited effectiveness of selection especially for quantitative characters of economic importance.

As a result of this, outstanding hybrid varieties could not be produced. Had there been use of improved inbreds (developed through population improvement) in the hybrid breeding programme, the hybrids would have been much more higher yielder than their contemporary hybrid varieties (Gardner cf. Rao and House, 1972). These ideas ultimately led to the discovery and use of an ingenious breeding system called recurrent selection. The system has several advantages over traditional hybrid breeding system :

(a) Production of hybrid varieties with substantially higher yield;

(b) maintaining sufficient amount of genetic variation;

(c) improving inbred lines to be used in hybrid breeding programme; and

(d) producing broad-based population varieties as a suitable alternative to hybrid varieties.

Recurrent selection, a method of population improvement (selection and intermating between the selects again and again) is a very wide term and includes several selection procedures. In broadest sense, any cylical scheme of plant selection (mass selection and its variants, simple recurrent selection, recurrent selection for gca or population test cross selection, recurrent selection for sca or inbred test cross selection, reciprocal recurrent selection or reciprocal half sib selection, etc.) aimed at to improve plant populations by increasing the proportion of desirable genes is recurrent selection. Sometimes, therefore, it is difficult to draw a sharp line between mass selection schemes and recurrent selection schemes. Further, there can be several populations with similar variance but with different means (Fig. 19.1). The identification and use of populations with highest mean would certainly facilitates production of best hybrids (Hallauer, 1981). The system, therefore, takes case of the drawbacks associated with traditional hybrid breeding system and can be an effective plant breeding scheme to move off the yield plateau of crops. Furthermore, population improvement methods can be employed in allogams as well as in autogams. The diallel selective mating system of Jensen (1970) described in Chapter 17, is a population improvement method for self-fertilizing crops. Of course, use of population improvement methods in autogams is comparatively more difficult.

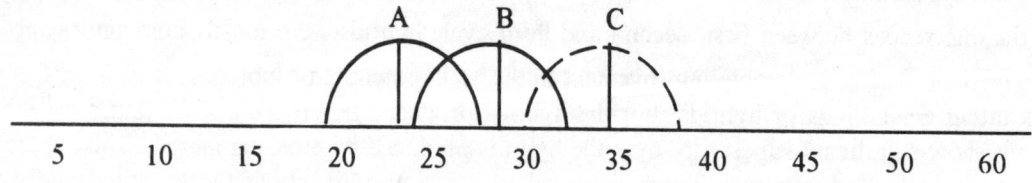

Figure. 19.1. Population with similar variance but with different means

Genetic Basis of Recurrent Selection

The two major forces which are applied during recurrent selection again and again are intermating and selection. The former force continuously generates genetic variability and the latter acts on this variability. If the base material is a random mating population (as is generally the case with open-pollinated varieties of cross-fertlizing crops), its genetic variance is expected to be ½ Dr + ¼ Hr (½D + ¼H if gene frequency is equal, where D and H are the additive genetic and dominance components of genetic variation, respectively, and Dr and Hr represent these two components respectively in a random mating population) if additive-dominance model is adequate. The knowledge about the nature and magnitudes of the different components of genetic variation greatly helps the plant breeder in formulating, shaping and finalizing his breeding programme. However, the assumption of the adequacy of additive-dominance model, especially for complex plant characters, by and large, appears to be unrealistic. In several cases, there is undisputed presence of epistasis and linkage. Their presence in the base material creates difficulty in knowing the real picture of the material regarding the relative magnitudes of additive and dominance components of genetic variation since the estimates of these two components will be biased to an unknown extent due to the presence of epistasis and linkage. The presence of linkage generally causes overestimation of dominance component and in several cases the apparent overdominance is actually pseudo-overdominance. However, linkages can be largely broken through intermating and the mean of the population can be shifted in the desired direction. The main purpose of using recurrent selection is, therefore, to break undesirable linkage blocks and accumulate desirable genes in the progeny.

If there is predominance of additive genetic variation in the base material, it is easy to improve characters through the use of simple selection procedures. On the contrary, if non-additive component is more important than the additive genetic component, the improvement will be difficult. In the first condition, undesirable alleles will be replaced by desirable alleles and ultimately favourable alleles will accumulate in the progeny. However, in the second condition, the accumulation of desirable genes will not be so fast as in the first condition. Therefore, the main job of recurrent selection is to increase the fixable component (additive variance plus additive × additive epistasis) of variance in the population and to isolate outstanding inbreds from such population or to use it in further breeding programmes.

Population Improvement Methods

Several intra-and inter-population methods have been suggested for improving plant populations. An improved population can be used in the following three ways :

(a) A source material for isolating outstanding inbreds to be used in the production of potential hybrid varieties;

(b) a source material for developing potential inbred lines with high general combining ability to be used in the production of synthetic or composite varieties; and

(c) the improved population can itself be used to evolve a population variety.

Mass selection and its variants : Mass selection is the oldest and operationally the simplest method aimed at to increase the frequency of desirable genes in the population. In fact, the non-heritable and unfixable components of variance are the main causes of reduction of gains through selection. The genetic gain through mass selection is determined by the magnitude of additive genetic component in the source material. This method becomes ineffective if the character to be improved has low narrow sense heritability since

(a) there is poor correspondence between the phenotype and genotype and thus difficulty in identifying superior genotype on the basis of phenotype,

(b) there is no control on the pollen parent as mass selection is carried out on female plants only,

(c) strict selection leads to reduction in population size followed by inbreeding depression, and

(d) no progeny tests are carried out.

However, mass selection is advantageous too since it is rapid and simple method and allows the plant breeder to handle large plant population.

Mass selection is more useful in cross-fertilizing crops than in self-fertilizers. The procedure of mass selection in cross-fertilizing crops involves retention of a large number of desirable individual plants (on phenotypic basis), bulking of harvested seeds without carrying out any progeny test and raising the mixture of seeds in next generation. The process is repeated till the desired level of improvement is achieved. Nevertheless, this method is mostly used to purify the existing improved varieties and not to evolve new varieties of crops.

In view of the drawbacks (low efficiency and less control on environment – soil and microclimate) associated with mass selection, some modifications have been suggested in this method to minimize the uncontrolled variation.

Modified or stratified mass selection or grid system : A modification in mass selection was suggested by Gardner (1961). Here, the plot (in isolation) is divided into small plots and best plants from each subplot are selected and retained for next generation. This modified method is also called as subplot system. Several plant breeders (Gardner, 1961, 1968, 1969; Lonnquist, 1967; Hallauer and Sears, 1972; Troyer and Brown, 1972; Verhalen *et al.* 1975) obtained impressive results by using this modified method.

Honeycomb system : Fasoulas (1973) suggested this modification. The sowing is done in hexagons and the best plant from each set is selected and retained.

The above mentioned two modifications of mass selection are restricted to the minimization of environmental variation only, that is, modification in the environmental design but no change in the breeding procedure. Some other modifications of this method which involve change in the breeding procedure (ear-to-row selection, modified ear-to-row selection, recurrent selection schemes, etc.) have also been suggested.

Ear-to-row selection : This modification in mass selection method was firstly suggested by Hopkins (1896). Here, ears are selected from the base material and are planted in rows alternating with remnant seed in the next season to assess the relative breeding value (that is open-pollinated progeny test). The progenies showing good performance on the basis of open-pollinated progeny test are carried forward. Each cycle of the method, therefore, requires at least two generations. The remnant seed to be used in the next cycle is the mixture of the high yielding progenies of the previous cycle. Since the breeding value of the selected ears is determined on the basis of open-pollinated progeny test, the trait (s) to be improved must have high narrow sense heritablity. Further, since the progenies are half-sibs, the method is also called as **half-sib family selection.**

Modified ear-to-row selection : To enhance the genuineness of ear-to-row selection, some modifications have been suggested in the method. In one of the modifications (Lonnquist, 1964), selected open-pollinated ears are planted in replicated trials in different environments, growing one replication at each site in isolation. Self pollination in ear rows in isolation block is totally restricted by detasseling the plants (in case of maize). Remnant seed of all entries is used to grow pollinator rows. Seeds of superior plants selected from ear-to-row plot of the isolation block across environments are bulked for using as base material for the next cycle.

Recurrent Selection Procedures

Although breeding schemes (without critical data) similar to recurrent selection procedure were long back suggested independently by Hayes and Garber in 1919 and East and Jones in 1920, Jenkins (1940) firstly published detailed description of a similar method on the basis of the results obtained from his maize experiments on early testing for general combining ability. However, it was due to the sincere efforts made by Hull in 1945 that the method could get its present name, recurrent selection. In the same year, he proposed recurrent selection for specific combining ability. In 1952, he discussed the aim for applying recurrent selection and also said that selection among genetically homogeneous group of plants (inbred lines, clones, etc.) is not recurrent selection. Three other types of recurrent selection, namely, simple recurrent selection, recurrent selection for general combining ability and reciprocal recurrent selection, were proposed and described by Sprague and Brimhall (1950), Lonnquist (1951) and Comstock *et al.* (1949), respectively. Recurrent selection procedures have also been described and discussed by many authors (Allard, 1960; Penny *et al.* 1963; Hallauer, 1981; and others). However, seveal modifications in recurrent selection schemes have been suggested from time to time according to the requirement of the breeder.

Simple recurrent selection : As has been mentioned earlier, this type of recurrent selection was experimentally demonstrated by Sprague and Brimhall (1950). The experimental evidence that simple recurrent selection can effectively change gene frequencies in a population, came from the studies carried by them for oil content in the kernel of maize. In this type of recurrent selection, a number of plants from the base material are self-pollinated. At maturity, selection is done for the characters to be improved. In the second crop season, selfed seed of potential plants is sown in an intercross block for all possible intercrosses among the plants. This completes the first cycle (involving two seasons) of the method. Composite of seed from intercrosses forms the base material for the second cycle. Similar to first cycle, a number of plants from the composite of seed are selfed and at maturity plants are selected for the characters under consideration. Selfed seed of potential plants is sown in intercross block for all possible combinations among plants. The material is then advanced to third cycle and these operations are repeated till desired improvement is achieved.

Sprague *et al.* (1952) provided additional information about the effectiveness of simple recurrent selection method using a synthetic variety (Stiff Stalk) as heterozygous base material. About 100 plants of the synthetic were selfed and oil percentage of seed from each ear was measured. Ten top ears of these plants (for oil percentage) were used to run two breeding programmes, namely, a recurrent selection series and a selfing series, parallely. For a reliable comparison between the two approaches for different characters, plot size, number of pollinations and intensity of selection were kept almost similar in the two programmes. In recurrent selection series, selfed seed of the ten top ears was grown in ear-to-row progenies in intercross block and all possible combinations were made. Seed from each combination was harvested separately to initiate next cycle. Same operations (selection of about 100 plants, their selfing and intermating among the progenies of top ten ears) as in the previous cycle, were done. However, to draw information for the degree of inbreeding per cylce, each intercross in the second cycle was grown separately instead of bulking the seed of all intercrosses. In selfing series, selfed generations were obtained from the same top ten ears selected from the original source material.

Recurrent selection caused changes in three population parameters, mean, range and standard error. The mean oil percentage showed a shift towards higher side, being minimum in the base material and maximum in second recurrent selection cycle. The mean of the 10 foundation ears selected from the base material and that of first recurrent selection cycle fell in between these two values (mean of the first cycle was more than that of foundation ears). Similarly, the range in oil content and the standard error had lower magnitude in the base material than in the first and second cycles. However, there was no significant difference between first and second cycles for these parameters. The comparison between recurrent selection series and selfing series led to the following conclusions :

(1) Recurrent selection was more effective (1.3 to 3 times) than selfing series;

(2) recurrent selection series was more variable than selfing series in terms of general phenotype and variation in oil content;

(3) recurrent selection did not reduce genetic variability as compared to the variability present in the base population; and

(4) in selfing series, the random fixation of genotypes was high due to a rapid increase in homozygosity, whereas such fixation of genotypes in recurrent selection series was generally low. However, if considerable fixation of genotypes is suspected in recurrent selection series, it can be minimized by

(i) allowing random mating in intercross population before reselection, and

(ii) growing intercrosses separately and maintaining pedigree records.

Furthermore, the effectiveness of selection will be highest if gene frequency is equal (u=v=.5) and lowest if gene frequency is zero or one.

Nevertheless, simple recurrent selection will be effective for those characters only which have high narrow sense heritablity. This method should not be recommended for complex characters (like yield) which generally have low heritability. However, depending upon the objective and need of the plant breeder, the method can be modified in a number of ways to make it more effective and purposeful.

Recurrent selection for general combining ability : This scheme of recurrent selection is also known as population test cross selection and requires three crop seasons for a cycle. The mthod involves the use of a heterozygous tester (a broad based related or unrelated population) and exploits mainly the additive component of genetic variation.

In the first crop season, a number of phenotypically desirable plants are selected from the population to be improved. These plants are then self-pollinated and simultaneously crossed to a heterozygous tester for testing the general combining ability of the selected plants. In the second crop season, progenies of all the top crosses are raised in replicated trials for evaluation. The selfed seed of parent plants (S_1 plants) which show high general combining ability, is grown in the intercross block in the third crop season. This completes one cycle of this recurrent selection scheme. The intercross block is harvested in bulk and the bulked seed (C_1 bulk) is used as the base material for the next cycle.

Plants of the source population may be crossed with the tester one generation later, that is, the selected plants of the base population are selfed and the S_1 plants are interplanted with the tester in isolation. Now the cycle would require four crop seasons (selection of promising plants at maturity and selfing, growing progenies of selfed plants and crossing with the tester, evaluation of test crosses in trials for general combining ability and raising selfed seed of promising parent plants in the intercross block). This modified method takes one additional crop season but allows rejection of undesirable S_1 lines at an early stage and makes the handling of the material more easy.

Since recurrent selection for general combining ability is a direct outgrowth of studies of early testing firstly carried out by Jenkins (1935), the method involves following two assumptions :

(i) Plants of an open-pollinated population show marked differences among them for combining ability; and

(ii) a plant sample selected on the basis of combining ability test is capable of producing more superior lines than that taken from the same population based on visual selection.

The idea of Jenkins got support from several other investigators (Sprague, 1946, 1952; Lonnquist, 1950; Wellhausen, 1952; and others). All these investigators demonstrated that early testing can help in detecting promising lines which are expected to produce inbred with high combining ability and thus is a good testing procedure particularly when yield is an important consideration. However, the procedure is of limited value where the genes governing the character under consideration have low frequency in the base population. On the other hand, Singleton and Nelson (1945), Richey (1945, 1947), Payne and Hayes (1949) and some other workers expressed different opinion about early testing procedure and considered visual selection as a better procedure (than early testing) for improving combining ability of inbreds in early generations. According to them, rejection of lines in early generations on the basis of top cross tests may result in an uncompensated loss of some promising lines. However, these conflicting arguments regarding the use of early testing could not stop several breeders from adopting this procedure.

Lonnquist (1950, 1951) applied recurrent selection for general combining ability to the open-pollinated Krug variety of maize for modifying combining ability and opined that top-cross combining ability can be modified by testing and selection in early generations. However, Mc Gill and Lonnquist (1955) obtained more interesting results regarding the effect of recurrent selection on genetic variability for combining ability in the same maize material. They compared four populations (variety Krug and three second cycle populations derived by recurrent selection) and noted that the genetic variability in Krug variety was significantly larger than in three other populations. Nevertheless, the three derived populations did not show significant differences for genetic variability. If the sufficient care is not taken to check the effect of inbreeding, the genetic variability in derived populations may decrease at a much higher rate than expected during recurrent selection.

Now it is amply clear that recurrent selection for general combining ability can definitely increase the frequency of desirable genes in the population and thus can be considered as an appropriate method for developing synthetic varieties as well as for developing reservoirs of germplasm to extract promising inbreds.

Recurrent selection for specific combining ability: As mentioned earlier, this scheme of recurrent selection was proposed by Hull (1945) and the scheme is also known as **inbred test cross selection**. As in case of recurreent selection for general combining ability (gca), a cycle here also requires three crop seasons (selection of promising plants from the base population at maturity and selfing as well as crossing with a suitable inbred tester, evaluating crosses in trials for specific combining ability, and growing selfed seed of parent plants showing high sca in the intercross block) except that the tester here is an inbred instead of an open- pollinated variety. The bulked seed of the intercross block is used as base material for the second cycle and so on.

The main purpose of proposing recurrent selection for sca by Hull was to take advantage of the non-linear component of heterosis which arises because of the non-linear interactios of genes at different loci or due to the interaction between alleles of the same gene (locus), or due to both these interactions. According to Hull, too much importance has been attached to selection within line and between lines in the standard breeding procedure and both time and effort can be saved by avoiding development of second cycle lines. However, Hull's view did not get support from many workers.

Reciprocal recurrent selection : Comstock *et al.* (1949) proposed reciprocal recurrent selection or reciprocal half sib selection for improving two genetically unrelated heterozygous and heterogeneous populations (say population 1 and population 2) simultaneously. In this scheme, population 1 is used as a tester for improving population 2 and reciprocally population 2 is used as a tester to improve population 1. In fact, it is a double recurrent selection for general combining ability method where the same two populations are used as testers for one another as well as the populations to be improved. In the same way, a number of phenotypically promising plants are selected from population 1. These plants are self-pollinated as well as crossed to plants from population 2. Similarly, a number of phenotypically desirable plants are selected from population 2. These plants are also self-pollinated as well as crossed to plants from population 1. In the second crop season, progenies of all the top crosses produced in populations 1 and 2 are separately raised in two replicated trials for evaluation and comparison. In the third crop season, selfed seeds of parent plants of two populations which show good combining ability are grown in two separate intercross blocks (one for each population) to make all possible intercrosses among progenies. The intercross block of each population is harvested in bulk. The two new populations (say 1 and 2) thus obtained are used as source populations to initiate the second cycle and so on. Finally, crosses between the plants of two groups (1 and 2) are made and commercial seed is produced from these crosses. However, to check the adverse effect of inbreeding (that is, to check the reduction in within group variability), a large number of plants should be selected in each generation. Also, all efforts should be made to avoid or minimize the hereditary relationship of all the plants selected in any generation to a few crosses made in the previous cycle of the scheme.

According to Comstock *et al.* reciprocal recurrent selection is not only useful in selecting for general combining ability but this method also provides opportunity in selecting for specific combining ability. However, some workers do not agree with them about the uefulness of this procedure. For example, Schnell (1961) emphatically opined that the use of this method cannot be made in selecting for specific combining ability.

Relative Efficiency of Recurrent Selection Procedures

Comstock *et al.* also compared relative efficiency of three recurrent selection procedures, namely, reciprocal recurrent selection, recurrent selection for gca and recurrent selection for sca, on theoretical basis, assuming absence of epistasis, linkage disequilibrium and multiple alleles. The comparisons were made considering three degrees of dominance, no dominance or incomplete dominance, full dominance and overdominance. In case of no dominance or incomplete dominance, reciprocal recurrent selection and recurrent selection for gca are equally efficient but recurrent for sca is inferior to both these proceures. However, all the procedures are equally efficient if there is full dominance. On the other hand, in case of overdominance, reciprocal recurrent selection and recurrent selection for sca are equally efficient but recurrent selection for gca is inferior to both these methods. Nevertheless as regards the three assumptions mentioned above, the third assumption (absence of multiple alleles) usually does not pose any serious problem in plant popultions since most of these populations have two-allelic system. Even most of the allopolyploids behave like normal diploids. But the first two assumptions (absence of epistasis and linkage disequilibrium) are not realistic since these assumptions are not fulfilled by many plant populations. However, if epistasis and repulsion phase linkages are present, the situation is similar to the presence of overdominance and therefore, under this situation, reciprocal recurrent selection and recurrent selection for sca are equally efficient and both are suprior to recurrent selection for gca. These comparisons made by Comstock *et al.*, therefore, indicate that in all situations, reciprocal recurrent selection is either equally efficient or superior to the other two recurrent selection procedures. On the contrary, simple recurrent selection method is inferior to all the three recurrent selection procedures discussed above.

CHAPTER 20

Development of Synthetics and Composites

As has been discussed in the previous chapter, recurrent selection procedure is important for the plant breeder from two main viewpoints : (1) Till today, recurrent selection is the most appropriate method for identifying outstanding inbreds which can be used for developing promising hybrids and synthetic varieties; and (2) the method helps in developing reservoirs of germplasm for extracting desirable inbred lines. Although promising single cross hybrids are expected to be highest yielder among all categories of plant varieties, population varieties (particularly synthetics where combining ability of inbreds has been tested) are superior to hybrid varieties in a number of ways :

1. Synthetic varieties have a broader genetic base than the hybrid varieties and thus have greater chances of adaptability and phenotypic stability over a range of environments.
2. These varieties have low degree of loss of vigour in subsequent generations as compared to hybrid varieties.
3. Synthetic varieties are more successful than hybrid varieties where pollination control is difficult.
4. These varieties have special advantage where enough quantity of hybrid seed cannot be produced. In some crops, replacement of hybrid seed is considerably low as off-season crop can be taken only in some pockets of the area under that crop.
5. Seed production in synthetic varieties is less costly.
6. Farmers can produce their own seed.
7. Where hybrid seed industryis costly and crop is grown on limited area only, synthetic varieties have an edge over hybrid varieties.
8. A synthetic variety can be further improved through population improvement.

Nevertheless, synthetic varieties have some demerits too. The biggest demerit of these varieties is their low yield potential than that of hybrid varieties (particularly single cross hybrids). This is mainly because the single cross hybrids are heterozygous and genetically homogeneous and thus can have advantage of both the additive and non-additive components of genetic variance, whereas synthetic varieties exploit mainly the additive component only based on gca testing. Second, manytimes, the breeder is forced to include some poor inbreds in his programme because the number of promising inbreds is generally low. The inclusion of such poorly performing inbreds adversely affects the yield potential of the synthetic variety. Third, the produce of single cross hybrids is highly uniform (in grain size, grain colour, etc.) and thus has higher marekt value than the produce of synthetics. Fourth, since single cross hybrids have more uniformity for characters like plant height and ear length, combine harvesting is easier in case of these hybrids than the synthetic varieties.

In view of the superiority of synthetic varieties over hybrid varieties, Hayes and Garber (1919) advocated the production and commercial utilization of synthetics. They emphasized that before the development of a synthetic variety, it is necessary to know the genetic worth in terms of yielding ability of all the F_1 crosses and only good combining inbreds should be used for producing a synthetic variety.

As regards the term synthetic variety, it has been used differently by different workers. According to some workers, synthetic variety is a wide term and its constituent genotypes (that is, parental genotypes crossed in all possible combinations to produce a synthetic variety) can be inbreds, populations derived through mass selection, clones and/or other types of materials. However, many other workers do not agree with this definition and, according to them, the terms 'synthetic variety' and 'composite variety' are different. There are three main differences between these two types of varieties :

1. The constituent genotypes of a synthetic variety are the inbred lines whose combining ability has been tested and which have proved good combiners in all combinations, whereas the constituent genotypes of a composite variety generally can be inbred lines whose combining ability has not been tested, mixture of inbred lines, hybrids, clones and open-pollinated varieties.

2. The yield potential of a synthetic variety can be predicted in advance on the basis of the mean performance of the parental inbreds and the mean performance of the hybrids produced in all possible combinations, whereas such early prediction for a composite variety is not possible.

3. A synthetic variety can be reconstituted by crossing the same set of parental inbreds in all possible combinations, whereas reconstitution of a composite variety is not possible.

However, several composite varieties developed by plant breeders of cross-pollinated crops, have their constitutent genotypes as inbreds whose combining ability has been very well tested. Such composite varieties, in real sense, are synthetic varieties. The producers of such composite varieties perhaps do not make a sharp distinction between a synthetic variety and a composite variety.

Development of Synthetic and Composite Varieties

The development of synthetic varieties usually requires three kinds of operations :

(1) Evaluation and identification of outstanding inbreds from the available lot : During this operation, general combining ability of the available inbred lines is tested to exploit mainly the additive component of genetic variation. Any appropriate biometrical procedure (mostly top cross method) can be used to tet the gca of the inbreds. The inbreds which show high gca are selected.

(2) **Method used for producing synthetic variety :** There are two ways of developing a synthetic variety :

(a) **Making crosses among the selected lines in all possible combinations :** The first step of this method involves crossing of the selected inbreds (that is, parents or Syn-0 generation) in all possible combinations in isolation to assure equal involvement of each of the parental inbreds in the formation of the synthetic. For producing synthetic variety (that is, Syn-1 generation), equal amount of seed from each combination (cross) is mixed and grown in isolation to allow open-pollination among all the crosses. The population is bulk harvested and the population obtained from this seed is Syn-1 generation.

(b) **Open-pollinating the selected lines :** This method is more simple than the method described above. Here, all possible crosses among the constituent inbreds are not made, rather equal quantity of seed from each of the parental inbreds is mixed and grown in an isolated plot to allow open-pollination among all the inbreds. In open-pollinated crops, if there are no sexual barriers, such a mixed planting of parental inbreds is expected to produce crosses in all possible combinations among the lines. The material is bulk harvested. The panmictic population raised from this seed is expected to perform similar to Syn-1 population of the method a.

Experimental results showed that the two methods described above are equally efficient as regards the genetic worths of the synthetic varieties produced using these methods. Therefore, the breeder can use any of the two methods for developing synthetic varieties.

(3) **Multiplication of synthetic seed :** Depending upon the crop, the Syn-1 panmictic population may be multiplied in isolation for one or more generations so that sufficient quantity of the seed may be made available to the farmers. In this way, Syn-2, Syn-3, etc. generations are synthesized. In most of the crops, Syn-2 seed is distributed among the farmers for cultivation. However, in some crops like sugarbeets, no multiplication of Syn-1 seed is done and the seed of this very generation is distributed among the farmers.

In fact, there is no basic difference in the production of synthetic and composite varieties. Since no test of combining ability of the constitutent genotypes of a composite variety is done and these parental genotypes may be inbreds, hybrids, open-pollinated population and/or mixture of genotypes, the method b described for the development of synthetic varietis is also a standard procedure for producing composite varieties.

The expected relative performance of different synthetic generations (Syn-1, Syn-2, Syn-3, etc.) may be given as follows :

Syn-1 > Syn-2 = Syn-3 = Syn-4 = ... = Syn-n

This expected performance of the synthetic generations indicates that Syn-2 generation is expected to be somewhat inferior to Syn-1 generation, but there will be no further deterioration in the yield and other charcteristics in the succeeding generations provided there is complete random mating among genotypes in each synthetic generation and no differential selection has been practiced. This statement is based on a very sound population genetics principle, the Hardy-Weinberg law. According to this law, a large panmictic population attains its zygotic equilibrium just after a single cycle of random mating irrespective of the genotype frequenceis in the parent population. Since Syn-2 generation is produced by random mating among all Syn-1 generation genotypes, no further derioration in yield and other characteristics is expected in Syn-3, Syn-4 and so forth.

Maintenance of Synthetics and Composites

Depending upon the floral structure and pollination behaviour of the crop, a synthetic or a composite variety is maintained by open-pollination in isolation or by random mating from hand pollinations among all the genotypes. To maintain the identity of a synthetic variety, the constituent inbreds are maintained and crossed in all possible combinations at regular intervals. However, for the maintenance of a composite variety, a large randomly taken sample of the seed is grown in an isolated area and open-pollination is allowed. A composite variety may be improved by adding some desirable gentotypes.

Natural and/or artificial selection may bring change in the genetic constitution of synthetic variety. This change may be towards positive side (that is, there is improvement in the performance of the synthetic) or towards negative side (that is, reduction in the yielding ability of the synthetic). Incomplete random mating or inbreeding due to smaller sample size or differential selection may also cause change in the performance of a synthetic variety.

Factors Affecting the Performance of Synthetic Varieties

Sewall Wright (1922) suggested a formula to predict the peformance of synthetic varieties. According to him, $1/p^{th}$ of the excess vigour present in the F_1 genration will be lost in the F_2 generation, where p is the number of parental lines comprising the synthetic. The formula may be given as follows :

$$\text{Syn-2 estimate} = \text{Syn-1 mean} - \frac{(\text{Syn-1 mean} - \text{Syn-0 mean})}{p}$$

Sewall Wright used the terms F_2, F_1, P for Syn-2, Syn-1 and Syn-0, respectively. Here, the estimate of Syn-2 will be determined by three factors, namely, mean of Syn-1, mean of Syn-0 and the number of inbred lines (p) comprising the synthetic. As has been discussed earlier, according to Hardy-Weinberg law, there will be no further deterioration in the vigour in Syn-3, Syn-4 and so on. The experimental proof in favour of the theoretical concept given by Wright (based on Hardy-Weinberg law) comes from the work done by Neal (1935) on single cross hybrids, three-way cross hybrids and double cross hybrids of maize. In Neal's experiment, the average loss of yield in F_2 generation for single cross, three-way cross and double cross hybrids (involoving 2, 3 and 4 inbreds in each cross, respectively) was 50 per cent, 33.3 per cent and 25 per cent, respectively, that is, $1/p^{th}$ of the excess yield of the F_1. Furthermore, there was no significant difference between the yields of F_2 generation and F_3 generation in case of single and three-way cross hybrids. Sprague and Jenkins (1943) also found similar results regarding the yield levels of F_2 and F_3 generations. Therefore, results of Neal's experiment clearly proved that there was a gratifyingly high agreement between Wright's theoretical expectations and the actual experimental results in two respects : (1) The average loss of excess vigour in F_2 generation was $1/p^{th}$, and (2) there was no further loss of vigour in F_3 and subsequent generations.

As has been discussed earlier, the three factors which can influence the performance of a synthetic variety are :

1. The number of constitutent inbreds (p);
2. the mean performance of the constituent inbreds (Syn-0); and
3. the mean performance of the F_1 hybrids produced by crossing the parental inbreds in all possible combinations (Syn-1).

   It means that the yield level of a synthetic variety can be increased by

1. increasing the number of parental inbreds,
2. increasing the mean yield of the parental inbreds, and/or
3. increasing the mean yield of the F_1 hybrids produced by crossing the inbreds.

That is, one or a combination of the above mentioned factors can be responsible for increase/decrease in the yield level of a synthetic variety. The increase in the number of parental inbreds, reduces the value of the component (Syn-1 mean – Syn-0 mean)/p of Wright's formula, and thus increases the value of Syn-2. Similarly, increase in the mean of the parental inbreds also reduces the value of the above component of Wright's formula and increases the estimate of Syn-2. The increase in mean of F_1 hybrids will also increase the value of Syn-2. Nevertheless, these three factors influencing the yield level of a synthetic variety are neither mutually exclusive nor are positively correlated. For example, increase in the number of parental lines means inclusion of some less yielding inbreds and thus reduction in the mean value of Syn-0 and *vice versa*. Also, increase in the number of parental inbreds will result in the inclusion of some poor combining inbreds which will adversely affect the mean of the F_1 hybrids. Therefore, to produce the best-yielding synthetic variety, there should be a balance among these three factors. Kinman and Sprague (1945) tried to know the relation between number of parental lines and theoretical performance of synthetic varieties of corn. They crossed 10 maize inbreds and produced 45 F_1 hybrids and F_2 generation of these hybrids. The inbreds were arranged in descending order on the basis of their mean combining ability to produce F_1 hybrids. Based on the F_1 mean and expected F_2 yield, they found that increase in the number of parental inbreds upto 6 was advisable to obtain optimum yield of a synthetic variety. Increasing the number of inbreds beyond 6 results in the reduction in the yield level of synthetic. However, such reduction in the synthetic yield can be checked by including high combining partially heterozygous parents. In this way, the optimum number of parents will be more than 6. Nevertheless, such synthetic varieties cannot be reconstituted. All these results may be summarized as follows :

(1) There was high agreement between Wright's theoretical expectations and actual experimental results.

(2) 6-line synthetic is expected to give optimum yield. However, this number of inbreds is greater than 6 if partially heterozygous lines are used to produce a synthetic variety.

(3) Inclusion of poor combining inbreds in the programme adversely affect the performance of synthetic.

(4) To obtain optimum yield of synthetic varieties, there should be a balance among the three factors determining the performance of synthetic varieties.

Synthetic (Composite) Varieties in Maize

Maximum work regarding development of synthetic (or composite) varieties has been done on maize crop. In India, Indian Council of Agricultural Research (ICAR), some agricultural research institutes and agricultural universities have done commendable work in this direction. Six maize composites, namely, Amber, Jawahar, Kisan, Sona, Vijay and Vikram, were released in 1967 at national level. Also, a large number of composite varieties (Chandan Safed-2 and Amber Pop in 1971, Makki Safed-1 in 1973, Chandan 1 and Rajendra Makka 1 in 1974, Chandan 3 and Ageti 76 in 1976, Moti in 1977, Diara composite, Tarun, Trikuta, Nishat and Mansar in 1978, and many others) were released by different states. In addition, three nutritionally superior opaque-2 composites, namely, Shakti, Rattan and Protina, were released in 1971 for commercial cultivation. These three composites have twice the level of the two essential amino acids (lysine and tryptophan) as compared to normal maize varieties.

Nevertheless, the use of short-term (partially heterozygous) inbreds whose combining ability has been tested in early generation, in the development of maize synthetic (composite) varieties would help the breeder in shortening the time of his breeding programme, increasing the number of constituent inbreds for producing best-yielding synthetic variety and broadending the genetic base of the variety. Such synthetic varieties generally possess higher adaptability and phenotypic stability than those produced by using completely homozygous inbred lines.

Synthetic (Composite) Varieties in Forage Crops

Like maize, most of the forage crops are cross-pollinated. But, whereas it is easy to develop synthetic varieties in maize, both the selfing and crossing are difficult in many forage species. For example, in case of lucerne, neither selfing nor crossing is easy because of cleistogamy of flowers and presence of self-incompatibility and somatoplastic sterility. Similarly, several other forage species having bisexual cleistogamous flowers, are higly cross-pollinated because of self-incompatibility or some other system or combination of systems of pollination control. In the beginning, therefore, mass selection was mostly used for improvement of such crops. But, mass selection method has its own limitations.

However, the forage breeders realized that

(1) if mass selection is to be used to improve forage crops, it should be based on progeny testing,
(2) the breeding principles investigated to accomplish improvement in maize, are also applicable to forage crops,
(3) some plants from non-inbred (open-pollinated) populations can more successfully transfer higher yielding potential to their progeny, and
(4) since progeny testing in forage crops is relatively more expensive and difficult, drastic elimination of undesirable plants before progeny testing is necessary.

Test of Combining Ability in Forage Species

Due to the presence of one or more pollination control systems in many forage species, both selfing and crossing of desired parent genotypes are difficult and the plant breeder manytimes has to depend on natural crossing for testing combining ability of these genotypes. Allard (1960) described four methods for testing the combining ability based on natural crossing :

(1) **Open-pollinated progeny test :** This test is based on the natural crossing between selected plants and other plants happen to occur in the breeding nursery. Here the progenies are derived from the outcrossed seed produced on selected plants or clones. If there is some degree of selfing in the

species, the test cross seed will be consisted of both the out-crossed seed and some selfed seed. This method is used for measuring general combining ability.

(2) Top-cross test : In strict sense, a top-cross means a cross between a selected inbred and a heterozygous tester. But, here the top-cross seed is obtained from selected plants or clones grown alternately with an open-pollinated variety (tester). The test cross seed contains seed of top-crosses to the tester, seed of intercrosses with other selected plants or clones, and selfed seed if selfing happens to occur. However, the proportion of top-cross seed can be increased by growing more plants of the tester variety. This test is also used for measuring general combining ability.

(3) Polycross test : A third test used for measuring general combining ability is polycross test. In this test, the progenies are derived from the seed of a selected line or clone grown with other selected lines or clones in the same breeding nursery. However, generally, the matings among the clones are not random. In some cases, the matings are highly non-random and thus hinder the production of a true polycross. To insure random mating among the clones, the breeding material should be replicated manytimes and/or should be grown at different sites at different dates.

(4) Single-cross test : Generally, all possible crosses among a number of clones are made in a diallel fashion. Since the test is based on natural crossing, either each pair of clones is grown in isolation or the inflorescences of each pair of clones are enclosed in the same crossing bag. This test can be used to measure general combining ability of clones as well as specific combining ability of various cross combinations.

D

Clone Breeding

CHAPTER 21

Development of Clones

The genetic structure of populations in cross-fertilizing plant species is different from that of populations in self-fertilizing species. Whereas a population in self-fertilizing species (barring pure line varieties) is consisted of different homozygous plants (genotypes), a population in cross-fertilizing species (barring inbred lines, single cross hybrids and clones) is made up of different heterozygous plants. Therefore, a population in self-fertilizing species is generally homozygous but genetically heterogeneous. On the contrary, a population in cross-fertilizing species is heterozygous as well as genetically heterogeneous.

Based on the type of reproduction in the species, the cross-fertilizing crops can be further divided into two major groups : (1) Sexually reproduced crop species and (2) asexually reproduced crop species. Maize, bajra, sunflower and castor are some examples of the first group. Some examples of the crops falling in the second group are sugarcane, potato, sweet potato, banana and a large number of fruit trees. The crops falling in the first group are propagated by seed and the breeding methods used to improve these crops have already been discussed in previous chapters. The crop species falling in the second category show variable behaviour with regard to flowering and seed setting. Some of these crops flower almost normally with satisfactory seed set, but some crops flower poorly or very poorly followed by poor seed set. There are some other crops which flower only under special conditions or do not flower at all. In this last group of crops, the vegetative parts of the plant are mostly the means of their propagation, that is, clones are developed to propagate the species. **A clone may be defined as a vegetatively propagated group of genetically identical heterozygous plants and can maintain its vigour and other characteristics for indefinite period**. The reasons for developing clones in these crops are : (a) Difficulties of flowering and seed setting in these species, and (b) to avoid the ill effects of Mendelian segregation and recombination.

In self-fertilizing crop species, heterosis can be fixed in homozygous condition by evolving homozygous varieties. But in sexually reproduced cross fertilizing crop species, heterosis cannot be fixed in heterozygous condition because of the effect of Mendelian segregation and recombination. On the other hand, vegetaively propagated crop plants provide an excellent opportunity of taking advantage of heterozygous condition for indefinite period without any deterioration in yield and other characteristics. These crops are reproduced through the vegetative parts of the plant and thus there is no union of gametes, no segregation and no recombination. Furthermore, in clones of vegetatively propagated species, heritability in broad sense has the same importance as has the narrow sense heritability. However, during improvement of clones through crossing and selection, narrow sense heritability is more important than broad sense heritability.

The three obvious advantages of asexual reproduction are :

1. Unspecified and genetically unstable mutants or any genotype irrespective of ploidy level and the degree of heterozygosity can be maintained indefinitely.

2. The genotypes being used in the crossing programme can be parallely maintained for comparison with the progenies and other materials.

3. No understanding of genetics is required for doing selection among clones.

Problems in Breeding of Vegetatively Propagated Crop

The plant breeder has to face a number of problems in breeding of vegetatively propagated crops. These problems are :

(1) Reduced flowering and reduced seed set : There are several reasons for reduced flowering and reduced seed set in these crops. For example, photo- and thermo-sensitivity of flower initiation is the main cause of reduced flowering and low seed set in sugarcane. Similarly, reduced seed set in potato generally results from incompatibility and male sterility involving a number of nuclear and cytoplasmic systems. Sometimes, complete or partial lethality is induced in potato due to the presence of a tetrasomic gene.

(2) Abnormal development of endosperm : Normal development of embryos and seeds is highly associated with the normal develoment of endosperm. In potato, the normal development of endosperm is directly related to a specific endosperm balance number (EBN) of the maternal and paternal contributions, that is, a 2:1 ratio of material and paternal EBN is necessary for the normal endosperm development (For details, see Poehlman and Sleper, 1995).

(3) Abnormal meiosis : Cultivated potato (*Solanum tuberosum*) is an autotetraploid species and thus there is regular formation of univalents and multivalents during meiosis. Therefore, it is virtually impossible to study the inheritance of important traits in this most produdctive and widely grown food crop. Similarly, cultivatd clones of sugarcane are mostly complex interspecific allopolyploids. The chromosomes are small and large in number. Depending upon the parents involved in the cross, the chromosome number may vary from clone to clone. There is abnormal meiosis followed by sterility due to the high polyploid number of chromosomes and faulty transmission of individual chromosomes. Therefore, selfing and crossing both are difficult in this crop. Furthermore, flowering in sugarcane is highly dependent on the environmental conditions. Similar occurrence of irregular meiosis and sterility is found in several other asexually reproduced species.

(4) Lengthy life cycle : A number of asexually reproduced species are perennials and thus the breeder has to wait for a long time to note complete observations about the breeding material.

(5) Transmission of diseases : Growing plants by vegetative parts (tubers, stem cuttings, etc.) is more risky from the point of view of transmission of diseases. On the other hand, crops reproduced by seeds have less chances of transmission of diseases from one generation to the next.

(6) Germplasm maintenance is difficult and expensive : The maintenance of vegetative parts of plant requires more space and it is not easy to maintain these parts for a long period of time.

Before we discuss the methods and techniques of breeding vegetatively propagated crops, we should know the characteristics of a clone. The important characteristics of a clone are :

1. Clones are mostly highly heterozygous and deteriorate drastically on inbreeding.

2. All plants of a clone are genotypically identical since a clone is maintained through vegetative propagation and does not face consequences of Mendelian segregation and recombination.

   All plants belonging to a pure line or an inbred are also genotypically identical or nearly identical (in an inbred), but these plants are homozygous or nearly homozygous (inbred).

3. Like those of a pure line, the plants belonging to a clone, do not constitute a Mendelian population.

4. Selection within a clone (like within pure line or within inbred) is ineffective since there are no genetic differences among the plants of a clone. The phenotypic differences among the plants of a clone are due to environment or genotype × enviornment interaction or due to both.

5. Barring mutations, mechanical mixture and/or sexual reproduction in clones, the clones are immortal. Some viral and/or bacterial diseases may also cause degeneration in the clone.

Methods of Breeding

As in sexually reproduced crops, the same three conventional breeding methods–introduction, selection and hybridization– are used to improve vegetatively propagated crops. However, the handling of breeding material in these crops is different from that practiced in sexually reproduced crops. The main cause of this difference is that the promising material can be selected and maintained at any stage of the breeding programme since there is no fear of loosing genetic integrity of the selected material in this group of plants. Nevertheless, in addition to the conventional breeding methods, some biotechnological techniques like haploid breeding (development of monoploids in *Solanum phureja* and protoplast fusion and development of dihaploids in *S. tuberosum*) and mutation breeding are being used to improve these crops.

Among the three conventional breeding methods, introduction and selection take advantage of the already existing variability, whereas hybridization creates new variability. Therefore, the main steps of the breeding programme will be :

(1) Indigenous and exotic collection of germplasm from different sources.

(2) Selection of promising clones from cultivated and wild clones.

(3) Hybridization among the clones to obtain new gene combinations.

(4) Selection of promising clones from the breeding material obtained after hybridization.

As has been discussed earlier, promising clone can be selected at any stage of the breeding programme and new variety can be evolved.

Sugarcane Breeding

Let us first discuss the methods of breeding used in sugarcane. In sugarcane, however, not only the performance of the first crop is important, but ratooning ability of the selected clone should also be high. The ratooning ability of the genotype/variety greatly affects the cost of production of the crop.

Collection of germplasm : Commercial as well as wild clones (from related species and genera) with a wide range of diversity in respect of different quantitative and qualitative traits, are collected and maintained by breeding centres for their use in future breeding programmes. Important traits in sugarcane are cane yield (mainly dependent on cane length and cane thickness), cane softness (relative proportion of juice and fibre content), sugar percentage, and resistance for drought, cold, salt, diseases and insects. Resistance for various traits is found in wild species like *S. spontaneum*. Indian Sugarcane Breeding Institute, Coimbatore has germplasm collection of more than 3500 clones.

Selection of promising clones : Selection of clones from commercial and wild germplasm and from segregating materials obtained after hybridization, is done for the following three purposes :

(1) For maintenance at the breeding station.

(2) For evolving new variety.

(3) For use in the further hybridization programmes.

The breeder should take extra care while selecting clones for the purpose of evolving a new variety. The procedure of selection for this purpose involves several seasons since the effect of environment and G×E interaction should be minimised if not completely eliminated. Therefore, for obtaining reliable results, it is necessary to grow the material at several locations and in several seasons. Selection done at single location and in a single year may be unreliable and misleading. The whole procedure of clonal selection should be as follows :

First season : A large number of desirable plants (few hundred to few thousand) are selected from the available genetically mixed plant material. Plants with obvious weaknesses should be avoided. However, even at this stage, selection for highly heritable characters will be effective.

Second season : Because of the limited quantity of the material of each selected clone and the huge number of plants selected during the first season, separate unreplicated rows of clones from selected plants are raised in the second season. At this stage, drastic rejection of undesirable plants must be accomplished since only limited number of clones can be accomodated in the next step which is demanding and costly. No laboratory test or genetical method is used at this stage, rather the selection is entirely based on eye evaluation of the breeder. Generally, 5-10 per cent (total 50-100) plants are selected based on important traits.

Third season : Plants selected at the end of the second season are grown in 2-row plots in replicated preliminary yield trials with standard checks. The breeding material is also raised in sick plots to test resistance for major diseases and insects. In addition, laboratory tests for sugar per cent and other quality traits are carried out. Only a few best performing plants are selected at the end of this step.

Fourth to sixth seasons : Selected plants are raised in replicated yield trials at several locations with standard checks. First and second ratoon crops are also raised for testing the ratooning ability of the selected clones. Very rigid evaluation of selected material is done for important traits including quality characters and for disease and insect resistance. The clone which proves better than the best check is finally selected to evolve new variety.

Seventh and eighth seasons : The clone selected at the end of the sixth season may be grown at farmers' field with the best local check. If the clone has significantly superior performance, its seed is increased and sent for release as a new variety.

Hybridization among clones : In sugarcane, all three types of crosses, namely, **intra-specific** or **within species** or **inter-varietal, inter-specific** and **inter-generic**, can be made to combine desirable characters from different sources. In intra-specific hybridization, crosses between different varieties or different clones of the same species are made (e.g., crosses between varieties or clones of *Saccharum officinarum*). In inter-specific hybridization, crosses between species of the same genus are made (e.g., *S. officinarum* × *S.barberi*, *S. officinarum* × *S. spontaneum*, *S. officinarum* × *S. sinense*). On the other hand, in inter-generic hybridization, crosses between different genera are made (e.g., *Saccharum* × *Zea*, *Saccharum* × *Erianthus*, *Saccharum* × *Selerostachya*, *Saccharum* × *Miscanthus*, *Saccharum* ×*Narenga*, *Saccharum* × *Imperata*, *Saccharum* × *Sorghum*). In vegetatively propagated crops like sugarcane and potato, the clones (belonging to the same species or different species) are highly heterozygous and thus the F_1 generation obtained after crossing such clones, is genetically heterozygous and heterogeneous. The three main steps of the hybridization and selection programme are : (1) choice of suitable parents, (2) production of F_1 generation, and (3) selection of promising clones.

Choice of parents : The extent of improvement through hybridization would depend on the type of parents used in the crossing programme. Two or more parents to be used in hybridization programme should contain all the desirable characters which the breeder wants to combine in his variety to be evolved, that is, the parents should be genetically diverse. The choice of parents would be easy for the breeder if he has thorough knowledge about the germplasm of the crop. The use of biometrical techniques like the estimation of general and specific combining abilities (gca and sca) of the parents and crosses can greatly help the plant breeder in this regard. With a sound knowledge of his breeding material regarding gca and sca, the plant breeder would be able to give a scientific basis of his results.

One or two cycles of inbreeding or sib-mating in the clones to be used in crossing programme may help the breeder in identifying better genotypes for hybridization programme. Such genotypes are expected to prove better parents than highly heterozygous clones.

Production of F_1 generation : Depending upon the availability of desirable genes in the clones to be used in the hybridization programme, seveal types of crosses are made in sugarcane :

1. Single or biparental crosses : In this type of crossing, two specific clones are crossed in isolation. The seed collected from the female parent will be hybrid seed if the female parent is self-sterile and a mixture of hybrid seed and selfed seed if the female parent is self-fertile. Therefore, to produce pure hybrid seed, it is necessary to identify self-sterile clones and use them as female parent. Hand pollination of self-sterile clone (parent) would further ensure the production of pure hybrid seed.

2. Area crosses : In this crossing system, a single male parent is used to pollinate several male sterile parents (clones) in isolation. The seed is collected from the clones used as female parents. To avoid the production of selfed seed, complete self-sterility of the female parents should be ensured.

3. Polycrosses : It is just like the first cycle of recurrent selection procedure. Here, natural cross pollination is allowed among selected clones in isolation. The seeds of different clones are harvested separately. However, only the female parent is known to the breeder in this method. If the breeder is interested to improve some common important character like cane yield, sugar content, or any other important trait, more cycle (s) of recurrent selection may be followed.

4. Diallel cross : If the selected clones are not large in number, the clones may be crossed in all possible combinations in a diallel fashion. This procedure will ensure equal participation of clones in the crossing programme. If necessary, hand pollination may be done to harvest sufficient quantity of the crossed seed.

Generally, a large number of crosses are attempted to ensure that at least some of the F_1 families will be superior to the existing clones/varieties. The selected few families can further be crossed in a diallel fashion to facilitate evaluation of the crosses in a better way.

Selection after hybridization : The selection procedure following hybridization is exactly similar to the selection procedure for evolving a new variety, described earlier in this chapter. The selection of promising clones involves at least 7-8 seasons.

Cultivated potato (*Salanum tuberosum*) is another important vegetatively propagated crop. Since cultivated potato is an autotetraploid, inheritance studies cannot be done easily in this species. For studying inheritance of important qualitative traits, dihaploids are produced in this species. Potato offers an excellent opportunity to the breeder to utilize diverse sources of germplasm since most of the wild relatives of cultivated potato can be easily crossed with this autotetraploid species. However, reduced seed set in potato results from complex nuclear and cytoplasmic genetic systems. Occurrence of male sterility at different stages (F_1 or F_2), complex control of incompatibility, presence of lethal tetrasomic gene, or abnormal development of endosperm may be responsible for reduced seed set in this important food crop. Sometimes, two or more phenomena occur simultaneously or successively.

Methods of breeding potato are almost the same as are used in sugarcane. However, crosses in this crop are mostly made by hand emasculation and hand pollination since the flowers are large and emasculation is easy. Of course, objectives of breeding different vegetatively propagated crops are different.

Mutation Breeding

In view of the problems faced by the breeder in improving vegetatively propagated crops, mutation breeding has a special significance in the breeding of these crops. Mutation breeding has been successful in developing new cultivars in ornamental plants. However, fruit plant breeders have not been very successful in this regard.

The major objective of induced mutagenesis in plants is to eliminate one or few specific defects of crop varieties which are otherwise high yielding and well adapted. Best examples of such mutagenesis come from sugarcane where quite a good number of attempts have been made to remove a number of defects (susceptibility to diseases like red rot, smut and *Helminthosporium* leaf spot, presence of leaf sheath spines, etc.). There are reports of the isolation of sugarcane smut disease resistant mutants (from variety CO 740), mutants resistant to red rot (from some commercial varieties), glabrous leaf sheath mutants (from variety 7717) and plants resistant to *Helminthosporium* leaf spot disease. Of course, till now there is no confirmed report of any official release of mutant varieties of sugarcane for commercial cultivation.

Similar efforts can be made to remove some specific defects of important and well adapted varieties of potato. More attention should be paid to eliminate the defects like susceptibility to late blight disease and frost and presence of red skin.

Use of Biotechnology

The use of biotechnological techniques (newer screening techniques for isolating disease resitance at cellular level, tissue culture techniques to induce numerical and structural chromosomal variation, somaclonal variation, somatic hybridization including protoplast fusion, etc.) appears to offer excellent opportunities to the breeders of vegetatively propagated crops. However, the breeders have to go a long way to see the practical application of these techniques in vegetatively propagated crops. Furthermore, the use of biotechnology in plant breeding should not be treated as an alternative to the conventional breeding methods, rather it is an adjunct to the conventional breeding system.

SECTION IV

Plant Breeding Techniques

CHAPTER 22

Mutation Breeding

A sudden heritable change in an organism is termed as mutation. The size of the genetic material undergoing such change varies greatly from the change related to a single nitrogenous base (transitions, transversions, etc.) causing gene mutation or point mutation to a change involving whole set of chromosomes (polyploidy and haploidy). In between these two extremes, several other types of changes like structural changes in chromosomes (duplications, deficiencies, inversions and translocations) and changes in the whole chromosome number (monosomics, trisomics, tetrasomics and nullisomics) may occur in an organism. Only gene mutations will be discussed in this chapter.

It is important to know that the presence of genetic variability in the base material is the basic requirement for achieving improvement in crop plants. More the genetic variability in the base population, more are the chances of improvement. However, mutation is the only process which can create new genes and thus can increase genetic variability in true sense. Of course, there are three basic sources of genetic variability : (1) Natural genetic variability present in the population; (2) variability released through recombinations; and (3) artificially induced variability through the use of mutagens (radiations, especially ionizing radiations, and chemicals).

Nevertheless, it should be kept in mind that both the spontaneous and induced mutations are usually undesirable, deleterious and sometimes lethal. Therefore, the technique of mutation breeding should be used only under specific circumstances like :

(i) If the crop to be improved is a vegetatively propagated one, mutation breeding can be preferably used to remove some specific defects. In these crops, unspecified and genetically unstable mutants can be maintained indefinitely.

(ii) In fruit trees and several other fruit crops which have lengthy life cycle, the use of conventional breeding methods is not easy. In these crops, mutation breeding can be used to improve some specific characters while retaining all the desirable characteristics of the old variety (like colour, flavour and taste of fruits).

(iii) In crops where conventional breeding methods can be successfully used, the use of mutation breeding should be done only when the natural variability of the species does not provide gene(s) for desired characteristic. One important example of this type is the low variability for Karnal bunt disease in wheat. The disease has caused serious damage and has threatened wheat consumption considerably. Induced mutagenesis has a good scope for increasing variability for this trait in wheat.

(iv) Mutation breeding has a good scope when a desired gene is tightly linked with undesirable gene(s).

(v) When recombination breeding reaches at a stage where the biological system becomes helpless in further increasing the productivity because of some sort of compensating mechanism, induced mutagenesis can work as a helping tool. There are several examples in crop plants where a ceiling is imposed on the yield level by the negative correlation between the two component traits of yield, that is, if we try to increase the value of one trait, the value of the other is automatically reduced and *vice versa*. In wheat, for example, grain number is negatively associated with grain weight. In such cases, the best solution is to keep the value of one trait constant and increase genetic variability for the other trait through induced mutagenesis. Successful attempts have been made in this regard by maintaining the mean grain number but inducing large variability for grain weight in wheat. Efforts can also be made by following a reverse process, that is, by maintaining mean grain weight but inducing variability for grain number.

(vi) When the breeder is willing to develop chimeras in an ornamental plant species, induced mutagenesis can be of great help in this regard. The development of chimeras in ornamentals is perhaps the most significant contribution made by mutation breeding.

(vii) Induced mutagenesis can be helpful in rectifying a simply inherited defect in an otherwise promising variety and maintaining the superior genetic background of the variety. Such a situation often arises when a highly developed variety succumbs to a new race of a pathogen.

(viii) When the breeder desires to change some biochemical pathway in any crop species, mutation breeding can make significant contribution. For example, increase in lysine content but maintaining the same total protein content in cereals, can considerably improve their quality.

(ix) When there is negative association between disease resistance and some quality trait, one can resort to mutation breeding. For example, level of disease resistance in sugarcane is negatively associated with juice percentage. Similarly, in some varieties of wheat, desirable resistance is associated with dark red grain colour.

The use of mutation breeding to remove a simply inherited defect of a variety possessing otherwise a superior combination of genes (the point vii above), has been especially advocated because this technique is not expected to disrupt a superior combination of genes present in the variety and can perhaps handle undesirable alterations in other traits more easily than standard hybridization methods. On the contrary, the use of traditional backcrossing (mainly recommeded for rectifying such defects of outstanding plant varieties) is not always followed by a complete reconstitution of genotype of the recurrent parent even after 5 or 6 backcrosses. However, the decision regarding the selection of one of these two methods (whether mutation breeding or backcross breeding) should be based on the following considerations :

(a) Relative ease with which the desired improvement can be achieved by the use of the two methods, and

(b) the proportion of useful mutants in the total number of mutants (useful plus deletrious) produced. However, it is not necessary that muation breeding will always prove to be a more efficient method for restoring the genetic background of the outstanding variety than backcross method. The use of marker genes, sometimes, makes the backcross breeding a quite efficient method for reconstituting the genotype of the recurrent parent.

Disadvantages of Mutation Breeding

The main disadvantages of mutation breeding are :

1. Mutations are mostly undesirable and deleterious, that is, the frequency of useful mutations is very low.

2. The breeder has to spend lot of energy for testing huge second generation materials. Further, for successful mutation breeding, efficient screening techniques should be available for the characteristics to be improved.

3. The field work required for mutation breeding is generally substantially greater than that required for traditional breeding.

4. The association of harmful side-effects with useful mutations in many cases, makes the duty of the plant breeder more difficult since mutation breeding in such situation requires more time, labour and expenditure.

5. Since most of the mutant alleles are recessive, such alleles fail to express themselves in heterozygous condition. The detection of such mutant alleles becomes more difficult in case of vegetatively propagated polyploid crops since mutation breeding in these crops requires application of greater doses of mutagens and growing of larger plant populations.

6. Most of the agriculturally important characters are governed by quantitative genes. However, mutation breeding for these characters is more difficult than for the characters governed by single or a few major genes. The detection and selection of useful mutations in case of quantitative characters are generally very difficult.

7. In some cases, mutant genes have pleiotropic effect. If pleiotropic effect is on the negative side, elimination of this effect involves lot of time and labour.

8. Manytimes, it becomes difficult to distinguish a mutant variety from the parent variety. Such a situation way create difficulty in the registration of the new variety.

9. An occurrence of mutation at DNA level is not always followed by a phenotypic expression of mutation in the individual. A number of events may take place between the reaction of mutagen with DNA and the ultimate phenotypic expression of mutation. Following factors may affect the realization of mutations :

(a) **Type of DNA repair :** Some mutational events at DNA level may go unscored because of normal DNA repair.

(b) **Diplontic selection :** Unaffected normal cells generally have a competitive advantage over mutated cells. Due to their faster growth, the normal cells restrict the mutant cells to a small sector only and exclude them from participation in the formation of gametes.This forced prevention of mutant cells by normal cells is termed as diplontic selection.

(c) **Haplontic selection :** When the competition between normal and mutant cells occurs at gametic stage and only normal gametes are functional, this is called haplontic selection.

(d) **Poor expression of mutant allele :** A mutant allele is generally highly affected by environmental conditions and manytimes fails to express itself.

History of Mutations

Although sudden changes were known to occur in plants and animals in the eighteenth and nineteenth centuries, Hugo de Vries (a Dutch botanist), at the turn of nineteenth century to twentieth century, gave the term "mutations" to "sudden heritable changes" or "discontinuous variations" (called as "sports" by Darwin). According to de Vries, mutations play a significant role in evolution. However, the rates of spontaneous mutations are so low that researchers manytimes ignore the naturally occurring mutations as a prominent cause of change in the variability of populations. But the discovery of artificial, radiation induced mutagenesis by Muller (1927) in *Drosophila* through X-rays and shortly afterwards by Stadler (1928) in maize and barley using the same radiation, proved beyond doubt that very high rates of mutations can be obtained both in plants and animals. As a result, many plant breeders became enthusiastic to use this technique for obtaining novel mutations as well as for increasing variability in metric traits. However, the plant breeders were disappointed on two accounts:

(1) The induced mutations, like spontaneous ones, were mostly undesirable and deleterious, and

(2) when the germplasms of plant species were rigorously screened (both at country and world levels) it was found that most of the desirable genes required for the improvement in important characteristics were already present in the germplasm.

Therefore, most plant breeders thought that it is better to use existing desirable variability of the germplasm rather than searching for induced mutations which are mostly haphazard in nature.

Nevertheless, despite this discouragement, some plant breeders continued their efforts in search of economically important induced mutations. The first success in this direction was perhaps achieved by Tollenaar (1934, 1938) in tobacco. The discovery of chemical mutagenesis by Auerbach and Robson (1947) increased the interest of plant breeders in this field of research. However, induction of several economically desirable mutants in barley (greater stiffness of straw, etc.) by Gustafsson (1947), became a source of great inspiration for many plant breeders. Auerbach (1976, 1978) gave further details of induced mutagenesis. Many other scientists made significant contributions in this direction. Some of them got practical success in inducing genes at specific loci. As a result, a number of varieties were released using mutatin breeding.

The attraction of plant breeders towards artificial mutagenesis was not without reasons. There were three reasons behind this attraction :

(1) The technique offers possibility of the addition of entirely new genes to the existing genetic variability of a species.

(2) It allows induction of all kinds of heritable changes (point mutations, segmental rearrangements of chromosomes like inversions and translocations, changes involving whole chromosomes and changes involving whole genomes) in a species. The technique, therefore, provides opportunity

for the movement of genetic material between different species and facilitates interspecific and intergeneric hybridizations.

(3) The new genes can be induced at specific loci. Such inductions can help in cutting short the time period of the release of a variety.

For a plant breeder, it is difficult to ignore these attractions.

Causes of Mutations (Forces Behind the Occurrence of Mutations)

Whether the mutations are spontanesous (naturally occurring) or artificially induced, two hypotheses have been put forward to explain the nature of mutations :

Target hypothesis of Lea : This hypothesis was suggested by Lea in 1946. According to Lea, purely mechanical forces operate to cause the process of mutation, that is, mutations occur due to the mechanical disruption of the hereditary material.

Indirect activation hypothesis : Under this hypothesis, the process of mutation is caused purely by chemical forces, that is, chemical alteration of the hereditary material is the cause of the occurrence of mutations.

Nevertheless, there are no reasons for believing that both the forces mentioned above cannot operate simulaneously in producing mutations. Therefore, there is every possibility that a combination of these two forces may be responsible for the production of mutations.

Molecular Nature of Gene Mutations (Mechanism of Mutations)

As we know, chromosmes are made up of nucleic acids and proteins. Now it has been proved beyond doubt that nucleic acids and not the proteins are the genetic material in all the living organisms. We also know that **gene is a segment of nucleic acid (mostly DNA and rarely RNA) and generally performs a specific biological function.** Further, **gene mutation is the result of an intra-genic change in the hereditary material. The mechanism of alteration within the structure of the gene is called as molecular basis of point mutation.**

However, the changes that occur in the gene at the level of nucleic acid molecule during mutagenesis are translated into changes in proteins or enzymes at molecular level, especially alterations in the kind and sequence of amino acids in a polypeptide chain. Therefore, molecular basis of gene mutations helps in understanding the genetic basis of gene-character or genotype-phenotype relationship since the expression of a character is determined by the type of the protein and the properties of a protein, in turn, are determined by the kind and sequence of amino acids present in that protein molecule. Thus, the change in the expression of the character is the result of the alteration in the structure of the gene governing that character.

A nucleic acid molecule is a chain of nucleotides. In turn, each nucleotide consists of a sugar molecule, a nitrogenous base and a phosphate group. The molecular nature of mutations can be explained on the basis of alterations in the base number, base sequence and base structure. There may be base insertion (base addition), base deletion, base substitution or alteration in the base sequence. These alterations can be broadly classified into the following two groups :

Alterations in base number : There may be an insertion (addition) or loss (deletion) of one or more nitrogenous bases in a DNA molecule. Both insertion and loss of base(s) involve breakage and reunion of the DNA molecule and cause mutations. Since one triplet (codon) of bases codes for one amino acid, the insertion or loss of three or multiple of three bases will result in addition or deletion of one or more amino acids in the protein molecule. However, since alteration in the base number (and consequently in amino acid number) is restricted to the added or deleted portion of the DNA molecule (or protein molecule), there will be no change in the DNA or protein molecule beyond the point of insersion or deletion. Such base alterations generally have little effect on the activity of the protein molecule and ultimately on the phenotype of the individual. On the other hand, if the number of inserted or lost bases is not three or a multiple of three, such addition or deletion would change the entire sequence of bases of the DNA molecule beyond the point of insertion or deletion, that is, there will be a shift in the reading frame of the codons beyond the insertion or deletion point. Such mutations are, therefore, called as **frame-shift mutations**. However, frame-shift mutations are generally deleterious since the proteins produced by these mutations are mostly non-functional.

The occurrence of inversion in a DNA molecule neither causes an addition or deletion of nitrogenous base(s) nor any change in the type of bases but results in the change of sequence of bases in the inverted region and a subsequent change in the sequence of amino acids in the protein molecule. The effect of such mutations generally depends on the size of the inverted region of the DNA molecule. Larger the size of the inverted portion more deleterious is the effect. Inversions, like additions and deletions of bases, involve breakage and reunion of the DNA molecule.

Though very rare, but there is possibility of breakage of a DNA molecule and reunion of the broken segment at some other place of the same molecule or there may be a cross-linking of DNA. Such structural changes in the DNA molecule would result in an alteration in the sequence of bases in the molecule and ultimately there will be a change in the sequence of amino acids in the protein molecule. Such alterations are generally deleterious and may lead to cell death (especially cross-linking of DNA).

Base replacements : The substitution of one nitrogenous base by another is called as base replacement. These substitutions are of two types : (1) Pyrimidine (one-ring bases) to pyrimidine (e.g., cytosine to thynime or *vice versa*) or purine (two-ring bases) to purine (e.g., adenine to guanine or *vice versa*) and (2) pyrimidine to purine (e.g., cytosine to adenine, cytosine to guanine, thymine to adenine and thymine to guanine) or purine to pyrimidine (e.g., adenine to cytosine, adenine to thymine, guanine to cytosine and guanine to thymine). The first type of substitutions are called as **transitions** where a pyrimidine base is replaced by another pyrimidine base or a purine base is replaced by another purine base. The second type of substitutions are termed as **transversions** where a pyrimidine base is replaced by a purine base or a purine base is replaced by a pyrimidine base.

Mutations involving single base replacement generally do not have large effect on the phenotype of the individual since in such a replacement only one codon (and subsequently only one amino acid) is affected. However, in a special case when the replacement results in the generation of a **non-functional** or **non-sense** codon, the polypeptide chain will terminate at the substitution point since a nonsense codon does not code for any amino acid. The mutations produced due to the generation of nonsence codens are termed as **non-sense mutations.** Such mutations generally have larger effects than the mutations produced by ordinary base substitutions since nonsense mutations lead to the production of incomplete and non-functional polypeptide chains.

Mutagenic Agents

Agents which are used for artificial mutagenesis are called as mutagenic agents or mutagens. These agents can be broadly classified into two groups : (1) Physical mutagens and (2) chemical mutagens.

Physical Mutagens

Physical mutagens include various kinds of ionizing and non-ionizing radiations. Among the ionizing radiations which induce **ionization (transfer of energy by rays to the electron of the atom, causing ejection of the electron from its orbit and making the atom positively charged)**, there are **particulate radiations** (like α-rays, β-rays, and neutrons) and **non-particulate radiations** (like X-rays and gamma rays). Ionizing radiations can also be grouped into densely ionizing radiations (like α-rays and neutrons) and sparsely ionizing radiations (like X-rays and gamma rays). Important example of the non-ionizing radiation is ultraviolet (UV) light which is a low energy radiation and cannot cause ionization. However, UV light is capable of causing **excitation (pushing the electron to a higher level orbit)**.

Chemical Mutagens

A large number of chemical mutagens are available. These chemicals include **alkylating agents** (ethylmethane sulphonate or EMS, methylmethane sulphonate or MMS, sulphur mustards, nitroso compounds, etc.), **acridine dyes, base analogues** and some **other chemicals** like nitrous acid and sodium azide.

As has been discussed above, a large number of physical and chemical mutagens are available to the plant breeder. Therefore, when making choice of a suitable mutagen, the breeder should be very careful. The following factors generally affect the choice of a mutagen :

(1) The effectiveness of mutagen to produce useful mutations
(2) The kind of the material to be treated
(3) Availability of the mutagen
(4) Reproducibility
(5) Persistence of the mutagen in the treated material
(6) Risk involved in the handling of the mutagen
(7) Economic consideration

Many plant breeders prefer chemical mutagens over physical mutagens since chemical agents produce less damaging effects (because they produce higher rate of gene mutations than some radiations which cause chromosomal disruptions) and their application is relatively simple. On the other hand, densely ionizing radiations like α-rays and neutrons (both fast and thermal) cause more chromosomal disruptions. However, the chemical mutagens have poor reproducibility, they and their derivatives persist in tissues of the material treated for longer time and their handling is relatively risky. Among chemical mutagens, ethylmethane sulphonate (EMS) is most widely applied for the induction of mutations. Since EMS is a powerful carcinogen, it should be used very carefully. The frequency and the spectrum of mutations, manytimes, depend on the kind and dose of mutagen used. Among physical mutagens, more sparsely ionizing radiations like X-rays and gamma rays have the advantage of good penetration and thus are very widely used. On the contrary, UV light does not have a good penetrating ability and is thus used to treat the tender plant materials (pollen, thin layer of *in vitro* cultured cells, etc.) only. But for the use of physical mutagens, we need costly equipments (X-ray machine and other apparatuses). Such equipments are not needed during chemical mutagenesis. Therefore, chemical mutagenesis is generally more economical than physical mutagenesis.

Mode of action of Physical Mutagens

Two radiations namely, X-rays and gamma rays are most widely used and thus occupy an important place among the physical mutagens. Both are sparsely ionizing electromagnetic radiations. The two radiations though are produced by different sources (X-rays are obtained from X-ray tubes and gamma rays from radioactive isotopes like cobalt-60 or caesium-137) both are high energy radiations (consist of small packets of energy called photons) and have similar physical properties and biological effects.

As discussed earlier, ionizing radiations cause ionization where an orbital electron of atom is hit by a ray photon. As a result of the transfer of photon energy (in the form of kinetic energy) to electron, the electron gets ejected and the atom becomes positively charged. Because of the presence of tremendous energy in the photoelectron (ejected electron), it can cause more ionizations. Both the X-rays and gamma rays are sparsely ionizing radiations since photon of these radiations produces only a few ionizations per micron of its straight path and thus has a low LET value (LET is the **linear energy transfer** of a photon, that is, quantity of energy deposited or lost by a photon per unit length of its path). On the contrary, photon of α-particles and neutrons is capable of producing several ionizations per micron of its path and thus has a high LET value. Accordingly, these radiations are called as densely ionizing radiations. The effects produced by low LET radiations are much more sensitive to environmental change than those produced by high LET radiations, that is, under changing environmental conditions, **the effects produced by densely ionizing radiations are more stable than those produced by sparsely ionizing radiations.**

The quantity RBE (relative biological effectiveness) is generally used to compare the efficiencies of different types of radiations. The value of RBE increases with an increase in the LET value upto the optimum point and thus RBE is generally a function of LET.

Ionizing radiations can cause biological damage either through **direct effect** or through **indirect effect**. In direct effect, radiation directly hits the DNA molecule so that the radiation energy is directly absorbed by DNA. Contrarily, in indirect effect, radiation initially hits some other molecules in the cell so that radiation energy is initially absorbed by these molecules and later on this energy is transferred to DNA molecule. However, under normal conditions, it is not easy to separate these two types of effects. In fact, the biological effects observed after ionizing radiation treatment are the result of the cumulative effect of these two types of radiation effects.

Ionizing radiations can cause breakage in one or both the strands of DNA and thus can disrupt the continuity of DNA molecule. The severity of the damage in the DNA molecule would generally depend upon the nature and the size of the break. If the break is small (involving one or few nucleotides) and is limited to one strand only, there is possibility of enzymatic repair of DNA and its normal structure may be restored. If DNA repair is normal, there will be no occurrence of mutation. But if DNA repair is faulty, mutation will occur due to some kind of base alteration. However, if there is peroxidation of the broken end in the presence of O_2, there is no possibility of enzymatic repair of DNA molecule and, under such situation, the breakage would ultimately result into strand deletion. The breakage in both the strands of DNA may result into the separation of fragments (if the breakage distance of two strands is 1-5 nucleotides). If breakage distance of the two strands is more than 5 nucleotides, the two breaks generally behave like two independent single strand breaks in the DNA molecule.

Depending upon the severity of disruption in the DNA structure, the breaks in DNA molecule may result into deletion, addition and/or substitution of base(s), cross-linking of DNA-DNA and DNA-proteins, and chromosomal mutations. Cross-linking of DNA mostly results in the death of cell. Therefore, ionizing radiations can induce any kind of change ranging from chemical alteration of a single base to chromosomal mutations. However, most of the changes brought about by X-rays and gamma rays treatment are gene mutations. On the other hand, α-particles and neutrans (densely ionizing radiations) are capable of inducing chromosomal mutations also.

Mercury vapour lamps or tubes are the source of UV light which is a non-ionizing radiation and is present in solar radiation. This radiation has a very poor penetrating capacity. UV light can act in two ways : (1) It can produce mutations by forming dimers of thymine, uracil and sometiems of cytosine in the same strand of DNA molecule, and (2) it can induce mutations by deaminating cytosine by adding a molecule of H_2O to the 5, 6 double bond of uracil and cytosine. Because of its poor penetrating ability, UV light, as a mutagenic agent, has a very limited use in higher organisms. In plants, its use is mostly limited to the irradiation of pollen. However, irradiation of pollen in most of the crop plants has its own difficulties because of the small quantity of pollen available and the limited duration of its viability. Therefore, UV light is not an important mutagenic agent for the plant breeder.

Mode of Action of Chemical Mutagens

Although the number of available chemical mutagens is very large, only a few chemical agents have been successfully used in the improvement of crop plants. Further, there are great variations in the mode of action of different groups of these chemicals. Most important of the chemical mutagens are the alkylating agents like ethylmethane sulphonate (EMS), nitrosoethyl urea and nitrosomethyl urea. During alkylation of DNA, an alkyl group is substituted for a hydrogen in the bases.

The alkylation of DNA may be related to : (1) Alkylation of phosphate groups of DNA, (2) alkylation of nitrogenous bases (particularly alkylation of guanine), or (3) depurination of deoxyribose sugar. During alkylation of the phosphate groups of DNA, unstable or semistable phosphate triesters are formed. Attached alkyl groups can affect the replication of DNA.The phosphate triester leads to breakage of DNA backbone if it is hydrolysed between sugar and the phosphate.

As regards the alkylation of bases, the most affected base by alkylating agents is guanine, especially sixth and seventh positions of this base. Alkylation of guanine can cause mispairing of bases and ultimately results in base substitution. For example, O^6 alkyl guanine is capable of pairing with thymine and thus causes transition. This replacement of cytosine by thymine will change a G:C pair to A:T pair.

If deoxyribose sugar is depurinated because of the separation of alkylated guanine from the sugar, the removal of guanine may cause breakage in the DNA backbone. However, if this gap is filled by some other base during replication of DNA, the alkylation may result in the substitution of base (transition or transversion).

Factors Affecting Mutagenesis Programme

Artificial induction of mutations using a physical or a chemical mutagen is termed as **mutagenesis**. Since the frequency of useful mutations is generally very low, a mutagenesis programme should be very well planned considering all important factors (objective of the programme, choice of the material and its part to be treated, choice of the mutagen and its dose, choice of characteristic for which an efficient screening technique is available, etc.) which can affect the success of the programme. **The experimenter should have a well defined objective in mind before starting a mutation breeding programme**. The experimenter should clearly know that whether the purpose of the programme is to remove defect from a promising variety or the purpose is to create new variability in a species for specific type of genes (e.g., alternative dwarfing genes in important cereal crops) or the programme is meant for some other purpose. Further, the knowledge about the character to be improved would definitely help the experimenter in deciding various steps of the mutagenesis programme. For example, the handling of the mutagen treated material would differ according to the number and kind of genes governing the character to be improved, that is, whether the character is under the control of one or a few major genes or it is polygenically controlled.

The choice of the variety to be treated would greatly depend upon the objective of the mutagenesis programme. For example, if the purpose is to eliminate a specific defect from a high yielding and well adapted variety, then the experimenter should naturally select the best variety available in the crop. On the other hand, if the experimenter is in search of some alternative specific genes, then some variety other than the best one can be selected for mutagenesis. As regards the plant part to be treated for mutagenesis, it is mostly decided according to the type of reproduction of the crop. **In sexually reproduced crops, the routine procedure is the treatment of dormant seeds.** Mutagenic treatment of seeds has some advantages over the treatment of other plant parts : (1) It is more convenient to control the environmental factors (temperature, moisture content, oxygen level, etc.) in seed treatment than in case of treatment of other plant parts, (2) the storage of seeds after treatment is relatively safer, and (3) a relatively much larger number of seeds is possible to treat at one time. However, the efficiency of the seed treatment can be increased by using water soaked seeds, especially at the time of DNA replication. **In vegetatively propagated crops, on the other hand, some other plant part (cuttings, suckers, etc.) is treated.**

As metnioned earlier, there is a large list of mutagenic agents available to the experimenter. Therefore, selection of the most effective mutagen is not easy. **The experimenter should not only know about the best mutagen, but the use of most appropriate dose of the mutagen is equally important.** Different mutagens have their own advantages and disadvantages. Also, a mutagen may be more effective in a particular biological system than in other system.

The success of mutagenesis programme also depends upon the type of character to be improved. The programme is expected to be more successful in improving a character for which efficient screening is available. As the availability of marker genes increases the efficiency of backcross method, in the same way, the development of good screening technique for a characteristic enhances the success of mutagenesis programme.

Furthermore, mutagenesis programme is likely to be more successful in case of self-fertizing crops than in cross-fertilizing crops which already have a **self-inbuilt-system of genetic variability**. Also, the indentification of mutants in generations after mutagenesis (M_1, M_2, etc.) and the production of selfed generations are easier in self-pollinated species than in cross-pollinated species.

As discussed earlier, the success of a mutation breeding programme is also affected by the type of the genetic control of the character to be improved. Improvement through mutation breeding is easier for a character if it is governed by oligogenes than for a character which is governed by polygenes. Both for maximizing the ability to detect small improvements and for maximization of the chance of producing big improvements, the size of mutation breeding experiments meant for improving quantitative characters is generally much large.

Procedure of Mutagenesis

After selecting a suitable mutagen and its optimum dose, the material (desired plant part) is treated with the mutagen. The first generation after treatment with a mutagen is called M_1 generation. The treated material (seed or any other plant part) is sown in the field to grow several hundred M_1 plants. Since the frequency of induced mutations is expected to be higher in the primary than in secondary tillers, a close planting should be done in the M_1 generation in case of cereals like wheat and barley to cut down the number of tillers. For knowing relative frequency of mutations in different tillers, single tiller selfing (2 to 3 tillers per plant) should be ensured and separate tiller harvesting should be done at maturity. However, in case of monoecious species like maize where there is sequential development of androecium and gynaeceum with a considerable gap between the development of the two reproductive parts (that is, the two parts develop from different tissues), the two organs are not always expected to share the same mutation. In such crop plants, selfing in M_1 generation does not serve any useful purpose since even after selfing of the plants the non-common mutant loci between the two reproductive organs will remain in heterozygous condition. In such plants, selfing should be done in the M_2 generation (that is, M_1 generation can be harvested in bulk).

Nevertheless, data on some types of mutations (dominant mutations, seed sterility, etc.) which indicate their appearance right in the first generation of mutagenesis, can be recorded in the M_1 generation.

In M_2 generation, progeny rows from separately harvested M_1 spikes or plants are raised for critical observation and detection of mutations. Due to selfing in M_1, the mutant alleles will be in homozygous condition in some plants and in case of a single gene trait such mutations can be detected at this stage. However, most of the mutations may not be useful for the experimenter. The plants indicating presence of mutations are separately harvested for growing their progeny rows in M_3.

However, in case of crops like maize where bulk harvesting is done at the M_1 stage, single plants instead of progeny rows are grown in M_2 generation. Several hundred plants are selfed at this stage and individual plant harvesting is done to raise progeny rows in M_3 generation.

In M_3 **generation,** progeny rows from the plants selected at M_2 stage are raised. Since most of the mutations are not useful, drastic rejection of plants or complete rows is done in M_3. Plants or complete rows showing similar mutations are bulk harvested.

At M_4 **stage,** a preliminary trial for evaluating the performance of M_3 selected lines with best available check variety is conducted.

In M_5-M_7 **generations,** promising M_4 lines with desirable mutations are tested in co-ordinated multilocation trials. One or a few lines showing significant superiority over the check(s) are recommended for release.

At M_8 **stage,** the seed of the outstanding mutant line(s) recommended for release at regional or national level is multiplied for distributing it among farmers.

The above described mutagenesis procedure, however, is mainly meant for improving oligogenic characters. Mutagenesis programme for improving polygenically controlled characters is relatively difficult. Here, both the production and detection of mutations are not easy. For improving **polygenic characters** mutagenized populations, especially at M_2 and M_3 stages, should be very large. Further, use of statistical parameters (mean, variance, etc.) is a must for evaluating populations varying for these characters. Here, single plants are uninformative and selection of superior types is based on the values of statistical parameters. At least five plants are selected from each M_2 family for recording observations. Only those M_2 families are selected which show high mean and high variance or normal mean and high variance. Separate bulking of each M_2 family is done. A large M_3 population is raised from each M_2 family so that genotype frequencies of the M_2 family is well represented in the M_3 generation. Single plants are selected in M_3 on the basis of high mean for the character under consideration. M_4 and subsequent generations are raised as described earlier.

Mutagenesis programme in **vegetatively propagated** plants needs special attention with regard to the choice of the mutagen as well as for handling the material after mutagenic treatment. The target of the mutagenic treatment in these plants is a differentiated bud containing generally three independent germ layers, that is, a multicellular complex tissue. Therefore, the mutagen to be used in these plants should have a high penetrating capacity. Ionizing radiations generally do not pose any problem, but the experimenter should be careful while using a chemical mutagen. Since induced mutations in this group of plants occur in diploid tissues as sectors only diplontic selection is the main factor influencing the realization of mutations. Further, since the cell(s) of any of the three germ layers can get mutated, the recovery of mutations would greatly depend upon the germ layer affected by the mutagen.

Utilization of Induced Mutations

Although the primary aim of artificial induction of mutations is to add more genes in the germplasm of a species, a mutagenesis programme is mostly started with a specific purpose in mind. A mutant may either be directly released as improved cultivar or may be used as a parent in hybridization/hybrid breeding programme.

Direct release as improved cultivar : In vegetatively propagated plants, especially those which are self-incompatible, desirable mutants are mostly used as improved clonal varieties. In this group of plants, it is very difficult to transfer desirable mutant genes from one variety to another because of high degree of heterozygosity and difficulty in inbreeding.

In seed propagated plants also, when mutation breeding is done to remove some specific defect from an outstanding variety, a mutant possessing superior combination of genes may be directly released as improved variety. There are several examples of such varieties.

As parent in hybridization/hybrid breeding programme : There can be several objectives of using mutants in hybridization programme. These objectives are:
(i) During mutagenesis, thousands of genes are exposed to the mutagen and, therefore, several desirable and undesirable mutations are likely to be induced after the mutagen treatment. Some of these mutations are able to express themselves, whereas some other are not realized. In such situation, the breeder would like **to eliminate undesirable associated effects from the mutant.** The best way to remove such effects is to backcross the mutant to the original variety and reselect side-effect-free or 'purified' mutant.

(ii) **To obtain a superior combination of genes,** a normal hybridization programme can be carried out by making mutant × mutant or mutant × other variety crosses.

(iii) As mentioned earlier, in some cases, mutant genes have **pleiotropic effect** with an association of desirable and undesirable effects. To minimize pleiotropic undesirable effect(s) in such cases, the mutated gene is transferred into different genetic backgrounds and the best combination is selected.

(iv) Induced mutations can be utilized **in the development of hybrid varieties** in the following two ways :

(a) The inbreds to be used in hybrid breeding programme may be improved through artificial mutagenesis, and

(b) **male sterility may be induced** in the inbred to be used as female parent in the production of hybrids.

(v) Artificial mutagenesis can be used **to transfer genes from alien chromosomes** and **to eliminate unwanted genetic material** after distant crosses. The best example of such transfer of gene comes from the classical work of Sears (1956) where a gene for leaf rust resistance was transferred from *Triticum umbellulatum* (a diploid wild grass species) to *T. aestivum* (hexaploid bread wheat). In this gene transfer, *T. turgidum* (tetraploid emmer wheat) was used as a bridging species. The whole procedure of transfer can be shown as follows :

T. turgidum × *T. umbellulatum* → Triploid hybrid → Doubling of chromosomes by colchicine treatment → Hexaploid hybrid × *T. aestivum* (Chinese Spring variety) → Repeated backcrossing with Chinese Spring and identification and selection of leaf rust resistant plants carrying one extra chromosome from *T. umbellulatum* → X-rays treament of resistant plants before flowering → Pollen from these plants was used to pollinate Chinese Spring → Substitution of single grass gene for portion of Chinese Spring chromosome without deleterious effects of grass chromosome.

However, some scientists do not put such transfer of genes under mutation breeding.

Some chemicals like colchicine can cause gross chromosomal changes and can alter the ploidy level of a species. Development of polyploids and haploids has, in some cases, played an important role in plant breeding.

Contribution of Mutation Breeding

Before 1964, there were only a few sporadic reports of crop varieties evolved through artificial mutagenesis. However, holding of an international symposium in this year by FAO and IAEA on **The Use of Induced Mutations in Plant Breeding** and the publication of its proceedings in 1965, provided a great stimulus to the plant breeders to work on this aspect more enthusiastically. As a result, there were 77 mutant varieties on the list in 1969. After this time, the number of mutant cultivars increased so rapidly that at another FAO/IAEA symposium in 1990, Micke *et al.* presented records of 1342 varieties of different seed and vegetatively propagated plant species developed in 48 countries of the world. China had developed maximum number of cultivars (264) followed by India (196), Netherlands (171), USSR (96), Japan (87), Germany (86) and USA (75). Most of these varieties (above 90 per cent) were, however, developed through the use of physical mutagens (X-rays or gamma rays). By now, the total number of mutant varieties should be more than 2000. Nevertheless, this number constitutes a very small portion of the total number of the plant varieties developed so far using both the conventional and non-conventional methods.

In India, more than 200 mutant varieties of different plant species have so far been developed. The number of mutant varieties developed in vegetatively propagated plants is almost equal to their number developed in seed propagated plants. Some important mutant varieties developed in India are Sharbati Sonora and NP 836 of wheat; Jagannath, Prabhavati, Gautam and CRM 13-324 of rice; BGM 408, BGM 417 and RS 11 of chickpea; TV 1, TAT 5 and TAT 7 of pigeonpea; TG 17 and TG 19A of groundnut; MCU 7 and MCU 10 of cotton; Aruna of castor; TAU 1 of blackgram; and CO 8152 and CO 8153 of sugarcane. Bhaba Atomic Research Centre (BARC) and some agricultural institutes/centres/universities have made excellent efforts in developing mutant varieties in India.

Many of the mutant varieties covered sizeable area after their release for cultivation and made valuable economic contributions. For example, the barley mutant variety **Pallas** became so popular in Denmark that about 40 per cent of barley acreage in 1966 in that country was under this single variety. Similarly, *Verticillium* wilt resistant mutants **Todds Micham** and **Murray Micham** of *Mentha piperetta* covered large acreage in 1970s in U.S.A. and saved the costly peppermint industry of that country from disaster. In China, rice, wheat and cotton mutants were grown on thousands of hectares. Mutant derived cultivars of barley constitute a major part of barley varieties grown in Czechoslovakia for malting purpose. Scientists in Pakistan have very successfully developed mutants of chickpea, mungbean and cotton. In India, castor mutant **Aruna** had covered large acreage in several parts of the country. Some pulse mutants developed at Bhaba Atomic Research Centre became very popular in Maharastra and were grown in large areas during the period 1985-92. For example, the mutant TAU-1 of blackgram covered about 90 per cent of the area under this crop in Maharastra in 1991-92.

Almost all types of plant characters have been improved through artificial mutagenesis. These characteristics include quality characters like protein content, amino acid composition and starch quality (maize, wheat, rice and barley), oil content (sunflower and soybean) and fatty acid composition (soybean, linseed and rapeseed), combining ability (maize and rice), male sterility (maize and rice), disease resistance (sugarcane, apple, pear, grape and peppermint), flowering time, self-compatibility or parthenocarpy (almond, cherry, citrus and grape), shape, colour and maturity of fruit (apple, blackberry, grapefruit or peach), size and architecture of plants, leaves and flowers, flower colour, flowering time and temperature requirement (ornamental plants) and yield (rice, barley, sugarcane, etc.). There is, therefore, clear indication that the advantage of mutation breeding is not restricted to oligogenically controlled characters only but a polygenically controlled complex trait like yield can also be improved by using this technique (e.g., rice, barley and sugarcane). Jagannath and CRM 13-324 mutants of rice and Co 8152 mutant of sugarcane have given significantly higher yields than their respective parent varieties.

Also, the technique of mutation breeding has good future prospects regarding its use for the advancement of genetics, in genetic engineering for localizing desirable genes, for providing useful genetic diversity in molecular gene banks, etc.

Mutation Breeding : Yes or No

Looking at the advantages and disadvantages of mutation breeding, this breeding technique should neither be overestimated nor underestimated. It should not be equated with the standard hybridization methods since the technique cannot replace conventional breeding procedures and thus should not be treated as a revolutionary tool in the hands of plant breeders. On the other hand, the technique should not be criticized just for the sake of criticism. As metnioned earlier, mutation breeding has its relevance in several areas and has made some significant contributions. In future also, this technique is expected to contribute a lot in some fields of research (molecular genetics, genetic engineering, etc.).

The main criticism of mutation breeding approach is that most of the mutations are undesirable and deleterious and thus have a little practical utility in crop improvement. But if we look at the proportion of useful recombinants to the total number of recombinants produced, the situation is not much different.

Also, the objectives of traditional hybridization breeding and mutation breeding are manytimes different. Whereas during hybridization breeding higher yield is the main objective in most of the programmes, the objective behind the mutation breeding programme is usually the removal of a specific defect from the old variety. Therefore, a variety developed through mutation breeding cannot compete with a variety evolved by traditional breeding, particularly when the procedure of identification and release of variety is lengthy and mainly based on higher yield.

CHAPTER 23

Polyploidy and Haploidy Breeding

The basic chromosome number or a set of chromosomes or a genome of a species (where there is only one chromosome of each kind) is denoted by x. Also, the gametic number (denoted by n) in a true diploid species (2n=2x) is equal to x. However, neither all plants are true diploids nor all gametes have the x number of chromosomes, that is, a plant may have more or less than the normal diploid number of chromosomes due to irregular meiosis or irregular mitosis. Some variations in chromosome number can even change the taxonomic status of the species.

All kinds of alterations in chromosome number come under a common term **heteroploidy**, that is, heteroploidy is the condition where an indivudual or a species carries chromosomes other than the basic diploid chromosome number (2x). Heteroploidy can be broadly divided into two parts : (1) Alterations (addition or reduction) involving whole set(s) of chromosomes, and (2) alterations involving one or more chromosomes but not the whole set so that the chromosome number is not the exact multiple of the basic chromosome number of the species. The first type of alterations are called as **euploidy** and the second as **aneuploidy**. There are further divisions and subdividions of euploidy and aneuploidy as shown in Figure 23.1. If there is addition of one or more whole sets of chromosomes in the normal diploid number, it is called **polyploidy**. On the contrary, if there is deletion of whole set of chromosomes from the diploid number, it is called as **haploidy**. Polyploidy can be of three types-autopolyploidy, allopolyploidy and segmental polyploidy. In **autopolyploidy**, the same chromosome set is repeated more than two times, that is, the chromosomes of the added set(s) are identical or similar to the chromosomes of the original species. In **allopolyploidy**, on the other hand, the chromosomes of the added set(s) are different from those of the original species and thus pairing does not occur between the added and the original chromosomes. Further, the chromosomes of the added set (s) may be neither completely homologous nor completely non-homologous to the original chromosomes of the species. Such a condition is termed as **segmental polyploidy** where the chromosomes are homeologous. Depending upon the number of sets of chromosomes present, polyploids may be triploids, tetraploids, and so forth.

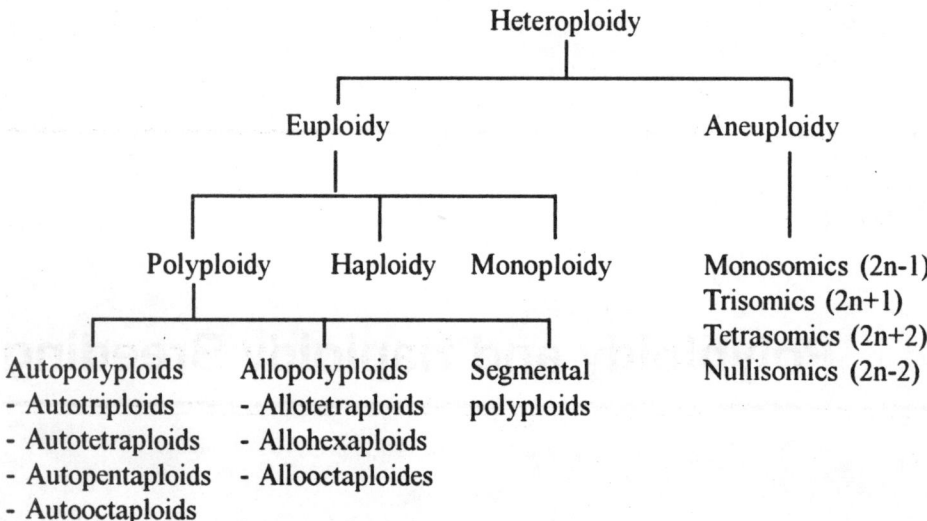

Figure 23.1. Classification of variations in chromosome number

Deletion of whole set (s) of chromosomes leads to the development of **haploids** and the condition is called as **haploidy**. A haploid may be a true haploid or **monoploid** having a single set of chromosomes (that is, having only one chromosome of each kind) or it may have two or more than two chromosomes of each kind. Haploids produced from polyploids are not true haploids.

Depending upon the chromosome(s) added or deleted, different kinds of aneuploid organisms may arise. If there is addition of a single chromosome in the diploid number, that is, 2n+1, the individual is called as **trisomic.** If there is addition of two chromosomes from different pairs, that is, 2n+1+1, the individual is known as **double trisomic.** However, if both the chromosomes of a pair are added to the diploid chromosome complement, that is, 2n+2, the individual is **tetrasomic.** Contrarily, if only one chromosome is missing from diploid chromosome number that is 2n-1, the organism is called as **monosomic.** If there is deletion of two chromosomes from different pairs, that is, 2n-1-1, the individual is **double monosomic.** But if both the chromosomes of a pair get deleted, that is, 2n-2, the individual is known as **nullisomic.** Nevertheless, both the addition and deletion of chromosomes may occur simultaneously. For example, an individual may have 2n+1-1 or 2n+2-2 chromosomes. These two individuals are called as **trisomic-monosomic** and **tetrasomic-nullisomic,** respectively. Furthermore, in addition to **primary trisomics** (where the additional chromosome is a normal chromosome of the same haploid chromosome complement), there may be **secondary trisomics** (where the extra chromosome has genetically identical arms, that is, an isochromosome) or **tertiary trisomics** (where the additional chromosome is a translocated chromosome).

Since aneuploids have unbalanced number of chromosomes, they are generally less vigourous than normal diploids, show irregular meiosis, are partly or highly sterile and genetically unstable.

There is formation of some univalents and multimatents at the time of meiosis. Cytological irregularities like non-disjunction of chromosomes and/or failure of unpaired chromosome(s) to reach either pole during meiosis are the main cause of their origin. The addition of chromosomes has less drastic effect on the viability of an aneuploid than the loss of chromosomes. Accordingly, tetrasomics are more viable than trisomics, monosomics and nullisomics, in order.

Aneuploids can be used as important genetic tools in a number of ways :

(1) Aneuploids are used for locating genes on specific chromosomes. This technique is relatively more accurate than that based on ordinary linkage.

(2) Trisomic-monosomic and tetrasomic-nullisomic studies facilitate substitution of specific chromosome(s), that is, these studies help in knowing whether the absence of any chromosome or pair of chromosomes can be compensated by the addition of any other chromosome or pair of chromosomes belonging to other genome. Tetrasomic-nullisomic studies for this purpose have been carried out in wheat.

(3) Balanced tertiary trisomics have been used for facilitating hybrid seed production in barley.

(4) Anenploids help in the production of substitution lines.

(5) Chromosomes involved in translocations can be identified with the help of aneuploids.

Polyploidy Breeding

Occurrence of more than two whole sets of chromosomes in a species is called as polyploidy. Polyploidy has several uses :

(1) Polyploidy has played a significant role in the creation of many plant species and is thus one of the main patterns of evolution (Chapter 2).

(2) Plant strains produced by polyploidy technique can be released immediately.

(3) The technique helps in overcoming hybrid sterility.

(4) The production of synthetic allopolyploids using the two parental species helps in knowing putative ancestors of the already existing allopolyploids, that is, the evolutionary history of naturally occurring allopolyploids can be traced and the degree of chromosome differentiation occurred in the diploid and polyploid species can be determined.

(5) Genetic variability of existing allopolyploids can be increased and thus their genetic base can be widened by producing allopolyploids afresh from their parental species.

(6) Polyploidy facilitates substitution or transfer of individual chromosomes or pairs of chromosomes.

(7) Polyploidy helps in transferring genes from related species.

(8) In segmental polyploids where the chromosomes are neither homologous nor non-homologous but are homeologous, there may be homozygosity within genomes but heterozygosity between genomes for the corresponding genetic regions of chromosomes (e.g. tetraploid and hexaploid wheats), that is, heterozygosity is fixed. Such heterosis is called as **homozygous genomic heterosis.** Segmental polyploids generally show higher adaptability as compared to their diploid counterparts.

(9) Polyploids are more vigorous or acceptable to consumers than their diploid counterparts. However, the optimum ploidy level may differ from species to species. For example autotriploids in

banana, watermelon and several other fruit species, tetraploids in potato, coffee, groundnut and alfalfa, hexaploids in wheat and sweet potato and octaploids in straw berry, have maximum acceptability/vigour. However, diploids show maximum vigour in maize and in blackberry the vigour is constant from 2x to 12x ploidy levels.

(10) Polyploidy provides more opportunity for mutations to occur and thus furnishes raw material for selection of desirable types.

(11) Autopolyploidy has been used as genetic bridge for interspecific hybridizations.

(12) Behaviour of chromosomes and chromatids at meiosis in autotetraploids helps in studying some phenomena like random chromosome assortment, random chromatid assortment and double reduction.

(13) Polyploidy facilitates the production of alien substitution lines, alien addition lines and aneuploids.

(14) Autotriploids behave abnormally at meiosis, produce non-functional gametes and thus there is induction of seedlessness in many fruit species (banana, watermelon, etc.). Autotriploidy also facilitates production of aneuploids.

(15) In some cases, autotetraploids show superiority in respect of quality over their diploid counterparts perhaps due to more number of genes in tetraploid than in diploid parent species.

(16) Autotetraploidy has also been found helpful in breaking the self-incompatibility barrier in some plant species.

Limitations of Polyploidy

It is true that polyploidy has played an important role in evolution, plant breeding and genetics but it has some drawbacks too which are as follows :

(1) It is not possible to predict the effects of polyploidy (both auto- and allopolyploidy) in advance and thus it is not always possible to combine together the desirable characteristics of two different species or genera. A number of attempts have been made in this direction (e.g. crosses between *Triticum* and *Secale, Raphanus* and *Brassica* and many other intergeneric and interspecific crosses) but except the development of *Triticale*, no success so far has been achieved.

(2) Several defects like cytological and genetical instability and low fertility are generally found associated with synthesized allopolyploids.

(3) The proportion of useful synthesized allopolyploids is very small.

(4) Improvement in synthesized allopolyploids is difficult as it requires extensive breeding at polyploidy level which involves lot of time and labour.

(5) Autopolyploids behave abnormally at meiosis and generally produce a few or no seeds. Therefore, most of these polyploids can be reproduced by vegetative means only.

(6) Autopolyploids grown by seeds are poor yielder than diploids due to high sterility followed by poor seed set shown by these polyploids,

(7) There is a very high positive correlation between the larger size of plant parts and the water content in autopolyploids. As a result, these polyploids do not show any improvement in the dry matter over the diploids.

(8) It is difficult to increase fertility in autotetraplaoids as these polyploids show complex segregation at meiosis.

Autopolyploidy

Autopolyploidy is also called as **simple polyploidy** as the autopolyploids arise because of the reduplication of the genomes of a single species. Due to abnormal segregation at meiosis and a high degree of sterility in autopolyploids, autopolyploidy has played a less important role in natural evolution than allopolyploidy. However, in many fruit crops (e.g. banana, watermelon, apple and pear), autotriploidy has played a significant role in producing seedless fruits. Similarly, the most important vegetable crop, potato is an autotetraploid. In sugar beets, triploidy is the optimum ploidy level.

As regards the inheritance in autopolyploids, they generally show complex segregation. Autotriploids show high degree of sterility due to irregular meiosis. Irregular meiosis leads to the production of non-functional gametes and ultimately to seedlessness.

Although autotetraploidy is considered to be more important to agriculture than other autopolyploidy levels, the cytology of autotetraploids is more complex. Since there are four chromosomes of each kind, quadrivalents are formed at the time of zygotene. The pairing occurs along the length of chromosomes in such a way that differnet regions of the same chromosome pair with different chromosomes. Crossing over takes place at pachytene. At diplotene, eight chromatids of the four chromosomes appear. As a result of the pairing of different regions of the same chromosome with different chromosomes, portions of different chromatids get attached to the same centromere. At this stage (diplotene), different types of configurations (bivalents, rings, chains, etc.) arise because of variations in the number and position of crossing over points (chiasmata). Except in case of bivalent formation where the two chromosomes of each bivalent go to the opposite poles, the adjacent chromosomes may go to the same pole or to opposite poles. Barring some exceptions, the disjunction of quadrivalents at anaphase I is two-by-two. However, at anaphase II of meiosis, the sister chromatids always go to opposite poles.

Depending upon the number of dominant/recessive alleles present, there can be five genotypes, namely, aaaa (nulliplex), Aaaa (simplex), AAaa (duplex), AAAa (triplex) and AAAA (quadriplex), at each locus of an autotetraploid. The gametic output in auto-tetraploids depends on three factors :

1. The genotype of the sporophyte;

2. the frequency of quadrivalents formed; and

3. the distance between the centromere and the locus in question which would determine the randomness of disjunction from quadrivalents.

If only bivalents are formed or quadrivalents are always formed and the separation of chromosomes at anaphase I of meiosis is two-by-two, only one kind of gametes will be produced by each nulliplex (aa type) and quadriplex (AA type) autotetraploids. However, as mentioned earlier, the gametic output in remaining three types of autotetraploids (simplex, duplex and triplex) will depend upon the regularity with which quadrivalents are formed and the distance between the centromere and the locus concerned.

Let us consider two extremes of gametic output in tetrasomic inheritance :

(1) **When only bivalents are formed or quadrivalents are always formed and the locus in question is inseparably linked to the centromere** : In this situation, the two chromatids carrying the gene in question always go to the same pole at anaphase I of meiosis and always go to opposite poles at

anaphase II, that is, the locus separates **reductionally** at I division and **equationally** at II division of meiosis so that sister chromatids never go to the same gamete. Such a behaviour of chromatids at meiosis is termed **random chromosome assortment**. The gametic ratios under random chromosome assortment for simplex, duplex and triplex individuals will be 1 Aa : 1aa, 1AA : 4 Aa : 1aa and 1 AA : 1 Aa, respectively.

(2) **When quadrivalents always form and there is 50 per cent crossing over between centromere and the locus in question :** Here, the locus may separate **reductionally** at first as well as at second division of meiosis, that is, the sister chromatids may go to the same gamete. Such a partition of chromatids at meiosis is called **random chromatid assortment**. When sister chromatids go to the same pole in the first as well as in the second division of meiosis, the phenomenon is called **double reduction**. The gametic ratios under random chromatid assortment for simplex, duplex and triplex individuals will be 1AA:12Aa:15aa, 3AA : 8Aa : 3aa and 15 AA : 12 Aa : 1aa, respectively (For details, see Allard 1960).

However, there can be intermediate situations between these two extremes (random chromosome assortment and random chromatid assortment) of tetrasomic inheritance owing to partial quadrivalent formation (that is, quadrivalent formation sometimes occurs and sometimes does not) and partial linkage between the centromere and the locus in question. The gametic ratios for simplex, duplex and triplex individuals and the frequency of double reduction will change accordingly.

Origin and Production of Autopolyploids

Autopolyploids may originate or can be produced in several ways :

(1) Autopolyploids may arise **spontaneously** due to cytological accidents. For example, misadventures in meiosis can cause formation of unreduced gametes with 2n chromosomes. Similarly, failure of mitosis may result in doubling of chromosomes in somatic cells. If doubling of chromosomes takes place at an early stage of plant development, such cells (with 4n chromosomes) may participate in the development of reproductive organs and ultimately gametes with 2n chromosomes are produced. If these gametes unite with other similar gametes at the time of fertilization, autotetraploids are produced. But union of these gametes with normal gametes (n) results in the origin of triploids. There are reports which indicate that at least in some cases the misadventures in meiosis are genetically controlled. Genes causing complete asynapsis or desynapsis have been identified.

(2) Autopolyploids can be artificially produced by **colchicine** ($C_{22} H_{25} O_6 N$) treatment. The source of this poisonous chemical are the corms or seeds of autumn crocus (*Colchicum autumnale*). Colchicine treatment is the simplest and the most effective method for doubling the chromosome number. The main function of colchicine is to block the formation of spindle fibres which are responsible for the movement of chromosomes/chromatids to the two poles at anaphase. All the chromatids/chromosomes present in the cell are surrounded by a single nuclear membrane (called **restitution of nucleus**). Thus, there is division of chromosomes without the division of cytoplasm. Colchicine treatment has been successfully applied in a large number of plant species (both monocots and dicots). Seeds, seedlings, growing shoot apices, etc. are treated with colchicine for chromosome doubling. The effect of colchicine, however, is not limited to chromosome doubling only but it can also be used for inducing mutations, haploidy, etc.

(3) There are some less effective polyploidizing **chemicals agents** (e.g. chloral hydrate, acenaphthene and sulfanilamide) and **physical agents** (radiations like X-rays and gamma rays, centrifugation and heat or cold treatments). However, their use for doubling the chromosome number has so far been very limited.

(4) In some plant species, **decapitation of plants** is a cause of chromosome doubling. There is development of callus with some polyploid cells at the cut end of the stem. The effectiveness of this technique can be increased by applying 1% IAA at the cut ends.

(5) *In vitro* cultured tissues manytimes have polyploid cells. Therefore, regeneration of plants from callus and suspension cultures may be used as a method for producing polyploids.

Role of Induced Autopolyploidy in Crop Improvement

In plant species where crops are not harvested for seed and vegetative propagation is possible, autopolyploidy has a special significance. Vegetatively propagated crops can be grown successfully irrespective of ploidy level. However, autopolyploids have been produced in seed propagated crops also. Good success has ben achieved in some crop species by producing triploids and tetraploid cultivars. In **watermelons**, for example, seedlessness provides a great ease in eating the fruit. In Japan, seedless triploid watermelons are successfully grown at commercial scale. Triploid cultivars are produced by crossing tetraploid (2n=44) as female parent with diploid (2n=22) as male parent. For ploidy level and some other reasons, reciprocal cross is unsuccessful. Diploid watermelons produce large number of fully viable seeds. Tetraploid watermelons also show a reasonable degree of genetic stability. Due to irregular meiosis, the triploids have very low fertility and are highly genetically unstable. Triploids generally produce white pseudoseeds of small size and sometiems large but empty seeds. However, there is much scope for improvement in triploid watermelon breeding particularly with regard to the improvement in genetic stability of tetraploids to be used as female parents, improvement in the fruit shape, and reduction in the cost of production of 3x watermelons.

In **sugar beets** where root is the commercial product, triploidy is the optimum ploidy level. In this root crop, triploids are superior to diploids (generally 10-15 per cent) in respect of root yield and sugar content per unit area and root size. On the contrary, tetraploids are inferior to diploids in these three respects. Triploids are produced by making crosses between tetraploids and diploids. Both types of crosses, 4x × 2x and 2x × 4x, are successful. Whereas diploids are higher yielder than tetraploids, the seed harvested from the tetraploid plants has much higher proportion of triploid seeds than has the seed harvested from diploid plants. Triploid sugarbeets are being grown at commercial scale in Japan and some parts of Europe. However, the efficiency of production of triploid sugar beet seed can be improved if suitable male sterile or incompatible lines are available to the breeder. Earlier, no procedure was available to sugar beet breeders to produce pure triploid populations and thus most of the commercially grown polyploids of this species were mixtures of diploids, triploids and tetraploids (called **amisoploids**).

In **tea**, since leaves are the commercial product, biomass yield is more important. A triploid tea clone (TV 29) released by Tea Research Association, India is superior to diploid cultivars in respect of biomass yield and tolerance to drought.

Autotetraploidy generally causes increase in the size of vegetative portions of plant like leaves, stems and roots. As a result, autotetraploids are mostly more vigorous than their diploid counterparts. However, this effect of autotetraploidy is not universal. For example, in sugar beets, autotetraploids produce smaller roots than their corresponding diploids. Nevertheless, production of autotetraploids in a large number of plant species (several forage crops and ornamental species) would be advantageous. For example, Pusa Giant Berseem, an autotetraploid variety of **berseem** (*Trifolium alexandrium*) gives 20-30 per cent more green fodder yield than diploid varieties of this forage crop.

Several autotetraploid varieties of ornamentals have been successfully produced. The objectives of the production of these varieties were mainly three : (a) Increasing flower size, (b) lengthening the blooming period of plant, and (c) improving the keeping quality of flowers.

Autotetraploids can also be produced in both vegetatively propagated and seed reproduced crop species for other purposes like adding disease resistance to commercially grown varieties (e.g., banana), improving quality characters (e.g. maize), overcoming self-incompatibility (e.g., tobacco, white clover and *Petunia*), and using as genetic bridge in distant crosses (e.g., interspecific crosses in *Brassica* and *Solanum*).

There are example of artificial production of autotetraploids in several plant species where crop is harvested for seed (e.g., rye, barley and jowar). But autotetraploidy has been most successfully induced in rye (*Secale cereale*). However, rye autotetraploids (2n=28) have their own advantages (increased kernel size and protein content and higher adaptability under adverse environmental conditions) and disadvantages (reduced tillering capacity, more plant height, difficulty in milling of large kernels, and problem in isolating autotetraploids from their corresponding diploid varieties).

A large number of attempts have been made to induce autopolyploidy in many plant species with different propagating systems. However, the success in terms of useful autopolyploids are not many. The past experiences of plant breeders regarding successes and failures of autopolyploidy programmes can be concluded as follows :

(1) Since most of the autotetraploids are cytologically and genetically unstable, autopolyploidy would have greater practical utility in crop species where some vegetative part of plant and not the seed is the commercial product.

(2) Cross pollination enhances gene recombination and thus helps in developing a self-variability generating system in plant species. Therefore, autopolyploidy would be more useful in cross-fertilizing species than in self-fertilizing species.

(3) Cell nucleus cannot accomodate unlimited number of chromosomes inside it. Presence of more chromosomes beyond its capacity can cause misdivision of nucleus. The production of useful autopolyploids is thus more successful in plant species which have low chromosome numbers.

These three generalizations would help in decision making about the production and utilization of autoploids in future crop improvement programmes. It should also be kept in mind that the performance of a species at diploid level has no correlation with its performance at autopolyploid level. An autopolyploid may be superior or inferior to its diploid counterpart. This situation can be found in the same crop species. For example, in sugar beets, triploids are higher yielders than diploids, whereas tetraploids are low yielders than their corresponding diploids.

Allopolyploidy and Segemental Polyploidy

Both the allopolyploids and segemental polyploids have genomes from two or more different species. Whereas the chromosomes of different genomes in an allopolyploid are non-homologous and thus do not pair at meiosis, the chromosomes belonging to different genomes in a segemental polyploid are homeologous and thus have some corresponding genetic regions. If a **diploidizing system** (as 5B system in polyploid wheats) has developed in a segmental polyploid, this polyploid behaves just like an allopolyploid. In the absence of diploidizing system in a segmental polyploid, there is multivalent/univalent formation because of partial pairing between the chromosomes belonging to different genomes. Segmental polyploids without diploidizing system generally do not survive except in a condition when they are propagated asexually.

It is considered that allopolyploidy has played a much greater role in evolution than autopolyploidy. A large number of important crop species like cotton, tobacco, mustard, oat, sugarcane and wheat are naturally evolved allopolyploids. Polyploid wheats are segmental polyploids with a diploidizing system and thus behave like true allopolyploids at meiosis. Allopolyploids show a cytological behaviour similar to that of diploids and thus are genetically stable. In addition to sexually reproduced allopolyploids (mentioned above), a number of apomictic allopolyploids are found in plants (*Solanum, Poa, Fritillaria, Parthenium, Tulipa*, etc.)

The main advantage of allopolyploidy is the combination of characteristics of two or more species in a single species (the main objective of hybridization). When two different genomes come together, they interact with each other and with the cytoplasm in which they are present. This interaction manytimes is complex and creates difficulty in the identification of putative ancestors of the allopolyploid. However, considerable differences are found from one allopolyloid to the other for this interaction. Even after the formation of an allopolyploid, chromosomal differentiation goes on occurring in nature and chromosomes of closely related species may show considerable differences. Mutations and structural changes in chromosomes are the main causes of chromosomal differentiation. The chromosomal differentiation in allopolyploid and in the diploid parental species may cause reduction in the homology of the chromosomes originally belonging to the same genome. Since degree of homology between the chromosomes of two species is one of the main criteria for tracing back the evolutionary history of an allopolyploid, chromosomal differentiation can create difficulty in conclusively knowing the real ancestory of an allopolyploid. Chromosomal differentiation in the polyploid species is mainly responsible for lack of resemblance between the newly synthesized allopolyploid (involving the same parental species) and the naturally occurring polyploid. Therefore, in addition to degree of homology of chromosomes, some biochemical tests like electrophoretic patterns of proteins and enzymes and chromosome banding patterns, should also be done to know the putative parents of an allopolyploid species.

Since allopolyploids contain two chromosomes of each kind and behave like diploids, they are also called **amphiploids** or **amphidiploids**. One simple and precise way of studying the cytology of allopolyploids is to produce synthetic amphidiploids by crossing the parental species of a naturally occurring allopolyploid and then doubling the chromosome number by colchicine treatment. However, such a synthesized amphidiploid may not be exactly similar to the naturally occurring allopolyploid because the latter might have undergone chromosomal differentiation and has faced consequences of natural selection since immemorial period of time. The fertility and stability (both the cytological and genetical) of synthesized amphidiploids is generally determined by the extent of homology between the chromosomes of the two species involved in the cross, that is, genetic divergence between the genomes of the two species. In case the two genomes are highly divergent, there would be normal bivalent formation and normal disjunction of chromosomes at meiosis. Such an amphidiploid generally shows high degree of fertility as well as a high degree of cytological and genetical stability.

Nevertheless, this relationship between the chromosome homology and the degree of fertility and stability of an amphidiploid should not be generalized. Sometimes, cytologically regular amphidiploids may show varying degrees of sterility perhaps due to some physiological reason. There are also examples among naturally occurring amphidiploids (e.g. tetraploid and hexaploid wheats) where the genomes of the parental species are less divergent in terms of chromosome homology (that is, chromosomes of different genomes are partly homologous and partly non-homologous), but behave like normal diploids (normal bivalent formation and normal disjunction of chromosomes) at meiosis due to the evolution of a device responsible for diploidization in the polyploid species. This shows that fertility and cytological stability of an allopolyploid are not always dependent upon the divergence of the genomes of the parental species.

Origin and Production of Allopolyploids

It is considered that naturally occurring allopolyploids and segmental polyploids have evolved through misadventures in meiosis either before or after the union of gametes of different species. Unreduced gametes may arise due to a pre-union meiotic irregularity and the union of unreduced gametes of two species at the time of fertilization, might have resulted in the spontaneous formation of an allopolyploid. Alternatively, there might be spontaneous occurrence of cross between the two species and then meiotic irregularity might have doubled the chromosome number (as is done for the artificial synthesis of amphidiploids where first a cross is made between two different species and the chromosome number of the sterile F_1 hybrid is doubled by colchicine treatment). It means that cytological accidents of one kind or another are the cause of spontaneous origin of allopolyploids. However, pre-union meiotic irregularity (the first possibility mentioned above) appears to be more likely possibility of the origin of allopolyploids. Possible evolutionary history of some important naturally occurring allopolyploid crop species (e.g., wheat, cotton, tobacco, mustard and groundnut) has been discussed in Chapter 2.

As regards the artificial production of allopolyploids, the procedure is short and simple. Cross is made between two different species whose desirable characteristics the experimenter wants to combine in a single genotype and then the chromosome number of the F_1 hybrid is doubled by colchicine treatment or by some other method. But,the simplicity of the procedure ends when the experimenter is not able to produce a useful amphidiploid. Unfortunately, man has not been very successful in producing new synthetic useful amphidiploids due to some cytological, genetical and physiological reasons.

Role of Induced Allopolyploidy in Crop Improvement

Allopolyploidy has several uses in crop improvement. These are :

(1) Creation of new plant species : Although the number of useful natural allopolyplolids is several times greater than the number of desirable natural autopolyploids, plant breeders have been considerably more successful in creating useful autopolyploids than evolving useful allopolyploids. The intergeneric allopolyploid, *Triticale* or *Triticosecale* (between *Triticum* and *Secale*) is perhaps the best example of man-made crop species. In this case, plant breeders have been able to combine useful characteristics of wheat (e.g. grain quality, yielding ability and photoperiod insensitivity) with the hardness of rye. Although three forms of *Triticale*, tetraploid (diploid wheat × rye), hexaploid (tetraploid wheat × rye) and octaploid (hexaploid wheat × rye) have been synthesized, hexaploids have been found most promising. It is perhaps that hexaploidy is the optimum ploidy level in this new species. Since D genome of wheat (found only in hexaploid wheat) contains some useful genes (particularly the genes for bread-making quality), efforts are now being made to transfer these useful genes of D genome to hexaploid *Triticale* rather than producing octaploid *Triticale*. The German scientist Rimpau was perhaps the first person to produce a fertile wheat × rye hybrid in 1888. Later on, wheat × rye hybrids were developed in several countries, namely, Sweden, Russia, Hungary, Canada, Mexico and America. *Triticale* is presently being grown at commercial scale in some countries of the world. Efforts are being made to remove some drawbacks (weak straw, shriveled grains and cytological instablity) of this man-made cereal.

Efforts have also been made to synthesize other intergeneric (*Triticum* × *Agropyron, Raphanus* × *Brassica, Solanum* × *Lycopersicon*, etc.) and interspecific (*Pennisctum purpureum* × *P. glaucum, Gossypium hirsutum* × *G. barbadense* , etc.) amphidiploids for combining desirable characteristics of parental species into a single genotype. However, interspecific amphidiploids have been developed with good success in some cases (e.g., *G. hirsutum* × *G. barbedense*). But the plant breeders have not succeeded in synthesizing useful intergeneric amphidiploids except triticale. For example, much discussed intergeneric amphidiploid *Raphanobrassica* developed by Karpechenko (1928) by crossing radish (*Raphanus sativus*) with common cabbage (*Brassica oleracea*) could not reach to an acceptable level till today. Similarly, plant breeders have not succeeded in developing an intergeneric amphidiploid between potato (*Solanum tuberosum*) and tomato (*Lycopersicon esculentum*) which could possess the useful characteristics of the two parental species, that is, same plant having potato tubers underground and tomato fruits above ground.

(2) Identification of ancestral parents of existing allopolyploids : Allopolyploidy has been helpful in knowing the evolutionary history of existing amphidiploids like tetraploid and hexaploid wheats, tetraploid cottons and tetraploid *Brassicas* on the basis of ploidy relationships between different plant species.

(3) Using as a genetic bridge : Allopolyploidy can be used as a genetic bridge between a wild species and a cultivated species and thus facilitates transfer of genes from related species. For example, for transferring lint strength genes from *Gossypium thurberi* (a wild diploid species of cotton) to the cultivated tetraploid *G. hirsutum*, first a cross was made between *G. arboreum* (cultivatd diploid species) and *G. thurberi* and the resulting amphidiploid was then crossed with *G. hirsutum*. Similarly, resistance genes have been transferred from related species to the cultivated species of tobacco (*Nicotiana tabacum*) by using amphidiploids as genetic bridges. The understanding of genomic relationships, between different species and different genera helps the breeder in overcoming interspecific sterility barriers and transferring genes from related species.

(4) Increasing genetic variability of existing amphidiploids : The genetic base of naturally occurring amphidiploids can be widened by synthesizing amphidiploids afresh from their parental species.

(5) Producing substitution lines involving individual chromosomes or pairs of chromosomes : These chromosomal substitutions would help in knowing that whether a chromosome or a pair of chromosomes of a genome is able to compensate the absence of an individual chromosome or a pair of chromosomes of a different genome. Such studies are especially important in segmental polyploids like tetraploid and hexaploid wheats. For this purpose, monosomic-trisomic and nullisomic-tetrasomic substitution lines have been developed in wheat.

Haploidy Breeding

A **haploid** plant contains gametic number (n) of chromosomes. A haploid plant produced from a diploid species is called as **monoploid** and contains basic chromosome number (x) so that gametic number is equal to basic chromosome number (n=x). Therefore, a monoploid has only one chromosome of each kind. In the same way, a haploid produced from a polyploid species is called **polyhaploid** (**dihaploid, trihaploid** and so forth produced from tetraploid, hexaploid and so on, respectively). Therefore, a polyhaploid plant has two or more chromosomes of each kind. If the chromosome number of a haploid has been doubled by colchicine treatment or by some other method, the haploid is called **doubled haploid** or **diploidized haploid**. A doubled haploid has two chromosomes of each kind.

Monoploid plants are weaker than diploids and show a very high degree of cytological and genetical instability. They behave abnormally at meiosis since there is absence of pairing between chromosomes and mostly unequal number of chromosomes (randomly) go to the two poles at anaphase I of meiosis. As a result, non-functional gametes are produced. However, dihaploids and doubled haploids show normal meiosis just like normal diploids.

Haploids have many uses :

(1) Haploidy facilitates production of completely homozygous lines (isogenic diploids) in a single step through chromosome diploidization. Such rapid development of homozygous lines has special significance in biennial and multi-year crops.

(2) Haploidy facilitates early detection and early isolation of mutants. Recessive mutant alleles which remain hidden in normal diploids and appear after selfing only, can be detected and isolated in haploids as easily as the dominant mutant alleles.

(3) The selective probability of recessive characteristics and recombinants is higher in haploids than in diploids. Gamete selection in a plant population is more efficient than zygote selection (that is, useful gametes are more frequent than useful zygotes).

(4) Since only one allele (either dominant or recessive) is present on each locus (hemizygous condition) in a monoploid, selection for dominant allele is easier in haploids than in diploids where selfing is necessasry to bring the dominant allele in homozygous condition.

(5) Promising dihaploids and doubled haploids can be used as parents in hybridization programme to utilize hybrid vigour.

(6) Haploidy facilitates transfer of cytoplasm from one line to another in a single step.

(7) Since dihaploids and doubled haploids behave like normal diploids, cytogenetic studies and breeding are easier in these haploids than in tetraploids (e.g., potato).

(8) Use of polyhaploids facilitates transfer of genes from the polyploid to diploid species.

(9) Diploidized haploids have a special significance where homozygosity of genes cannot be achieved easily due to the presence of self-incompatibility.

(10) Doubled haploids can be very effectively used in recurrent selection programmes aimed at to improve plant populations.

(11) An important future use of doubled haploids will be in physiological and biochemical experiments which require uniform standard plant materials.

(12) Doubled haploids can play an important role in developing cytogenetic and cytoplasmic stocks in future.

Therefore, in addition to its importance in basic cytogenetical research, haploidy offers some unique opportunities in the improvement of crop plants.

Origin and Production of Haploids

Haploid plants spontaneously occur in a large number of plant species (maize, rice, wheat, barley, cotton, tobacco, etc.) but at a very low frequency. Theoretically, haploid plants may originate from the haploid male gamete (that is, through **androgenesis**), from any haploid cell of the female gametophyte without fertilization (more likely through **parthenogenesis**) or due to accidental hybridization in nature between different plant species belonging to the same genus or different genera followed by parthenogenetic development of embryo or elimination of complete set of chromosomes of one of the two species (that is, through **genome elimination**). In maize, they have a parthenogenetic origin, that is, the egg cell develops into a haploid plant without the occurrence of fertilization.

Haploid plants are generally easily identifiable as they are weaker than diploid plants. However, in induced haploidy, use of markers can speed up their identification.

A number of methods have been used for the artificial production of haploids and probable uses of haploidy in basic cytogenetical research and in crop improvement have been discussed by many workers (Bourgin, 1978; Chaleff and Parsons, 1978; Choo and Kannenberg, 1978; Kihara, 1979; Choo, 1980; Collins and Genovesi, 1982; Chaleff, 1983; Baenziger *et al.*, 1984; Bajaj, 1990; and others). Haploids can be produced using **physical agents** (Pandey and Phung, 1982), **chemical agents** (Lacadena, 1974) or by attempting **interspecific or intergeneric crosses** (Kasha and Kao, 1970). In addition, some **genetic systems** like haploid initiator gene in barley (Hagberg and Hagberg, 1981), indeterminant gametophyte gene (Kermicle, 1969) or stock 6 (Coe, 1959) in maize, can enhance the production of haploids. Simultaneous presence of genetic factors and alien cytoplasms can also be a cause of the production of haploids. However, the speed of production of haploids can be increased considerably by using *in vitro* techniques. The application of these techniques is preferred over the use of conventional methods by plant breeders since these techniques are relatively more efficient, less cumbersome and involve less labour (Bajaj, 1990). Following three *in vitro* techniques have been used to produce haploids :

(1) Excised anther/pollen culture technique.

(2) Excised ovary/ovule culture technique.

(3) Genome elimination technique (generally known as *bulbosum* technique)

Of the three techniques mentioned above, the first and third techniques are most widely used.

Anther culture technique : In this technique, the flower buds which have anthers containing uninucleate microspores are selected for excision of anthers. If necessary, flower buds can be surface sterilized and washed with distilled water. Anthers of appropriate age are cultured by dissecting and placing them on a solid agar medium in petri dishes or by floating them on a liquid medium. Depending upon the genotype and the medium used, plantlets will emerge from the anthers in 4-8 weeks. The development of plantlets from anthers may be either **direct** through embryo development (that is, microspores directly give rise to embryos and then embryos develop into plantlets) or **indirect** through callus formation (that is, first there is callus development and then the callus tissue gives rise to plantlets after putting it in a regeneration medium). When the plantlets have attained a height of about 5 cm, they are removed from culture, freed from agar and transferred to small pots containing sterile soil. To protect the plantlets from desiccation and sudden shocks, they are generally covered with glass beakers and kept in a humid greenhouse with proper light facilities. The beakers are removed after one week but the plantlets are kept in the same small pots for about another 15 days. Ultimately, they are transferred to large pots and kept in these pots till they attain maturity.

The chromosome number of haploid plants can be doubled preferably by colchicine treatment or by stem-segment culture.

The success of producing haploid plants by anther culture technique would depend upon the following factors :

(1) Knowledge of experimenter about the most appropriate stage of anthers to be cultured,

(2) the degree of care with which buds have been selected (there should be no injury to buds during their excision and buds should be taken from healthy plants grown under proper light conditions),

(3) knowledge of experimenter about the culture media and nutritional requirements of the plant species used (which may vary from species to specis and in some cases, from genotype to genotype of the same species),

(4) the genotype used for producing haploids, and

(5) pretreatment of plants and buds selected for producing haploids with thermal shocks, etc. for enhancing the frequency of haploids.

Ovule culture technique : Excised ovules are generally cultured for enabling the young embryos to have their normal development. Alternatively, egg cell may develop into an embryo, without fertilization (parthenogenetically) and gives rise to a haploid plant. The ovule culture technique has been used to develop haploid plants in rice (Zhou and Yang, 1990) and several other plant species.

Genome elimination technique : Kasha and Kao (1970) suggested a novel technique called **'bulbosum method'** for the artificial production of haploid plants. They used it to produce barley haploids. In this method, the cultivated diploid barley (*Hordeum vulgare*) is crossed (generally as female parent) to *H. bulbosum* (a wild diploid barley with 2n=2x chromosome number). The reciprocal cross is also successful. To enhance the percentage of seed set, the emasculated spikes are treated with hormone before pollination and also 1-2 days after crossing. The process of fertilization is normal, that is, there is a normal union between the male gamete and the egg cell. But as the growth in the developing embryo progresses, the **bulbosum** chromosomes start eliminating during mitotic divisions till the whole set is lost and only the *vulgare* chromosomes remain in cells. About 2 weeks after pollination, the immature embryos are dissected out and cultured on agar containing a nutrient solution. The plantlets arising from these embryos are mostly monoploids and rarely diploid hybrids. The chromosome number of these haploid plantlets can be doubled by colchicine treatment or by any other method and then they are grown under controlled environmental conditions.

Based on the points mentioned below, Kasha and Kao (1970) advocated that these haploids were caused by genome elimination and not by parthenogenesis :

(1) Some cells of developing embryo had more than seven chromosomes indicating incomplete genome elimination.

(2) The reciprocal cross (that is, *H. bulbosum* × *H. vulgare*) was also successful.

(3) The cross between diploid *H. vulgare* and tetraploid *H. bulbosum* (as male parent) was equally successful in terms of seed set percentage.

(4) The percentage of seed set in induced pathenogenesis experiments is considerably lower than in the present case.

According to Ho and Kasha (1975), this elimination is controlled by genes.

The phenomenon of genome elimination has also been observed in several intergeneric crosses like *Triticum aestivum* × *Hordeum bulbosum*, *H. vulgare* × *T. aestivum*, *H. vulgare* × *Secale cereale*, *T. aestivum* × *Zea mays*, *Triticum* × *Agropyron*, *Agropyron* × *Triticum* and *Triticum* × *Elymus*.

Genome elimination technique is superior to anther culture technique in several ways :

(1) This technique is very efficient and can be used for the large scale production of haploids since the frequency of haploid formation is very high in this method (Jensen, 1974, 1977).

(2) This technique, unlike anther culture technique where success is based on the genotype used, is genotype-independent.

(3) This technique is better than anther culture in terms of genetic purity since the plants obtained after using this technique are all haploids, whereas the plants obtained after using anther culture include non-haploid (aneuploid) plants.

Role of Induced Haploidy in Crop Improvement

The importance of haploidy in basic cytogenetical studies and practical plant breeding has already been discussed earlier in this chapter (uses of haploidy). Here, we will discuss mainly the theoretical and practical aspects of doubled haploid breeding. Narrowing of gap between these two aspects would certainly help in increasing the prospects of doubled haploid breeding.

Expectations from Doubled Haploid System

Many workers (Baenziger *et al.*, 1984; Bajaj, 1990; and others) have emphasized the significance of doubled haploids in crop improvement. No doubt, homozygous lines can be produced by conventional inbreeding, but this process is time consuming as it takes several crop seasons/plant generations to develop completely or almost completely homozygous lines. On the contrary, production of doubled haploids is a matter of weeks. Further, development of inbred lines in several plant species like alfalfa, berseem and sunflowr is a difficult task because of the presence of self-incompatibility, somatoplastic sterility, male sterility, protandry, protogyny, or any other device enforcing cross pollination. The inclusion of doubled haploid system in breeding programmes is not only helpful in reducing the time needed for the production of inbred lines but also it introduces the potential to enhance the efficiency of selection and evaluation in advanced generations. The inclusion of this haploid system in pedigree breeding programme, for example, can reduce the duration of the scheme by several crop seasons/years. The duration of single seed descent (or multiple seed descent) scheme which is largely similar to doubled haploid breeding can also be reduced by several years. Both these methods of breeding (single seed descent and doubled haploid breeding) have the same objective, the rapid production of homozygous lines followed by selection of promising lines and their evaluation in yield trials. In recurrent selection schemes and in the development of synthetics and composites aimed at to increase the frequency of desirable alleles and the proportion of additive genetic variance in populations, double haploids can play a great role for the exploitation of heterosis. The use of doubled haploids in recurrent selection schemes has a greater significance when the trait under improvement is highly affected by the change in environment (like drought tolerance) or by gene dominance. Doubled haploid system can also be used in diallel selective mating systems (Jensen, 1970) to develop provising doubled haploids and release them as commercial cultivars. Therefore, doubled haploids can play an important role in the breeding of both the self-pollinated and cross-pollinated crop species. However, the success of this system in the breeding of any crop species would depend mainly on the efficiency in the production and isolation of haploids and in the doubling of chromosome number in that species.

Production of doubled haploids helps in the detection and isolation of both the dominant and recessive mutants with equal efficiency. Doubled haploids have their future use in physiological studies, biochemical selection experiments, linkage analyses, somatic cell genetics, statistical genetics and in the conservation of genetic resources.

Achievements of Doubled Haploid Breeding

The *in vitro* production of doubled haploids has resulted in the release of cultivars in several crop species (barley, wheat, rice, etc.). Perhaps **Mingo**, a spring barley, was the first cultivar developed through doubled haploid breeding. This variety, the highest yielding barley in Ontario that time, was licensed for sale in 1979. Another variety of barley which was also released at that time was **Rodeo**. The release of Mingo barley was considered as the most dramatic demonstration of doubled haploid breeding (Ho and Jones, 1980). According to Kasha and Reinbergs (1981), another doubled haploid barley, **Gwylan**, was ready for marketing in New Zealand.

A number of doubled haploid varieties of wheat and rice have been released in China (Hu, 1986; Loo and Xu, 1986). These varieties were developed through anther culture technique. Another popular wheat variety **Florin** was released in France (de Buyser *et al.*, 1987). This variety was also developed using anther culture method. Anther culture technique has been used at large scale in China to develop large number of wheat and rice varieties and strains. One of the best examples is winter wheat vareity **Jinghua No. 1** developed through this method. Among the 81 rice varieties and strains developed in that country, **Hua Yu No. 1** proved to be very promising in terms of yielding ability, wide adaptability, resistance to bacterial blight and tolerance to salt (Hu, 1985). Similarly, a new rice variety **Xin Xiu** possesses a very wide adaptability. Another important doubled haploid rice cultivar which was released in Texas is **Texmont**, a high yielding and early maturing rice (Bollich, 1990). Another use of anther-derived doubled haploids has been made for the improvement of nutritional quality of rice. *In vitro* production of doubled haploids has also been practically used for the rapid development of maize pure lines. For example, **Qun Jua**, a selected pure line developed through anther culture, took only one year for its development.

Many attempts have been and are being made in other countries including United States of America towards the development of useful doubled haploids.

In view of the information given above, it can be safely said that doubled haploid breeding has become a reality for several crop species and has a bright future. However, doubled haploidy should not be considered as panacea for all plant breeding problems and, like other breeding methods, must be used most judiciously. In fact, for optimizing the utilization of haploid techniques in plant breeding programmes, doubled haploidy system should satisfy the following requirements :

(1) Large scale production of haploids should be possible,

(2) an efficient technique should be available for doubling the chromosome number,

(3) it should be possible to develop haploids in large numbers from a random sample of gametes, and

(4) the production procedures should involve relatively less time, cost and other facilities.

Nevertheless, since success of haploidy breeding may vary from species to species and genotype to genotype for the same species, it is advisable to use this breeding system in those cases only which fulfil the above mentioned requirements.

CHAPTER 24

Distant Hybridization

A cross between two distant species of the same genus is termed as interspecific hybridization, for example, a cross between *Lycopersicon esculentum* and *L. pimpinellifolium*. Similarly, a cross between two species belong to different genera is called as intergeneric hybridization, for example, a cross between *Raphanus sativus* and *Brassica oleracea*. These both kinds of crosses (interspecific and intergeneric) are known as wide crosses or distant crosses, that is, distant hybridization includes both the interspecific and intergeneric hybridizations.

As has been discussed in Chapter 2, distant hybridization has been occurring in nature and some very valuable crop species have evolved because of these spontaneously occurring wide crosses. Hexaploid and tetraploid wheats (*Triticum aestivum*, *T. durum*, etc.), tetraploid cottons (*Gossypium hirsutum* and *G. barbadense*), tetraploid tobacco (*Nicotiana tabacum*) and amphidiploid *Brassicas* (*B. carinata*, *B. juncea* and *B. napus*) are some examples of such crops. Man has also made a number of attempts to make inter-specific and intergeneric crosses and in some cases he has achieved a good success. Perhaps the most striking example where man has succeeded in producing a new (man-made) crop species was to attempt crosses between wheat (both hexaploid and tetraploid) and rye (*Secale cereale*) and the evolution of *Triticale*.

Objectives of Distant Hybridization and Role in Plant Breeding

The objective of distant hybridization though greatly depends upon the aim of the plant breeder, the main objective of this kind of hybridization is to broaden the genetic base of crop species/varieties. On a broad basis, the objectives of distant hybridization may be classified as follows :

1. Academic interest : Crossing between varieties of the same species is a routine method followed by plant breeders and geneticists. Such crosses, in general, are very easy to attempt and, barring some situations (presence of self incompatibility and male sterility systems), do not present any problem in the production of subsequent generations. However, sometimes, plant breeder or geneticist may attempt wide crosses just to meet their curiosity about the results of such crosses, that is, what will happen if such and such species are crossed together. Most of the attempts made in this direction in the early periods of plant breeding probably had this basis. The first artificial plant hybrid produced by Thomas Fairchild in 1717 (See Roberts, 1965) between carnation (*Dianthus caryophyllus*) and sweetwilliam (*D. barbatus*) and a large number of distant hybrids produced in subsequent years by many others did not have a considerable agricultural value. Even now, sometimes, wide crosses are attempted just for the academic interest.

' **2. Defect elimination or deficiency improvement :** One of the most important objectives of attempting distant crosses, is to improve a crop species by transferring to it one or a few characters from other cultivated or wild species belonging to the same or other genus. Or, to phrase it otherwise, the distant hybridization is used manytimes for eliminating some defect (disease susceptibility, etc.) of a species which otherwise possesses all desirable characteristics. For improving deficiency in a species, it is crossed with a species or genus in which the deficient character is present in highly developed form. Such transfer of character(s) may be accomplished by transferring small chromosome segments carrying desirable genes from donor species to the recepient species or by producing alien substitution monosome (a single chromosome and not a pair, of the recepient species is substituted by a single chromosome of the donor species), alien substitution lines (one chromosome pair of the recepient species is substituted by one chromosome pair of donor species), alien addition monosome (in addition to its normal somatic chromosome number, the recepient species has one chromosome and not a pair, of donor species), alien addition lines (the recepient species, in addition to its normal somatic number, has one chromosome pair of donor species) or by some biotechnological technique (RFLPs, RAPDs, etc.).

(a) **Transfer of small chromosome segments :** The transfer of small chromosome segment has been mainly used to transfer genes for disease resistance from one species to another. Sears (1956) perhaps demonstrated the most elegant method for the transfer of disease resistance from one species to another by introducing small segments of chromosome from the alien species. He was able to successfully transfer genes for leaf rust resistance from *Aegilops unbellulatum* (a diploid wild grass having U genome) to *Triticum aestivum* (common wheat possessing A, B and D genomes) in spite of the fact that the seeds produced by this cross are not germinable when *Ae. umbellulatum* is crossed as male or female parent with *T. aestivum*. In view of the difficulty, an amphidiploid between *T. turgidum* (tetraploid emmer) and *Ae. umbellulatum* was produced and used as a bridge for combining the genomes of these two species (*T. aestivum* and *Ae. umbellulatum*). The F_1 hybrid was backcrossed to common wheat twice and a plant possessing resistance to leaf rust and having a single added chromosome from *Ae. umbellulatum* was obtained. However, this added chromosome, in addition to the genes for leaf rust resistance, had some deleterious genes having adverse effect on plant characteristics and pollen viability. The progeny of this plant was raised. Some plants from this progeny were X-rayed before the occurrence of meiosis and wheat plants were pollinated by this irradiated pollen. Forty out of total 6091 plants gave clearcut evidence of translocation (at least 17 different translocations) of *Aegilops* chromosome segments carrying rust resistance to a wheat chromosome.

There are some other examples of this kind of translocation of small chromosome segments from one species to another. Elliot (1957) and Knott (1961) were able to transfer genes for stem rust resistance from *Agropyron elongatum* to common wheat by irradiating the hybrid material (carrying rust resistance genes) involving these two species and translocating *Agropyron* chromosome segments to wheat chromosomes. Similarly, Driscoll and Jensen (1964) transferred rye (*Secale cereale*) chromosome segments carrying genes for both leaf rust and powdery mildew resistance to common wheat. Efforts are also being made to transfer genes for stem resistance from *A. elongatum* and rye to *durum* wheat (Prabhakar Rao, 1978). Both kinds of translocations (rye-*durum* and *Agropyron-durum*) were obtained, but because of the difficulty in the transmission of these translocations through male gametes, no alien translocation homozygotes in *durum* background could be obtained.

In addition to the production of translocations by X-rays or gamma rays treatment, the transfer of resistance genes (or other desirable genes) from one species to another may also be accomplished by other methods. For example, if spontaneous crossing over occurs between the foreign chromosome and a chromosome of the recepient species, alien substitution monosomes may be produced and desirable gene(s) may be easily transferred to the chromosome of the recepient species. Another method of such transfers is by promoting pairing between the foreign chromosome and a chromosome of the recepient species if spontaneous crossing over does not take place between these chromosomes because of infrequent pairing. In common wheat, there is a gene on the chromosome 5 of B genome which induced diploidization in this species by suppressing pairing between homeologous chromosomes belonging to different genomes. By using a wheat line which is nullisomic for 5B or which lacks the homeologous pairing suppressor on 5 B chromosomes, the pairing between the foreign chromosome and a chromosome of the recepient species may be promoted. However, this technique should be used only by those who have suitable genetic stocks with them and are well acquainted with the genetic control of chromosome pairing.

Several other reports of transfers involving small chromosome segments or single genes for disease resistance in wheat from its related species or genera are available in the literature. Friebe *et al.* (2001) gave a long list of wheat-alien translocations. Some of these translocations obtained through homeologous recombination (e.g., T2B/2G#1 carrying S 36/Pm 6 genes from *Triticum timopheevii*), through irradiation (e.g., 7 DS. 7DL-7Ae#1L carrying Lr 19/Sr 25 genes from *Agropyron elongatum* and 6 AS.6 AL–6 Ae#1L carrying Sr 26 from *A. elongatum*) or through spontaneous translocation (e.g., T3DS.3DL-3Ae#1L carrying Sr 24/Lr 24 genes from *A. elongatum* and T1BL. IR#1S carrying Lr 26/Sr 31/Yr 9/Pm 8 genes from *Secale cereale*), have made significant contribution towards development of new wheat varieties and some of these genes are still effective in wheat growing belt of India. Nevertheless, the 1B/IR translocation (a simple transfer between rye and wheat in the Soviet Union cultivar **Kavkaz**) carries a number of useful genes from rye and has contributed towards high yield, multiple disease resistance and adaptation to marginal environments. Veery wheat lines, resulting from the cross (Kavkaz/Buho's'// Kalyansona/Blue Bird) with Kavkaz, and their progenies like Kauz, Attila, Pastor and Baviacora, all carry the 1B/IR translocation. Though there are some concerns about their milling and baking properties, the Veery lines and progenies involving Veery lines in their pedigrees are being grown on estimated 40 million hectares under a number of different names (**Anadolu** in Turkey, **Annapurna 1** and **NL 459** in Nepal, **Arganda** and **Carthaya** in Spain, **Cordillera 3** in Paraguay, **Dashen** in Ethiopia, **Estanzuela** in Uruguay, **Gamtoos** in South Africa, **Genaro F 81**, **Seri M 82** and Ures T 81 in Mexico, **Lima 1** in Portugal, **Loerie** and **Seric** in Zambia, **Millaleau Inia** and **Nobo Inia** in Chile, **Pakistan 81** and **Pirsabak 85** in Pakistan, **R 253** in Kenya, **Rusape** in Zimbabwe, Viri in Tanzania, **DWR 162, GW 190, HD 2160, HPW 42, HS 207, HUW 206, K 8804, MACS 2496, PBW 343, PBW 373, VL 719, VL 738, WH 533** and **WH 542** in India) and occupy more than 50 per cent wheat growing area in developing countries (Skovmand *et al.*, 1997).

Transfer of desirable genes through small chromosome segments is not only confined to the transfer of disease resistance genes, but some other desirable genes have also been transferred from one species to another by this technique. In cotton, Saunders (1965) transferred genes for stem and leaf hairiness (a desirable character for insect resistance) from *Gossypium raimondii* (2n=26) to *G. hirsutum* (2n=52). For this, chromosome number of the hybrid produced by crossing these two species was doubled and the allohexaploid thus produced was crossed to *G. hirsutum* for obtaining pentaploid. Again, the pentaploid was crossed to *G. hirsutum* and by doing selfing in the progeny for some generations, the gene for hairiness was transferred to a *hirsutum* chromosome. Similarly, genes for fibre strength were transferred from *G. thurberi* (2n=26) to *G. hirsutum* by crossing the allotetraploid of *G. thurberi* × *G. arboreum* (2n=26) with *G. hirsutum*. The transfer of black arm resistance from *G. barbadense* (2n=52) to *G. hirsutum* by crossing the two species and then backcrossing the hybrid with *hirsutum* is an example of such a gene transfer where interspecific hybrid is sufficiently fertile and easily crossable to the recepient species. The chromosomes of these two species possess enough degree of homology required for recombination between the donor and the recepient chromosomes.

Brassica napus (n=19, A and C genomes) is a natural amphidiploid combining basic sets of *B. campestris* (A genome) and *B. oleracea* (C genome) and is a self-compatible species. For producing *napus* hybrids, self-incompatibility alleles have been transferred from *B. campestris* to this amphidiploid.

Cultivated tobacco (*Nicotiana tabacum*, 2n=48) which is a hybrid polyploid from *N. sylvestris* (2n=24) and *N. tomentosa* (2n=24), has low adaptability. Genes for high adaptability have been transferred to this amphidiploid from *N. sylvestris*, one of its putative parents.

Low coumarin content is a desirable character in sweet clover (*Melilotus*). This character cannot be directly transferred from *M. dentata* to a better adapted and more drought tolerant species, *M. officinalis* since the two species are not crossable. However, Smith (1948) was able to transfer this character from *M. dentata* to *M. alba*. Webster (1955) transferred this character from low coumarin *M. alba* to *M. officinalis* by producing fertile hybrids between these species with the help of embryo culture technique.

However, one should not get very much fascinated with this method. The method has some serious drawbacks also. It is quite likely that an undesirable gene may be linked with the desirable gene under transfer. In fact, a large number of small segment transfers could not be commercially exploited because of this linkage between an undesirable gene and the desirable gene.

(b) Alien substitution and alien addition lines : Alien substitutions and alien additions also have been used to transfer one or a few desirable genes from one species to another. These lines can be produced by making cross between the recepient species and donor species. The F_1 hybrid which is usually sterile is treated with colchicine to double its chromosome number. The amphidiploid thus produced is backcrossed, as female parent, to the recepient species and again a hybrid is produced which will contain normal diploid complement of the recepient species and single chromosome set from the donor species. When this hybrid is again backcrossed to the recepient species, the progeny

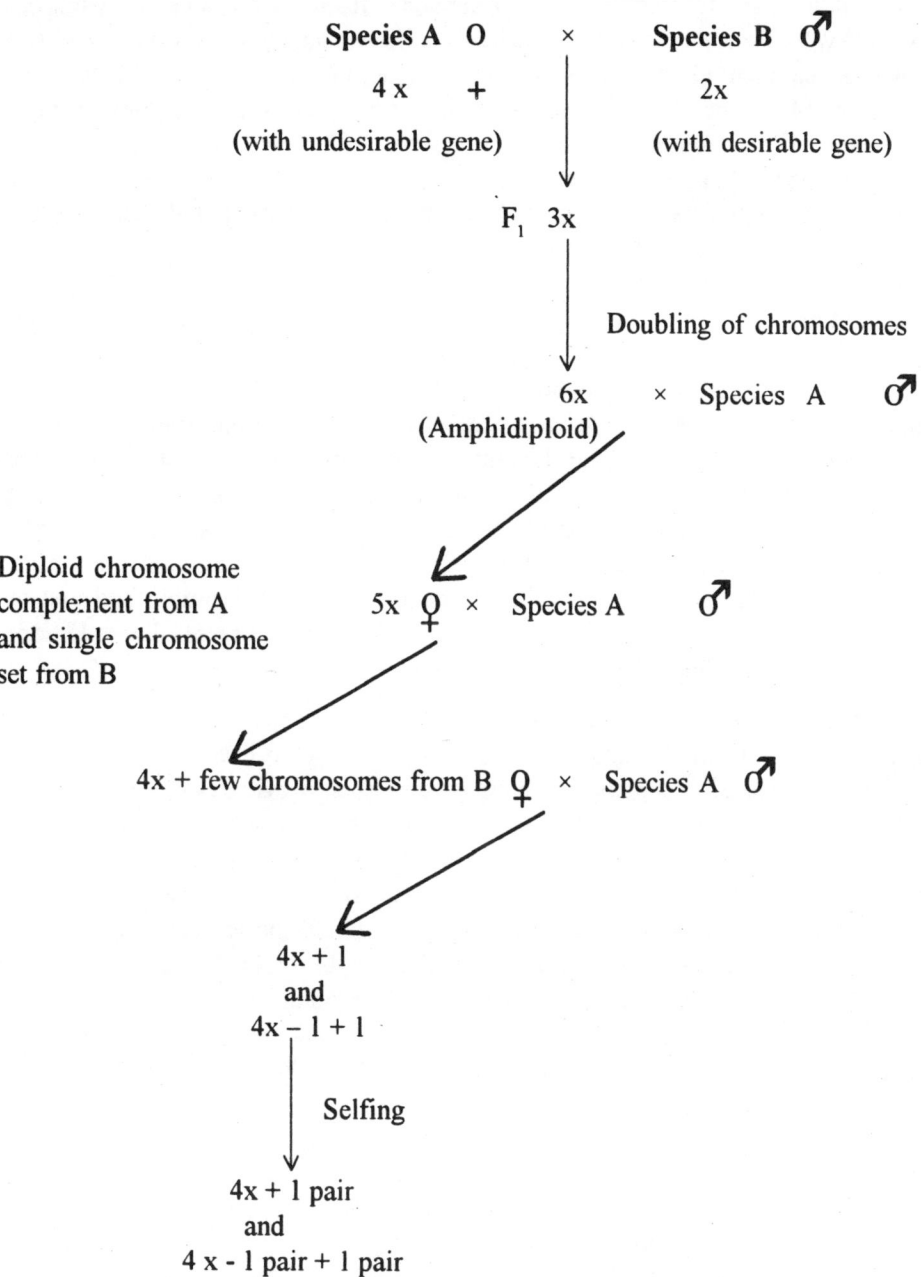

Figure 24.1. Production of alien addition lines and alien substitution lines

will have normal somatic complement of the recepient species and a few chromosomes from donor species. Plants with character under transfer are selected and again backcrossed to the recepient species and alien addition monosomes (4x+1 chromosome from donor) and alien substitution monosomes (4x-1+1) with desirable character are produced. On selfing, these addition and substitution monosomes will produce alien addition lines (4x+1 pair from donor) and alien substitution lines (4x-1 pair of recepient +1 pair of donor), respectively. A schematic representation of this is given in Figure 24.1. However, the production of substitution lines and addition lines is mainly confined to the polyploid species since both alien substitutions and alien additions are not effectively viable in case of diploid species. As regards the relative desirability of the alien additions and alien substitutions, the former show less undesirable effects than the latter.

Alien substitution lines and alien addition lines have been produced in a number of crop plants (tobacco, wheat, etc.) and desirable genes have been transferred from one species to another. The gene for mosaic resistance in tobacco was transferred from *Nicotiana glutinosa* to *N. tabacum* by developing both the alien substitution lines and alien addition lines. The resistant substitution line contained 23 pairs of chromosomes of *N. tabacum* and the pair of H chromosomes (carrying gene for mosaic resistance) from *N. glutinosa* and thus the somatic chromosome number of the line was 48. The somatic chromosome number of the resistant addition line on the other hand was 50, that is, normal somatic complement of *N. tabacum* (4x=48) plus the pair of H chromosomes from *N. glutinosa*.

The alien chromosome transferred to the recepient species may, in addition to the desirable gene under transfer, have some undesirable gene(s). Such an occurrence of desirable and undesirable genes on the same chromosome (which is quite frequent) restricts the use of alien additions and alien substitutions in plant breeding.

(c) **Introducing cytoplasmic male sterility :** In addition to the interspecific and intergeneric transfers of nuclear genes through different techniques, wide crosses have also been used for transferring male sterile cytoplasm in some crop plants. The transfer of such a cytoplasm, sometimes, is felt necessary for the production of male sterile lines in a species where desirable cytoplasm is not available. Since the plant breeder would like to have this cytoplasm in the genetic background of the recepient species (to which the cytoplasm is to be transferred), simple successive backcross method is usually followed for the transfer of cytoplasm. First, the donor species (carrying male sterile cytoplasm) and the recepient species are crossed together. The F₁ hybrid is backcrossed to the recepient parent and backcross progeny is produced. Plants from the segregating backcross progeny are again crossed to the recepient species. This process is repeated till the genotype of the recepient species is completely reconstituted. Generally, 5 to 6 backcrosses are required for the complete reconstitution of the genotype of the recepient species.

Male sterile cytoplasms have been transferred to common wheat, *durum* wheat, tobacco, cotton, etc. from other species. In case of common wheat (*T. aestivum*), male sterile cytoplasms have been transferred from *T. timopheevii* and *Ae. caudata*. To *durum* wheat, the male sterile cytoplasm has been transferred from *Ae. ovata*. In tobacco, *N. tabacum* was crossed with *N. megalosiphon*, *N. debneyi* and *N. bigelovi* for transferring male sterile cytoplasm from these three species of the same genus. Similarly, in cotton, *G. hirsutum* × *G. anomalum* and *G. hirsutum* × *G. arboreum* crosses were used to transfer male sterile cytoplasm from *anomalum* and *arboreum* to *hirsutum*.

These all kinds of transfers (small chromosome segment, complete chromosomes and cytoplasm) may be put under a common heading, **introgressive hybridization**, which means transfer of certain characters from one to another species or genus without impairing its taxonomic status. Introgressive hybridization has played an important role in the evolution of some crop plants. For example, crosses between maize (*Zea mays*) and gamagrass (*Tripsacum* sp.) occurred spontaneously and some genes for disease resistance got transferred from the latter genus to the former.

(3) Evolving new varieties : When the desirable characters in different species or genera to be crossed are almost equally distributed or the number of characters to be transferred is so large that it is not easy to transfer them all together, efforts are made to combine all desirable characters of the two species/genera and evolve new varieties. Backcross method, usually does not help in such cases. Rather, pedigree selection is used to handle the segregating progeny and selection is practiced for all the desirable characters. However, limited backcrossing with the species carrying more number of desirable characters, at some intermediate stage of the programme, may help in quickly concentrating the characters of the species in the progeny. If variety is to be evolved as a hybrid, as in case of cross-fertilized crops, selfing after crossing is not needed. But, the high productiveness of the hybrid and its seed production at commercial scale should be ensured. In case of vegetatively propagated species, one can select any genotype with any number of chromosomes and with any level of heterozygosity. The vigour of the variety can be maintained for indefinite period.

One of the best examples of evolving a popular variety by distant hybridization is perhaps the development of well known cotton hybrid variety Varalakshmi. The hybrid which is being grown over a large area (about 50,000 hectares) was developed at the University of Agricultural Sciences, Banglore by crossing *G. hirsutum* and *G. barbadense* and possesses desirable characteristics of both the parent species (high yielding capacity of *hirsutum* and better quality staple of *barbadense*). In rice, the variety Co 31 has been developed through pedigree selection from the segregating material of the cross between cultivated rice (*Oryza sativa*) and wild rice (*O. perennis*).

With a view to combine desirable characteristics of different species of the genus *Saccharum* (*S. officinarum, S. spontaneum*, etc.), the present day commercial varieties of sugarcane have been developed by following complex interspecific hybridization among these species. Whereas *S. officinarum* has contributed a number of agronomically important characters (softness of cane, high juice percentage, less fibre content, etc.) to these complex interspecific hybrid varieties, the disease and insect resistance characteristics have come from the wild species, *S. spontaneum* (kans). The elephant grass-bajra hybrids possess desirable characteristics of both the parent species, that is, high forage production capacity of elephant grass (*Pennisetum purpureum*) and drought tolerance and seed production characteristics of bajra (*P. glaucum*).

Perhaps, the best authenticated examples regarding the development of cultivated varieties through interspecific hybridization come from garden strawberry. Crossing and recrossing between *Fragaria virginiana* and *F. chiloensis* and selection from the segregating generations of this cross for a long period of time have resulted in the development of varieties which have fruitfulness, high quality of flesh and excellent flavour similar to the former and large fruit size similar to the latter species. Similarly, peach-plum hybrids and moong (greengram) - mash (blackgram) hybrids which combine desirable characters of the two parent species, are on the way.

(4) Evolving new species : There are a large number of naturally occurring amphidiploids (many crop plants and fruit trees). These amphidiploids probably evolved due to meiotic irregularities either before or after the union of gametes of two different species. A pre-union meiotic irregularity may give rise to unreduced gametes in both the species and a post-union irregularity may cause doubling in the chromosome number of the sterile F_1 hybrid. Both kinds of irregularities result in the development of an amphidiploid.

In some cases, however, artificially synthesized amphidiploids have been produced as new man-made species. Since most of the interspecific and intergeneric F_1 hybrids are sterile, doubling of chromosome number is required to develop fertile allopolyploids. This is usually done by treating the sterile F_1 hybrids with colchicine. *Triticale* which has been developed by crossing wheat with rye, is the most famous example of a man-made plant species. Rimpau in 1888 in Germany (see Rimpau, 1891) for the first time, produced this new cereal. Two kinds of triticales, octaploid (obtained by crossing *T. aestivum* with *S. cereale*) and hexaploid (obtained by crossing *T. turgidum* with *Secale cereale*), have been developed. The hexaploid triticales have performed better than octaploid triticales probably because the 6x is the optimum ploidy level in triticales. Triticales possess good characteristics of both the parent genera (high yield potential and quality from wheat and hardiness from rye) and are being cultivated as a commercial crop in some countries (Canada, Hungary, Brazil, etc.) where wheat cannot be successfully grown because of acidic soils or very low temperature during the crop season. In the beginning, triticales had some very serious defects like shrivelling of grains and poor fertility. These defects, however, have been removed to a considerable extent from the present day triticales.

The development of *Raphanobrassica* by the Russian scientist Karpechenko (1928) by crossing radish (*Raphanus sativus*) with cabbage (*Brassica aleracea*) is another effort in this direction. The idea behind the making of this intergeneric cross was to develop a species which could have root like radish and leaves like cabbage. Unfortunately, the amphidiploid had a combination of undesirable characteristics of the two genera, that is, root like cabbage and leaves like radish. However, one should not get disappointed after getting this undesirable combination. It is just possible that sometime in future one may obtain a desirable combination in the form of a transgressive segregant. For this, intensive studies with regard to selection from different segregating materials of this cross are required.

Another example of man-made species is the development of *Festulolium* (*Festuca* × *Lolium*).By crossing ryegrass (*Lolium perenne*) with fescue species (*Festuca arundinacea, F. pratensis* and *F. rubra*), amphidiploids have been developed. These amphidiploids though possess certain favourable properties of the parents, still need considerable improvement with regard to their cytological stability and fertility.

Efforts have also been made to combine edible parts of potato (*Solanum tuberosum*) and tomato (*Lycopersicon esculentum*). But, the efforts to develop **pomato** have so far been confined to an academic interest only.

(5) Helping basic research : Distant hybridization is one of the main patterns of evolution in cultivated species and has resulted in the evolution of a large number of valuable crop plants and fruit trees (wheat, cotton, tobacco, pears, plums, cherries, grapes, etc.). It is greatly helpful in determining evolutionary relationships among different species of a genus. Production of substitution and addition lines not only helps is transferring desirable characters from one species to another, but also allows one to do inheritance and evolutionary studies. By producing nulli-tetrasomic series in amphidiploids, one can, on the basis of cytological (pairing behaviour), genetical and biochemical (protein banding pattern, etc.) studies, know whether a particular chromosome pair of one genome is able to compensate a pair of chromosomes belonging to another genome. Such studies have been carried out in wheat.

There are geographical, cytological (Gerstel, 1953; Gerstel and Sarvella, 1956), genetical (Phillips, 1963) and biochemical (Cherry *et al.*, 1970) proofs available which clearly show that amphidiploid (tetraploid) cottons have originated from hybrids between *G. herbaceum* and *G. rainmondii* and that the origin of these amphidiploids has been monophyletic (Phillips, 1966).

The polyploids like wheat where the chromosomes of different genomes are homeologous, offer a unique system of intergenomic heterosis at corresponding loci of different genomes (homozygosity within genomes but heterozygosity between genomes).

Triticale is perhaps the species which provides best material for doing basic type of research regarding inheritance of characters, pairing behaviour of chromosomes, DNA replication (particularly delayed replication of DNA in heterochromatic parts of rye chromosomes) and substitution and addition of chromosomes. Delayed replication of DNA in heterochromatic parts of rye chromosomes is the main cause of cytological instability in triticales.

Furthermore, wide crosses may give some unexpected results. Some genes change their mutability when transferred to a new genetic background. In cotton, for example, the gene for petal spot shows a considerable increase in its mutation rate when transferred from *G. hirsutum* to *G. barbadense*.

Limitations of Distant Hybridization

The importance of distant hybridization need not to be emphasized further. This kind of hybridization is used to combine diverse characters particularly when variability within a species has exhausted. However, wide crosses show varying degrees of success depending upon the kinds of species involved in the cross. In general, more taxonomically closely related are the species, more are the chances of crossability between them. In some cases, crossing between the two species is quite easy and desirable genes from one species to another may be transferred almost as easily as between the varieties of the same species. For example, crosses between maize and teosinte (*Zea maxicana*, formerly *Euchlaena maxicana*) are very easy. The two species have the same chromosome number (2n=20) and their chromosomes differ for some knob patterns only. The F_1 hybrid between the two species is robust and fertile. In tomatoes, the two species *Lycopersicon esculeutum* and *L. pimpinellifolium* belonging to the same subgenus *Eulycopersicon* (red fruited) are easily crossable. This cross is as easy as the intervarietal crosses within *L. esculentum* (except that there is a very slight reduction in the chiasma frequency and pollen fertility) and does not present any difficulty at the stage of hybridization, gene recombination in the hybrid or the production of progeny by the hybrid (Rick and Butler, 1956).

There are examples where crosses can be made with some difficulty or crosses are attempted in some specific ways. In tomatoes, the crosses between the species of two subgenera, *Eulycopersicon* and *Eriopersicon* (wild, green fruited) are successful only when species of the latter subgenus are used as male, that is, reciprocal crosses are not successful. Similarly, their F_1 hybrid and *L. esculentum* are crossable only if the latter is used as a female parent. On the contrary, the crosses between the F_1 hybrid and the species of *Eriopersicon* succeed only when the latter is used as male parent. Similar difficulties are faced when crosses are made between blackgram (*Vigna mungo*) and green gram (*V. radiata*).

On the other hand, there are a number of cases where crosses are not at all successful because of one barrier or the other. For example, *G. davidsoni* is not crossable to a number of *Gossypium* species like *G. hirsutum*, *G. barbadense*, *G. thurberi* and *G. stocksi*. The control of hybrid inviability may vary from very simple genetic mechanisms to very complex mechanisms like genotypic disharmonies because of large differences in the genes and their organization in the chromosomes of the two parent species. The criterion of taxonomic relationship between the species to be crossed also does not hold always true. Crossability between *Brassicas* present the most striking example of the failure of this criterion. The species *B. oleracea* is not crossable to its closely related species *B. napus* (rape) and *B. rapa* (turnip) but is crossable to *Raphanus sativus* (radish), a species of different genus.

The somatic hybridization also does not appear to be of much help in the case of distant hybridization. If successful, it can only help in producing a hybrid possessing the chromosome number equal to that of the two parents. But, the production of a hybrid does not guarantee the initiation of flowering in the hybrid nor it rules out the possibility of occurrence of hybrid sterility even though the hybrid flowers.

Genetic Disability and Reproductive Incapacity

The genetic differentiation between two species may occur at different stages of the separation of gene pools. If crosses are made between separated groups, depending on the stage of genetic differentiation (not morphological differentiation and ecological preference), two kinds of properties can be observed in addition to normal segregation and recombination of genes. If the differentiation (genetic) is at an earlier stage, genetic disability may be found in the F_2 and later generations. However, the worth of the new combination as compared to that of old ones, would depend on the kind of genetic systems present in the parental species. If the genetic systems in the parental species are very well balanced, one should not expect new combinations superior to the old ones. But if the reverse is true, new combinations may be far superior to parental combinations.

In case the genetic differentiation is at an advanced stage, the hybrid may show reproductive incapacity or sterility in the F_1 generation. However, as has been discussed earlier, the degree of hybrid sterility may vary from zero (hybrid is fully fertile) to 100 per cent.

The limitations of distant hybridization may be classified into two broad groups : (A) barriers to distant hybridization and (B) hybrid sterility.

Barriers to Distant Hybridization

There may be some general barriers (geographic isolation, failure of flowering due to differences in the day length, flowering of parental species at different times, etc.) which may not allow the pollen to reach the stigma of the parent species used as female. Even though the pollen reaches the stigma, there are a number of barriers which could adversely affect the success of distant hybridization. These barriers are as under :

(1) Barriers to zygote formation : In wide crosses, the zygote may fail to form either due to the failure of pollen germination on the stigma, slow pollen tube growth, bursting of pollen tube in the stylar region, short pollen tube, thick pollen tube or pollen tube reaches the embryosac but there is no fertilization. Because of the cross incompatibility or some other genetic or physiological mechanism operating in some distantly related species, the pollen does not at all germinate on the stigma of the foreign species. For similar reasons, pollen tubes in some cases grow slowly in the styles of other species. One major reason of the failure of interspecific hybridization in Datura is the bursting of pollen tubes in the styles of other species. Hybridization between maize (as female) and gamagrass (*Tripsacum*) is not usually successful because the pollen tubes of gamagrass fail to travel the whole length of maize styles and thus no fertilization. A similar situation of short pollen tubes and long styles has been observed in some species of Nicotiana. The polyploids have usually thick pollen tubes than diploids. The crosses between polyploid (used as male) and diploid species sometimes do not succeed because the thick pollen tubes fail to grow in the ordinary styles of diploids. Rarely, a pollen tube may reach the embryosac but fails to burst.

(2) **Barriers to zygote development** : Even though fertilization does take place between the two species, difficulties may be encountered during the zygote development and in some cases it may completely fail to develop. A number of factors like lethality caused by single genes, general incompatibility of parental genotypes, disharmonies between the endosperm and the developing embryo, chromosome elimination and incompatibility between genes and cytoplasm may completely or partially arrest the development of embryo. In *Crepis*, for example, the species *C. tectorum* carries a lethal gene l (its normal allele is L) which though has no deleterious effect on *C. tectorum* itself, causes death of *tectorum* hybrids with *C. capillaris*, *C. bursifolia* and *C. leantodontoides* at an early stage of their development. Similarly, the survival of hybrids between diploid wheat and *Ae. umbellulatum* would depend on whether the latter species carries the gene l (hybrid is viable), L^e (causing early death) or L^l (causing late death).

Interspecific hybridization in cotton when either *G. gossypoides* or *G. davidsoni* is used as one of the parents, presents a bit different picture. In such crosses, genotypic disharmony rather than the action of specific lethal genes, probably causes the death of embryo. Whereas hybrids of *davidsoni* with many other *Gossypium* species (*G. thurberi*, *G. hirsutum*. etc.) die in the cotyledonary stage or even earlier, some hybrids of *gossypoides* with other species may survive upto anthesis stage (of course, some hybrids die in the early embryonic stages). In some other cotton interspecific hybrids, the hybrid inviability is controlled by simple genetic mechanisms rather than by complex genetic unbalance between the parents.

Gradual to drastic elimination of chromosomes from zygotes has been observed in some cases of distant hybridization. The development of zygote in such cases is usually normal but chromosome elimination can lead to meiotic irregularities in the later stages of the hybrid development. In exceptional cases, this elimination is so drastic that all the chromosomes contributed by one of the parental species are eliminated and chromosome complement of the other parent only remains. In such extreme cases, since the plants obtained possess chromosome complement of one parental species only, they no more remain interspecific hybrids. Interspecific crosses in *Hordeum* and *Nicotiana* show chromosome elimination. An elimination of complete chromosome complement takes place when *Hordeum bulbosum* is crossed, as female parent, with *H. vulgare* or *Triticum aestivum*. In both cases, there is complete elimination of *bulbosum* chromosome complement. This is, however, a novel technique for the production of haploids.

There are some examples where there is poor or abnormal development of endosperm in distant hybrids leading to its abortion at an early stage of the development of embryo or at a later stage. If endosperm aborts at the early stage of the embryo development, it will prevent further growth of the embryo and no viable seeds will be obtained. In the hybrids of *H. bulbosum* and *H. vulgare*, the endosperm abortion occurs at an early stage. On the contrary, in the hybrids of *Secale* with *Triticum*, the endosperm aborts at a quite later stage resulting in the development of some viable seeds.

The incompatibility between a specific cytoplasm and the genome of some other species, in general, leads to male sterility or hybrid weakness. In a few cases, however, such an interaction between cytoplasm and nuclear genes may adversely affect or completely prevent the development of embryo.

(3) Barriers to hybrid seedling development : In some cases, the death of distant hybrids occurs after the embryo has fully developed. The occurrence of death may vary from seedling development to anthesis period and, in some cases, the plants may die even after the completion of anthesis. In cotton, some hybrids of *G. gossypoides* with other species of *Gossypium* die even after the occurrence of meiosis. The cause of death of these hybrids is not yet fully known. Perhaps some complex genetic mechanism (lethal genes in combination with genotypic disharmony between the parental genomes) operates for hybrid inviability. Similarly, the gene L^1 carried by *Ae. umbellulatum* causes the death of *Ae. umbellulatum* × 2x wheat hybrids at a later stage. Some interspecific crosses in *Melilotus* lead to albinistic defects in the hybrid plants which fail to survive.

Overcoming Barriers to Distant Hybridization

The obstacles in distant hybridization may be overcome by taking some general precautions before, after and at the time of hybridization and by using some special techniques which facilitate this kind of hybridization. With regard to general precautions, the species used as female parent should be healthy. Diseased plants or those attacked by insect pests may fail to produce fully developed crossed seeds. Since the success in wide crosses depends on the degree of seed setting, all efforts should be made to enhance seed setting by selecting better buds (early buds usually set better) for emasculation and pollination and removing all unwanted buds. The special techniques which can facilitate distant hybridization are as under :

(1) Reciprocal crosses : In some cases, reciprocal crosses may be attempted to overcome barriers to distant hybridization. When a diploid species is used as female and polyploid species as male, the thick pollen tubes of the polyploid usually do not grow properly in the stylar tissues of the diploid. The use of parent with higher ploidy level as female will overcome this difficulty. The use of gamagrass as female parent in *Tripsacum* × *Zea mays* crosses overcomes the barrier of too-long styles of maize.

(2) Use of better nicking varieties : The crosses between *Crepis tectorum* and other species of *Crepis* are successful and produce normal hybrid plants if *tectorum* varieties carrying non-lethal gene L are used in these crosses. On the other hand, if *tectorum* variety carries lethal gene l, the hybrids die at the early stage of their development. A similar situation exists in the varieties of *Ae. umbellulatum*. If the *umbellulatum* variety used in intergeneric crosses with diploid wheat, carries non-lethal gene l, the progeny is normal. The degree of necrosis in some interspecific wheat hybrids depends on the varieties used in the crossing programme.

(3) Use of nutrient culture : If embryo fails to develop because of endosperm, the embryo can be removed from young seed before the endosperm aborts and grown in nutrient culture. Embryo culture technique has proved very helpful in a number of cases where crosses were not successful. Crosses between *Secale cereale* and *Hordeum jubatum* and different species of *Linum* (*L. perenne* × *L. austriacum*) have been made successful by using this technique.

(4) Grafting hybrid seedlings : The F$_1$ hybrid seedlings of *Melilotus alba* and *M. dentata* are albinistic defectives and eventually die. If these hybrid seedlings are grafted on the healthy seedlings of one of the parental species, they reach normal maturity and produce viable seeds.

(5) Shortening the style artificially : If the styles of the seed parent are too long as compared to those of the species used as male parent, the crosses may be made effective by cutting off a portion of the styles. Pollen germination and tube growth may be enhanced by applying the stigmatic paste of the female parent or agar with sugar and gelatin to the cut styles. This method has been found successful in producing interspecific hybrids in potato (Swaminathan, 1955) and hybrids of maize with teosinte, *Tripsacum* and *Sorghum*.

(6) Early pollination : Early pollination increases the time of pollen tube growth. In some cross incompatible species, bud pollination may help in making them cross compatible. This technique is also used to overcome self-incompatibility as in case of some *Brassica* species.

(7) Use of growth regulators : Application of NAA and GA in low concentration to the pistil of the seed parent facilitates pollen tube growth in some interspecific crosses. GA treatment has been found to increase pollen tube growth in case of autotetraploid barley ($\overset{+}{Q}$) × rye crosses.

(8) Mixed pollinations : In cotton, pollinating the seed parent with a few of its own pollen grains and then with an excess amount of pollen from the species used as male, helps in increasing boll retention and obtaining both the crossed (small in size and shrivelled) and the selfed seeds.

(9) Stigmatic paste from male parent : For providing a favourable environment for the pollen to germinate on the stigma of other species, a paste from the stigma of the male parent is prepared and applied to the stigma of the seed parent.

(10) Use of self anther paste : Application of paste prepared from the mature anthers of the seed parent to the stigma before pollination, provides a stimulus to the stigma and facilitates germination of foreign pollen.

(11) Smearing stigmatic secretions : The stigmas of the seed parent can be smeared with the stigmatic secretion of fresh flowers from male parent to faciliate pollen germination and tube growth.

(12) Test tube pollination : To overcome all barriers regarding failure of male gametes to reach the ovules of the female parent (no pollen germination, slow pollen tube growth, bursting of pollen tubes in the style, shorter pollen tubes, etc.), a suspension of pollen of the male parent can be prepared and injected into the ovary of the female parent. This technique, however, cannot be conveniently applied to the species with small flowers.

(13) Vegetative rapproachment : In this method, grafted plants (with stock and scion of different species) which set freely, are used in successive generations for wide crosses for overcoming interspecific cross incompatibility. The method has been used in jute for producing hybrids between cross incompatible species (*Coschorus capsularis* and *C. olitorius*).

(14) Chromosome doubling : In many cases, distant hybridization fails because of the differences in the ploidy levels of the two parent species. This barrier, in some cases, can be overcome by doubling the chromosome number of the species having lower ploidy level. After doubling the chromosome number of *Lotus tenuis* (2n=12), this species has been successfully crossed with *L. corniculatus* (2n=24).

(15) Use of bridging species : There are a large number of cases where two species are not directly crossable. However, some distant hybridizations have been made effective by using a third species or an amphidiploid as a bridge between the two uncrossable species. An important example of this type may be cited in *Nicotiana*. The F_1 hybrid between *N. glutinosa* (n=12, resistant to tobacco mosaic) and *N. tabacum* (n=24, susceptible to mosaic) is sterile. But the amphidiploid (*N. digluta*, n=36) produced by crossing these two species is easily crossable to *N. tabacum.* In this way, genes for tobacco mosaic resistance have been transferred from *N. glutinosa* to *N. tabacum.* In cotton, the species *G. thurberi* which has genes for strong fibre, is not directly crossable to *G. hirsutum* which is deficient in this character. For transferring this character to *hirsutum*, an amphidiploid of *G. thurberi* with *G. arboreum* was produced and was then crossed with *G. hirsutum*. Similarly, for transferring leaf rust resistance from *Ae. umbellulatum* to *T. aestivum*, amphidiploid of *Ae. umbellulatum* with *T. turgidum* was used as a bridge.

(16) Attempting crosses in different environments : The success of distant hybridization may also be influenced by environment in some cases. It is thus advisable that crosses may be attempted under different environmental conditions as far as possible.

The use of above mentioned techniques has substantially lowered the existing barriers to distant hybridization.

Hybrid Sterility

Genetic disharmonies between the genomes of the parental species are the basic reason for hybrid inviability as well as for hybrid sterility. The difference between the two metabolic disturbances (hybrid inviability and hybrid sterility) relates to the stage of life cycle at which the genetic unbalance affects the hybrid. If the genetic disharmony expresses itself before the occurrence of gametogenesis in the F_1 hybrid, it will result in hybrid inviability. On the other hand, the expression of disharmony after the occurrence of gametogenesis in the hybrid will result in the hybrid sterility.

The degree of hybrid sterility may vary depending upon the parental species used in the hybridization. In some cases of interspecific hybridizations, the feritility of the F_1 hybrid is almost normal, e.g., crosses between *Lycopersicon esculentum* and *L. pimpinellifolium*. In some cases, distant hybridization leads to partial hybrid sterility, e.g., *L. esculentum* × *L. perunianum* and *L. esculentum* × *L. chilense*. On the other extreme, F_1 hybrid produced by crossing sugarcane with maize at Coimbatore is fully sterile and being maintained through vegetative propagation.

On a broad basis, sterility may be of three types : (1) chromosomal, (2) genic, and (3) cytoplasmic.

Chromosomal sterility : Differences in the number and structure of chromosomes in the parents which lead to a number of meiotic abnormalities, are responsible for this kind of sterility. The abnormalities may be in the form of reduced pairing (or no pairing in extreme cases), loops at pachytene because of deficiencies and duplications, rings and chains at first metaphase because of translocations and bridge and fragments at first anaphase because of inversions. All these abnormalities usually lead to abnormal distribution of chromosomes in the daughter nuclei and thus production of all or many non-functional gametes. The daughter nuclei produced from the hybrids having structural chromosomal differences, in general, have less than full chromosome complement because of disorganized disjunction of chromosomes. However, if hybrid sterility is due solely to the chromosomal sterility, the hybrid can be made fertile by doubling its chromosome number and producing an amphidiploid.

In some cases, the structural differences in the chromosomes of two species are so small that they are not able to affect chromosome pairing materially and thus are not detectable. The difficulty in the detection of such too small but important structural differences is more pronounced at metaphase I (the stage at which the chromosome pairing has been analysed in most of the distant hybrids) when the chromosomes show high degree of contraction and have only a few association points. Stebbins (1950, 1958) has called such minute structural differences as 'cryptic structural changes' and the phenomenon of sterility due to these structural changes as 'cryptic structural hybridity'. A peculiar behaviour has been shown by some cotton interspecific hybrids. The F_1 of *G. hirsutum* and *G. barbadense* is fully fertile, but there is degeneration in F_2 since non-parental F_2 segregants are weak and sterile and most of the F_2 plants resemble to one or the other parent, that is, F_2 population reverts back to the two parental species. This hybrid breakdown, according to Stephens (1950), is attributable to cryptic structural hybridity.

Genic sterility : The doubling of chromosomes may not always make a hybrid fertile, that is, in spite of normal pairing and normal disjunction of chromosomes the hybrid may be sterile. There could be some genes (and not the chromosomes) responsible for hybrid sterility. Such genes have now been identified in some plant species. The sterility caused by genes is termed as genic sterility. F_1 hybrid of *Setaria italica* (foxtail millet, 2n=18) with *S. viridis* (wild, 2n=18) show normal pairing of chromosomes. But only 30% pollen and 50% ovules are fertile. This sterility seems to be due to the presence of sterility genes.

However, in some cases, it becomes difficult to make clearcut distinction between genic sterility and cryptic strcutural hybridity particularly when the structural differences in the chromosomes of the two parents are very minute. Behaviour of F_1 subspecific hybrids of rice may be cited as an example for this. In rice, the hybrids produced by crossing two subspecies *indica* and *japonica* though show normal chromosome pairing, the develpment of both the male and female gametophytes is inhibited by complementary recessive alleles responsible for sterility (Oka, 1955). But, Mashima and Uchiyamada (1955) did not agree with the explanation given by Oka. They said that structural differentiation has already taken place between chromosomes of *indica* and *japonica* since autotetraphoids derived from a single subspecies have higher number of quadivalents and smaller number of bivalents than induced subspecific tetraploids and thus cryptic structural differences rather than sterility genes are responsible for sterility in these intersubspecific hybrids. The main difference between crytic structural hybridity and genic sterility that doubling of chromosomes in the former case will overcome sterility, vanishes because amphidiploid may become sterile because of multivalent formation like in genic sterility. The only possible generalization in such cases is that probably genotypic disharmonies and minute structural changes in the chromosomes, singly or jointly, are responsible for hybrid sterility.

Cytoplasmic sterility : Reciprocal differences with regard to sterility in interspecific hybrids indicate that cytoplasm may sometimes be the cause of hybrid sterility. Such a situation has been reported in *Oenothera*. This kind of hybrid sterility, however, can be overcome by attempting reciprocal crosses.

Segregation in Distant Hybrids

Following generalizations can be made with regard to the segregation in distant hybrids :

(1) Since the species used as parents in distant hybridization are usually highly genetically diverse and produce extremely heterozygous F_1 hybrids, a tremendous diversity of types appear in the F_2 and subsequent generations of such crosses. Many of the segregants may be entirely different from the parental combinations and their occurrence cannot be explained on the basis of morphological or physiological characteristics of the parental species.

(2) The meiosis in distant hybrids is mostly irregular due to reduced pairing (or no pairing at all) followed by abnormal segregation of chromosomes at first anaphase. Even though the pairing is almost regular, recombination may not occur normally. Additionally, many of the recombinant gametes and zygotes may not survive because of their less adaptive value. In both the situations, classical Mendelian ratios for different characters are not obtained.

(3) The two species used in the wide crosses are usually those which have established themselves according to their respective environments after facing consequences of natural selection for immemorial time. When attempts are made to disturb such well developed (balanced) genetic systems, usually a great resistance is displayed by specific gene complexes and, as a result, only those gametes and zygotes which have gene combinations similar to parental species, show normal develoment. Rest of the segregants are usually weak, less adapted and sterile. Due to such elimination at gametic and zygotic levels, in addition to linkage and pleiotropy, only a part of the total possible recombinants is recovered. Therefore, segregating generations of distant hybrids represent a limited range of variability rather than the total possible variability.

(4) Chromosomal and genic disharmonies do not influence the female gametogenesis and male gametogenesis equally. The latter is affected more by these disharmonies than the former. Therefore, whenever the distant hybrids are propagated by backcrossing, the F_1 is used as female and one of the parents as male. It has also been observed in some cases (e.g., hybrid of *T. vulgare* × *T. timopheevii*) that when the F_1 is pollen fertile and is propagated through sefling, the progeny is not like a true F_2 but resembles more to a backcross progeny. Two reasons may be attributed to this similarity of selfed progeny to backcross progeny : (a) The male gametes possessing genes from both parents have very low transmission (rather fail to survive) and only those gametes which have gene combinations similar to one or the other parent, will be normally transmitted; (b) Since the gametes having gene combinations similar to one or the other parent have a balanced genetic system (like parents), they are favoured more by natural selection. Both the situations will lead to the production of progeny like backcross types.

Distant Hybridization and Quantitative Variation

It would be worthwhile to know that whether distant hybridization can be used for broadening quantitative variation. Some scientists may have very high expectations in this regard. However, the value of wide crosses in broadening quantitative variation must not be exaggerated. This aspect of distant hybridization has been discussed in detail by Schnell (1978). He defined wide crosses as 'crosses between genetically very divergent forms within a species or even beyond the limits of species' and discussed problems faced in the use of quantitative variation. He made it clear that the consequences of an increase in the genetic divergence between parents would depend on the method of breeding the varieties, that is, whether one has to evolve a variety through hybrid breeding (as in case of cross-fertilized crops) or through pure line breeding (as in case of self-fertilized crops). Based on the results obtained in maize, it can be safely concluded that the magnitude of heterosis in maize increases with an increase in the genetic diversity of parents to a particular degree of the latter and after that degree it starts declining. The stage at which this decline in heterosis starts is certainly reached before the sterility barriers between parents appear. In fact, the variance of the resulting population will depend mainly on the genetic variance within the initial populations and very little on the genetic diversity between them.

The use of distant hybridization for creating useful quantitative variation may not be free of problems. In distant crosses, usually, one parent is a well adapted variety, while the other parent may be some wild species possessing a number of undesirable characters with a few desirable ones and thus giving rise to a number of undesirable recombinations.

Use of distant hybridization in self-fertilized species will ultimately result in the development of homozygous lines showing increased heritable variance. The mean of these lines, however, is expected to be around the mean of the performances of the two parents. Further, selection opportunities are more if one parent has a very high value and the other parent a low value of the character under consideration. Under such a situation, the high value of the character in better parent will be compensated by the low value of the poor parent. Thus, presence of large heritable variance alone is also not of much help. In distant hybridization also, as in other methods, the genetic progress in a metric trait will depend on variance as well as on mean.

However, some fascinating biotechnological techniques are now being used to provide new opportunities to the plant breeder by considerably reducing the time period of his crop improvement programmes. These techniques have provided unconventional and rapid means of transferring genes across sexual barriers and thus widened the scope of hybridization between unrelated species. Such a creation of new crop plants cannot be achieved by conventional breeding procedures (Chapter 25).

SECTION V

Recent Trends in Plant Breeding

Biotechnology and Plant Breeding

The biotechnology subject is not so new as it is often supposed. Knowingly or unknowingly, man has been making use of biotechnological processes for his day to day's needs since times immemorial. For example, the process of fermentation using naturally occurring microorganisms has a prehistoric origin. Some important biotechnological developments took place in early years of the second decade of the twentieth century (conversion of starch into butanol and acetone), 1940s (production of penicillin at commercial scale) and in early 1960s (microbes aided mining of uranium). However, a rapid progress in biotechnology research started only after 1970. Enormous work done on this subject in 1970s and 1980s clearly proved that some seemingly impossible things can be made possible through the use of biotechnology. For example, transfer of genes from one species to other unrelated species or from microorganisms to plants or from animal kingdom to plant kingdom and *vice versa* is now possible through genetic engineering, a component of bitechnology. Such transfer of genes across genetic barriers is not possible by conventional plant breeding methods. But, even today, some pessimistic plant breeders consider biotechnology as a white elephant. However, whether this subject is a white elephant or a golden calf (according to some over-optimistic biotechnologists having exaggerated claims) will be decided in due course of time.

Biotechnology has been variously defined. However, the definition, **the controlled use of biological agents, such as, microorganisms or cellular components for beneficial use**, given by U.S. National Science Foundation, is perhaps the most acceptable definition of biotechnology. In fact, biotechnology is a hybrid subject arising from a complex interaction between biology and technology. This interaction is based on a two-way-effect relationship where both the components of biotechnology (biology and technology) influence each other.

Biotechnology has a very wide range of its **usage** :

(1) A sufficiently large plant population can be grown in a small space through tissue culture technique. This allows rapid detection of mutant/resistant/tolerant plants.

(2) Embryos obtained after interspecific or intergeneric hybridizations can be rescued by embryo culture technique. This technique, therefore, helps in circumventing barriers of distant hybridization.

(3) Haploid plants can be produced through anther culture and ovule culture techniques. Production of haploids facilitates detection and isolation of dominant and recessive mutants with equal efficiency and helps in producing homozygous lines in a single step. Production of haploid plants has several other advantages (Chapter 23).

(4) Meristem culture helps in rapid multiplication of clones and early identification of promising genotypes of fruit and forest trees.

(5) Biotechnology helps in isolation of promising and stable somaclonal variants.

(6) Genetic engineering facilitates transfer of genes from any species to any other species, that is, gene transfers between different species, between different genera or between different kingdoms and thus helps in developing transgenic or genetically modified plants.

(7) Biotechnology has created a new commercial environment due to its important role in basic as well as applied research in molecular biology and microbiology.

(8) Recombinant DNA technology facilitates cloning of random DNA or specific genes.

(9) Molecular markers like RFLPs (restriction fragment length polymorphisms) and RAPDs (random amplified polymorphic DNAs), have been used for preparing chromosome maps in several organisms. RFLPs also have an evolutionary significance as these polymorphisms can be studied in related species to know phylogenetic relationship between different species. Furthermore, RFLPs can facilitate strain identification for PVRP (Plant Variety Right Protection), marking of parental genes in new lines, and mapping of genes affecting quantitative characters.

(10) Biotechnology can be used to check environmental pollution by controlling plant diseases and insect pests biologically (using viruses, bacteria, etc.), treating sewage efficiently, degrading petroleum, etc. In this way, biotechnology can protect the environment from insecticides, pesticides, etc.

(11) Biotechnology is playing a significant role in medical science as it is being used to synthesize vaccines of different types, to identify parents/criminals through DNA finger-printing and to produce monoclonal antibodies, DNA probes for disease diagnosis, valuable drugs (human insulin, human interferon, etc.), etc.

(12) *In vitro* fertilization facilitates production of test tube babies in humans and induction of superovulation and embryo splitting in farm animals.

(13) Like plants, transgenic animals with desired characteristics can be created.

(14) Biotechnology has several industrial uses. It helps in producing important enzymes (α-amylase, proteases, etc.), antibiotics (penicillin, streptomycin, etc.), immunotoxins and several useful chemical compounds (glycerine, lactic acid, ethanol, etc.). It also helps in transforming less useful chemical compounds into more useful ones.

(15) *In vitro* conservation of genetic stocks or *cryopreservation* (preservation of cells/tissues/small organs in liquid nitrogen at very low temperature) has several advantages (facilitates preservation for long period of time, requires very less space, minimum risk in labelling, preserved material is ideal for germplasm exchange, material remains free from diseases, insects, etc., and the preserved material is the best source of nucleus seed) over ordinary method of conserving genetic stocks.

In view of the role played by biotechnology in different areas like plant sciences (particularly in plant breeding for expanding and sustaining crop production and productivity), animal sciences, medical sciences, industry and environmental protection, this subject has been divided into **plant biotechnology, animal biotechnology, medical biotechnology, industrial biotechnology** and **environmental biotechnology**. As a result of its multidirectional contribution, biotechnology is providing new opportunities for employment and trade. Looking at the scope of the present book, our discussion onward will be mainly restricted to plant biotechnology.

Plant Biotechnology

The traditional plant breeding methods (based on the principles of Mendelian genetics) supplemented by the knowledge increased in other disciplines/phenomena (biometrical genetics, plant pathology, entomology, mutagenesis, polyploidy, male sterility, incompatibility, heterosis, etc.) have played the main role in the evolution of high yielding, fertilizer responsive, disease/insect resistant, and photoinsensitive varieties of crop plants. As a result, there has been a phenomenal increase in the production and productivity of many crops. But this whole crop improvement was restricted to the limits of sexual reproduction. However, some biotechnology techniques have no sexual barriers and thus can facilitate gene transfers beyond the level of sexual reproduction. Plant biotechnology, therefore, can serve as a useful supplement to the conventional plant breeding methodology.

The two components of plant biotechnology, **plant tissue culture** and **plant genetic engineering**, will be discussed separately.

Plant Tissue Culture (*in vitro* Techniques)

Several techniques like embryo culture, meristem culture, anther culture, and somatic cell hybridization, come under plant tissue culture. These techniques are used for regeneration of functional plants from cells, tissues or plant organs. Since cells, tissues, embryos, etc. are generally grown in test tubes on artificial nutrient media, these techniques of tissue culture are also called as *in vitro* techniques. Each of these techniques is used for a specific purpose. However, the success of these techniques would vary from species to species and from genotype to genotype for the same species. Manytimes, specific materials require specific procedures to be followed for regeneration of functional plants. Therefore, the knowledge of the experimenter about the nutritional requirements of different genotypes/species would greatly affect the success of the programme.

Perhaps the idea of plant tissue culture technique originated from Haberlandt's (1902) prediction that each plant cell, like zygote, is capable of developing into a complete plant. This ability of plant cells to develop into functional plants is known as **totipotency**. Some useful work in this direction was done in France and U.S.A. in the third and fourth decades of twentieth century. Braun (1959) succeeded in regenerating first plant from a mature plant cell. However, the credit for the artificial production of first haploid plant goes to the Indian scientists Guha and Maheshwari (1964). They produced haploid plants of **Datura** through anther culture technique. Anther culture is now being used as a routine procedure for haploid production in many countries of the world and a number of doubled haploids are being grown as crop cultivars (Chapter 23).

In general, multicellular tissue fragments obtained from any part of living plant are used to initiate *in vitro* culture or tissue culture. These tissue fragments are called **explants**. Depending upon the objective of the experimenter, the explants may be from any plant part (embryo, hypocotyl, cotyledon, meristem, stem, etc.).

The explants are generally surface sterilized with a suitable disinfectant (1% calcium hypochlorite, 5% solution of commercially available bleach, etc.) to make them free from fungi and bacteria. Then the disinfectant is removed by reinsing the explants 2-3 times with sterilized distilled water.

After surface sterilization, explants are cultured on a liquid or solified by agar medium. Several nutrient media are available to be used for different purposes/genotypes. The explants, if cultured on an appropriate medium, start dividing after some time and within a few weeks give rise to a **callus**, an undifferentiated mass of cells, or directly to plantlets. Salts, sugar and vitamins are the common contents of a nutrient medium. In addition to these common contents, plant hormones (e.g. auxins and cytokinins) may be a part of the medium for controlling cell division and cell growth. The auxin : cytokinin ratio generally determines the initiation of root or shoot primordia. If this ratio is high, there will be root initiation before the emergence of shoot buds, whereas a low ratio would favour initiation of shoot buds and suppression of root initiation. However, if there is an intermediate ratio, the undifferentiated callus would continue to grow without initiation of shoot and root primordia. Several refinements have been suggested in the preparation of culture media for successful and wider applicability of tissue culture.

The formation of plantlets (the ultimate objective of the breeder) from explant may take place through different routes. One more common route is that the explant first gives rise to callus and then callus gives rise to embryos and ultimately embryos develop into plantlets. Another route may be that embryos directly arise from explant and then plantlets are formed from these embryos. The formation of plantlet may also be through adventitious shoots or axillary buds. The regeneration of various plant organs (shoot, root, etc.) is called **organogenesis**. Regeneration of plantlets from explants has been successfully accomplished in many plant species (barley, wheat, rice, maize, tobacco, potato, sugarcane, etc.)

Nevertheless, the experimenter should be careful about the mortality of plantlets. At early stage, the plantlets may be in a **heterotrophic** state (unable to synthesize their own organic food) and may show a high degree of mortality. The best solution of this problem of mortality is that the plantlets should come to an **autotrophic** state so that they are able to synthesize their own food material. Another problem with the growing plantlets is that, due to inadequate development of their root system, they may have high water loss and thus should be protected from desiccation. However, right from the stage of the excision of explants to the formation of self-dependent plantlets, the experimental material (cultured cells, organs and growing plantlets) should be protected from fungal and bacterial infection. For this, the culturing of tissues and organs should be done under an **aseptic** environment. After the plantlets have achieved sufficient hardening, they are transferred to greenhouse for a few weeks and ultimately to the field.

Embryo Culture and Ovule Culture

In vitro culturing of immature embryo on an appropriate nutrient medium for obtaining seedling, is termed as embryo culture. The first step of this process is the aseptic excision of an immature embryo from a developing seed. The excised embryo is then cultured on a suitable nutrient medium for obtaining a seedling. To provide a more favourable chemical and physical environment for growth and development of developing hybrid embryos, it is advisable to culture the ovules (ovule culture) instead of culturing embryos (embryo culture). Aseptic excision of ovules at the initial stage of the developing embryos inside them, is done and then the ovules are culture on an appropriate nutrient medium. Embryo culture (or ovule culture) has several uses :

(1) Embryo culture is very commonly used for **circumventing post-zygotic barriers to distant hybridization**. In distant hybridization, genomes come from two different species or two different genera and may have varying degrees of incompatibility between them. The zygotes of distant hybrids mostly fail to develop into mature embryos/seeds due to an inadequate supply of metabolites to the developing embryo by the endosperm. Hybrid embryos may collapse or abort at any stage of their development. However, in such a situation, a wide cross may be made successful by protecting embryo through embryo culture technique. The embryo culture, therefore, is a kind of **embryo rescue technique**. The stage at which the embryo would abort or collapse is generally determined by the degree of incompatibility between the embryo and the endosperm. It is easier to culture mature or nearly mature embryos since they are largely autotrophic and thus need simple nutrient medium. On the contrary, embryos at a few-celled stage of development (i.e., proembryos) are largely heterotrophic and thus need complex nutrient medium. However, due to a remarkable advancement in the knowledge about the nutrient media, it is now possible to successfully culture quite young embryos. The embryo culture technique is quite old and is being widely used for the recovery of distant hybrids.

(2) In some cases, the seeds of wide crosses are dormant and take relatively long time to germinate. However, the hybrid embryos mostly do not have the problem of dormancy. In such cases, embryo culture can be used **to circumvent the seed dormancy barriers.** Also, in some plant species, seed dormancy is an acute problem and the period of dormancy may vary from several weeks to several years. Seed dormancy is still more problematic in case of some fruit trees where generation period is quite long. The use of embryo culture in such cases can be helpful in considerably **reducing the generation period.**

(3) Embryo culture technique can facilate increased production of haploid plants. For example, in genome elimination technique of haploid production, genome(s) of only one species remain, whereas the chromosomes of the other species are progressively eliminated (e.g., *Hordeum vulgare* × *H. bulbosum*, *H. vulgare* × *Triticum aestivum* crosses, etc.). After the elimination of chromosomes of one species, embryos may be cultured in an appropriate medium to produce haploid plantlets.

(4) Almost complete absence of endosperm in the seed, that is, seed without stored food, causes incomplete development of embryo. The occurrence of such stored food deficiency is a common characteristic of orchid seeds where the embryos are almost naked. In such cases, embryos may be culutred on an appropriate medium and normal plants can be produced. Thus, embryo culture technique may play an important role in the **propagation of orchids**.

(5) Ovule culture facilates *in vitro* fertilization in cases where the growth and penetration of pollen tubes are not normal.

Anther Culture or Pollen Culture

As the name of this technique indicates, the explants used in anther culture are anthers or pollen. The procedure of anther culture and its uses are described in Chapter 23.

Meristem Culture

Meristem culture is the **culturing of apical meristems** (preferably of shoot apical meristems) for rapid multiplication of already existing shoot apical meristem and regeneration of roots. It has been observed that plant genotypes produced through meristem culture are genetically more homogeneous (i.e., have less number of genetic variants) than those produced through the culturing of more mature tissues. The size of the explant in meristem culture largely depends on the objective of the experimenter. If the objective is the rapid vegetative propagation of a heterozygous genotype, the explant should contain one or few leaf primordia in addition to the meristematic portion. Such explants have three advantages over the smaller explants containing only the meristematic region : (1) Their handling is easier, (2) their excision is less time consuming, and (3) their survival rate is higher than that of smaller explants. However, if the objective is to propagate disease-free genetic stocks, the explant should contain meristematic region with the least part of its adjacent tissue. To obtain more than one potential plantlet from one explant, each shoot tip may be divided into several small pieces. Also, sub-culturing of axillary buds produced on the shoot tip explants, helps in increasing the number of plantlets. The explants are cultured on a suitable nutrient medium to produce plantlets. The plantlets are then cultured on a rooting medium and are ultimately shifted to soil.

Meristem culture may be utilized in several ways :

(1) Meristem culture is perhaps the best *in vitro* technique **to produce genetically homogeneous genotypes** because of the productin of fewer genetic variants in this technique. Therefore, the technique helps in maintaining the genetic identity of the plants propagated.

(2) Meristem culture may be used for **cloning** or **clonal propagation** or **micropropagation** of vegetatively propagated heterozygous genotypes. The main objective of *in vitro* clonal propagation is to obtain large number of vegetative progeny through a quick regeneration of specific genotypes. Clonal propagation has several uses like quick regeneration of self-incompatible genotype, rapid multiplication of male sterile line to be used in a hybrid breeding programme, increasing breeding materials at an earlier stage of the breeding programme in crop plants with poor seed setting, regeneration of a few plants of a distant hybrid, rapid multiplication of a few propagules of a germplasm line, regeneration of few disease-free plants and increasing number of genotypes for germplasm exchange.

Since thousands of plantlets can be produced from a single genotype through micropropagation, the technique has widespread potential in ornamentals, horticultural crops and in crop species like potato and sugarcane. However, in crop species where seed can be produced in abundance without any difficulty or in crop species grown over widespread area, meristem culture is of no practical utility.

(3) Meristem culture has an important role in **propagating disease-free genetic stocks**. The vegetative parts of the plant have highest density of virus and other plant pathogens, whereas the seeds and newly developing meristematic tissues are completely or almost completely free from these pathogens. Therefore, in the seed-propagated plant species, these pathogens have no chance or least chance of their transmission to the next plant generation. But, in vegetatively propagated plant species, the pathogens have every chance of their transmission to the next plant generation through vegetative propagules. However, regeneration of plantlets in large numbers through meristem culture can facilitate elimination of pathogens from propagating stocks. Of course, the degree of recovery of disease-free stock would depend on the size of explant excised. The smaller explants containing meristematic region and least portion of its adjacent tissue, are more useful for this purpose. Meristem culture may be utilized for protecting promising clones from plant pathogens and for producing disease-free breeding lines and foundation seed stocks in vegetatively propagated crop species. Such a use of this technique has been made in some crop species (sugarcane, potato, etc.) for producing virus free breeding lines and foundation seed stocks.

Of course, there are other methods which may be used for eliminating plant pathogens from propagating stocks of vegetatively propagated plant materials. Any one of the three kinds of treatment, cold treatment (**freeze therapy** or **cryotherapy**), heat treatment (**thermotherapy**) or chemical treatment (**chemotherapy**), can be used for this purpose. Cold treatment has an edge over heat treatment since heat treatment adversely affects the survival of plants and is unable to eliminate all types of pathogens from plant materials. Also, the defence mechanisms of surviving plants are weakened due to heat treatment and these plants become susceptible to other pathogens. In case, the shoot apical meristem is itself infected with some pathogens, a combination of meristem culture and heat treatment may be more effective than the use of any one of these methods.

(4) Meristem culture can be effectively used for **cryopreservation** or **freeze preservation** of germplasm. This technique is relatively more effective where there is a rapid decline in the viability of plant propagules (tubers, etc.). Also, cryopreservation requires considerably less space than the ordinary preservation of germplasm.

(5) Meristem culture is becoming increasingly important for its use in germplasm exchange and distribution of promising clones among farmers. It is considerably easy to exchange seedlings in small containers like test tubes than exchanging bulky vegetative propagules.

Somaclonal Variation

Transient (epigenetic) or genetic variation observed among plants regenerated through cell culture is known as somaclonal variation. It is only the latter (genetic) which is heritable and is thus important for the plant breeder. Therefore, the practical utility of somaclonal variation would depend on the proportion of the useful genetic variation in total somaclonal variation. In fact, in the early stage of tissue culture technology it was presumed by a large number of scientists working in this field that the plants regenerated from this technique would be exact duplicate copies of the parent plant used for culture. This was perhaps the main reason that somaclonal variation was ignored for long time. However, as soon as plant regeneration became a routine procedure, many scientists noticed the presence of variation among regenerated plants from the same parent plant. Such plants were then more critically examined and ultimately the significance of somaclonal variation in plant breeding programmes was realized. Since this variation has a **somatic origin**, it was called somaclonal variation.

The exact origin of somaclonal variation is not yet fully known. However, the probable causes of its origin are :

(1) Presence of genetic variation in parent plant itself, that is, pre-existing genetic variation in donor plant,

(2) gene or point mutations of both dominant and recessive types,

(3) structural aberrations of chromosomes (deficiencies, duplications, inversions and translocations),

(4) numerical changes in chromosomes (aneuploidy and polyploidy).

(5) symmetric and asymmetric crossing overs during mitosis,

(6) gene amplification (increased number of copies of a gene per genome),

(7) activation of transposable elements, and

(8) production of intracellular mutagens during *in vitro* growth.

It means all kinds of mutations can occur in somatic cell cultures. Further, two or more factors causing somaclonal variation can operate simultaneously.

Although the application of mutagenic treatment increases the frequency of somaclonal variants, use of mutagens for the recovery of somaclonal variants is generally avoided because of two reasons: (1) Mutagenic treatments manytimes have undesirable side-effects, and (2) mutagenic agents are generally costly. A comparison of somaclonal variations with induced mutations indicates that the former type of variation has superiority over the latter type in several ways :

(1) The frequency of beneficial mutations is relatively high in somaclonal variations than their frequency in induced mutagenesis.

(2) A particular deficiency (disease susceptibility, etc.) of the genotype can be corrected more efficiently through somaclonal variations than through the induction of mutations.

(3) Since it is possible to grow a huge number of cells in a limited space in cell and tissue culture experiments, screening for resistance to abiotic stress conditions (tolerance to salt, low temperature, resistance to herbicides, etc.) and plant regeneration from resistant cells are easier at cellular level than at plant level.

(4) Unlike induced mutations, somaclonal variants are not generally associated with undesirable features.

(5) Somaclonal lines showed considerable superiority over mutagen-treated populations for several quantitative characters in some studies carried out to compare the effectiveness and usefulness of somaclonal variants and induced mutations. Also, somaclonal lines showed improvement simultaneously for several characters.

(6) Unlike induced mutations, somaclonal variants are free from chimera formation.

(7) It has been possible to recover new genes through somaclonal mutants.

(8) Somaclonal variation can facilitate reduction in time required for the release of a new variety.

However, cell culture technique has some drawbacks too :

(1) Selection for complexly inherited characters like yield is not effective at cellular level.

(2) *In vitro* studies require costly equipments and laboratory facilities.

(3) A part of the somaclonal variation is epigenetic and thus unstable. Only genetic part of somaclonal variation is useful for the plant breeder.

Similar to the effect of mutagens (both physical and chemical), somaclonal mutations cannot be used to alter specific characters. However, both the qualitative and quantitative characters may be affected by somaclonal variation. Somaclonal variation has been utilized to correct some specific defects in sugarcane and potato. Some high yielding and disease resistant somaclonal varieties of sugarcane are being grown at comercial scale in India and Hawaii. However, the advantage of somaclonal variation may be taken more effectively if the cell culture technique is used as a supplement to conventional plant breeding methodology.

Somatic Cell Hybridization (Protoplast Fusion)

A cell-wall-less cell is called as protoplast. The fusion of protoplasts of two different species is called somatic cell hybridization or somatic cell fusion. The hybrids produced through protoplast fusion are called **somatic hybrids**.The cell walls are removed with the help of a mixture of some cell wall degrading enzymes and then the fusion of protoplasts is effected by some appropriate fusigenic agent called **fusogen**. Although the success in hybridizing somatic cells was achieved in animals earlier than in plants, but because of the possibility of regeneration of hybrid cells into complete functional plants, protoplast fusion is becoming increasingly important in plant breeding. The main objective behind somatic hybridization is to combine characteristics of two sexually incompatible species, that is, to circumvent sexual barriers to distant hybridization. The transfer of characteristic (s) from one species to another may involve,

    (i) Whole genome or whole nucleus,

    (ii) one or few chromosomes,

    (iii) one or few nuclear genes, or

    (iv) cytoplasmic organelles like mitochondria and chloroplasts.

However, the plant breeder would like to make a meaningful use of somatic hybrids and, therefore, would like to transfer a few nuclear genes or one or few mitochondia or chloroplast borne characteristics (e.g., male sterility genes residing in mitochondria or triazine resistance which is chloroplast borne) rather than transferring whole genome or whole chromosome(s) from the donor species/parent. As a result, attempts are being made to produce **cybrids** (cytoplasmic hybrids) and to obtain **variable combinations of nuclear and cytoplasmic genomes**. Because the nuclear division is independent of organalle replication, mitochondria and chloroplasts show a random assortment in the progeny hybrid cells. Further, since mitochondria and chloroplasts segregate independently, it is possible to produce lines with different chloroplast characteristics with the same type of mitochondria. For example, if there is a fusion of male sterile protoplast with a triazine (a herbicide) resistant protoplast, it is possible to recover both the triazine resistant and triazine sensitive male sterile lines with the help of markers. The protoplast fusion technique, therefore, has opened new avenues for obtaining variable combinations of nuclear and cytoplasmic genomes and also for recovering variable combinations of mitochondria and chloroplast borne characteristics.

Protoplast fusion may be **spontaneous** or **induced**. Sometimes, there is spontaneous fusion of some chloroplasts (lying in close proximity) at the time of their isolation for culture. Such fusion of chloroplasts of the same species gives rise to **homokaryons** or **homokaryocytes**. It appears that spontaneous fusion of chloroplasts largely depends on the age of leaf. It has been observed that more protoplast fusions take place during isolation of protoplasts from young leaves. Induced protoplast fusion is a multi-stage process and will be discussed in detail.

The process of induced protoplast fusion may be divided into following four stages : (1) Isolation of protoplasts from two parent species; (2) fusion of protoplasts; (3) identification and isolation of fused hybrid protoplasts; and (4) regeneration of fertile hybrid plants.

Isolation of protoplasts from two parent species : It is possible to isolate protoplasts from almost all plant parts. However, isolation of protoplasts from the mesophyll of young leaves has given best results particularly in dicotyledons. Mesophyll tissues are taken after surface sterilization and peeling off the lower epidermis of the leaf. These tissues are then treated with an appropriate mixture of cell wall-degrading enzymes in an incubator to produce naked protoplasts. The naked protoplasts are purified after washing with an appropriate washing medium.

Fusion of protoplasts : Induced fusion of purified chloroplasts of the parents to be hybridized requires a suitable fusogen. Several techniques like **treatment with calcium ions at high pH, Na NO$_3$ treatment, polyethyl glycol (PEG) treatment** and **electrofusion technique**, have been used for the fusion of plant protoplasts. In the first three techniques, the protoplasts of the two parents are mixed and centrifuged with the fusogen. All the three techniques are non-selective as they cause fusion of any two or more protoplasts and generally produce **homokaryons** (fusion between protoplasts of the same parent), **heterokaryons** (fusion between protoplasts of two different parents) and **mixture of protoplasts** of the two parents. However, the PEG treatment is preferred over the remaining two treatments because (1) its results can be highly reproduced, (2) there is low cytotoxicity, and (3) fusion between any two protoplasts (between two species, between the genera and even between plant and animal kingdoms) is possible by this technique. In electrofusion technique, first the purified protoplasts are lined up between two electrodes by applying a potential difference. Then, the fusion between protoplasts is induced by applying extremely short pulse of high voltage. The technique is relatively more selective, more desirable and less drastic.

Identification and isolation of fused hybrid protoplasts : This step of the process of somatic hybridization is critical and bit difficult because of the following reasons :

(1) The proportion of heterokaryons (hybrid protoplasts) in the total protoplast population (heterokaryons + homokaryons + unfused protoplasts of the two parents) is generally very small (0.5-10.0 per cent).

(2) Different plant materials generally require different techniques for the identification and isolation of hybrid protoplasts.

(3) Due to a number of reasons (formation of more than two-nucleate protoplast hybrids, formation of asymmetric hybrid protoplasts because of varying rates of DNA replication in two parents or because of variable rates of cell division in two parents and occurrence of somaclonal variation during cell culture), only a few hybrid protoplasts have the exact total number of chromosomes of the parents. For the identification of true somatic hybrids, cytological studies are required.

Several procedures like growing protoplasts on **genotype-specific medium** where only hybrid protoplasts and none of the parental protoplasts can grow (e.g., *Nicotiana*), identifying hybrid protoplasts on the basis of **green colour of callus** (e.g., *Petunia*), labelling chloroplasts of two parents by **different fluorescent agents** (e.g., *Nicotiana*), and culturing the whole protoplast population and then identifying the hybrid protoplasts on the basis of chromosome number, callus morphology and banding patterns, are being used for the identification and isolation of true somatic hybrids.

Regeneration of fertile hybrid plants : After the isolation of hybrid protoplasts, the next step is to regenerate functional plants from these hybrids so that they may be exploited in crop improvement. However, successful regeneration of functional plants from hybrid protoplasts is a difficult task due to **somatic incompatibility** (inability of somatic combinations to produce fertile plants). In fact, the formation of true **symmetric hybrids** (somatic hybrids possessing full somatic complements of both the parents) is very rare due to the loss of the entire genome or several to many chromosomes of one of the two parents. This loss of chromosomes causes formation of **asymmetric hybrids** (somatic hybrids containing chromosome complement of one parent and some or no chromosomes of other parent).

There are some other difficulties in regeneration of fertile plants. For example, even if a symmetric hybrid is formed, there may be loss of chromosomes of one parent during subsequent mitotic divisions. It has been observed that even symmetric hybrids produced from distantly related sexually incompatible parents show a very high degree of sterility.

As mentioned earlier, the division of nucleus is independent of the replication of mt-DNA (mitochondria DNA) and cp-DNA (chloroplast DNA). Also, the replication of mt-DNA is independent of the replication of cp-DNA. Since the somatic hybrids contain cytoplasms of both the parents in addition to nuclear genome(s), it is quite likely that there is presence of recombinant organelles (chloroplast and mitochondria) in the progeny of these hybrids.

Production of Cybrids or Cytoplasmic Hybrids

A somatic hybrid containing nuclear genome of one of the two parents and cytoplasmic genomes of both the parents is known as a **cybrid** or **cytoplasmic hybrid.** There are four probable routes of cybrid production :

(1) When normal protoplast of one parent fuses with a nucleus-deficient (i.e., enucleate) protoplast of other parent, that is, fusion of normal protoplast with **cytoplast**,

(2) when normal protoplast of one parent fuses with the protoplast of other parent containing non-functional or non-viable (i.e., inactivated) nucleus,

(3) elimination of one of the two nuclei of somatic hybrid, or

(4) progressive elimination of chromosomes of one parent from the somatic hybrid.

Irradiations or high speed centrifugation may be used to produce cytoplasts (enucleated protoplasts).

Uses of Somatic Hybridization

Somatic hybrids and cybrids may be used for various purposes. If the regeneration of fertile plants is possible, symmetric hybrids provide a rare opportunity of creating, new plant species. These somatic hybrids also facilitate interspecific transfer of genes. Asymmetric or deficient somatic hybrids can be used for interspecific gene or cytoplasm transfers. Transfer of genes or cytoplasm from one species to another is easier through asymmetric hybrids than through symmetric hybrids.

Cybrids can be used to transfer both the nuclear genes and plasma genes, to recover hybrids from sexually incompatible parents, to obtain variable combinations of nuclear and cytoplasmic genomes and to produce progeny with recombinant organelles.

Scientists in France have been able to combine two different characters into a breeding line of *Brassica napus* through somatic hybridization. Lines containing intergeneric hybrid combinations of mitochondria and chloroplasts (mitochondria of one genus and chloroplast of another genus) have also been developed using this technique. Perhaps the maximum use of somatic hybridization technique has been done in different genera (*Nicotiana, Solanum, Lycopersicon, Petunia,* etc.) of Solanaceae family.

Plant Genetic Engineering

Genetic engineering or **recombinant DNA technology** (also referred to as **transformation**) **is the transfer of a specific DNA segment of an organism to another through an agent (vector) and incorporation of this DNA segment into the host genome so that the foreign DNA becomes an integral part of the host DNA and replicates like normal genetic material in succeeding sexual generations.** If this DNA transfer involves a plant species, it is termed as **plant genetic engineering**. The DNA or gene so transferred is called **trans-DNA** or **transgene** and the plant containing such genetically transferred foreign gene is known as **transgenic plant, transformed plant, genetically transformed plant** or **genetically modified plant.**

Avery, McLeod and McCarty, for the first time in 1944, demonstrated that DNA is the substance responsible for inducing transformation in bacteria. This discovery created a lot of interest among the molecular biologists for further research in this field. In the early 1970s, molecular biologists were able to prove that DNA could be cut and spliced and transferred from one organism to another. This advancement in biotechnological research opened new avenues for the practical application of this technology for bringing improvement in crop plant species.

The early 1980s was the time of very active research in the field of biotechnology and also of the beginning of valuable and eye-opening results of the untiring efforts of the biotechnologists. It was this time that the novel system of genetic transformation (plasmid-mediated gene transfer) became available with the discovery of T_i and R_i plasmids of *Agrobacterium tumefaciens* and *Agrobacterium rhizogenes*, respectively. The development of first transgenic plant in 1983, provided extra stimulus and enthusiasm to the scientists working in this field for continuing and accelerating their efforts in this direction. Due to such efforts made in 1980s and 1990s, transgenic plants resistant/tolerant to diseases, insect pests, herbicides, etc. of more than 50 plant species have been successfully produced. In the beginning, more emphasis was given to the development of transgenics of **tissue culture-friendly** dicotyledonous plant species, but now there are significant breakthroughs in transformation of so-called **tissue culture-shy** (recalcitrant to *Agrobacterium* strategy) monocotyledons (rice, maize, etc.).

Plant Transformation Process

Important steps of plant transformation process (or recombinant DNA technology) are : (1) Gene identification and isolation; and (2) gene transfer, integration and expression.

Gene identification and isolation : Following techniques can be used to identify and isolate a particular gene :

(i) **Using synthesized DNA as probe to identify the gene from genomic libraries** – The DNA, to be used as probe, may be either (a) synthesized from the mRNA of the gene capable of producing large quantity of mRNA or (b) on the basis of the amino acid sequence in the product protein of the gene.

(ii) **On the basis of transposable elements.**

(iii) Another technique which is being used in **Drosophila** is based on **chromosome walking**.

The techniques (a) and (b) falling under (i) mentioned above have been most extensively used for identifying a particular gene. **Molecular probes** (small DNA or RNA segments) facilitate identification and isolation of specific DNA sequences from an organism as they (probes) are capable of recognizing complementary sequences in DNA or RNA molecules.

Gene transfer integration and expression : The techniques used for gene transfer may be classified into two broad groups : (i) Vector-mediated gene transfer and (ii) direct gene transfer.

Vector-mediated gene transfer : In this technique, a **plasmid (an extrachromosomal, self-replicating, circular DNA molecule)** or a **virus** is used as vector. The T$_i$ plasmid found in *Agrobacterium tumefaciens* (a pathogenic soil bacterium causing tumors, crown galls, in many dicotyledons) has been used most extensively as a vector (a delivery system). The R$_i$ **plasmid** found in *Agrobacterium rhizogenes* (causing hairy root disease in several dicotyledons and monocotyledons) has also been used as a vector but not so extensively as T$_i$ plasmid. The T$_i$ and R$_i$ plasmids are also called **tumor-inducing** and **root-inducing** plasmids, respectively. Each of these plasmids has a **T-region** or **T-DNA** with a border on both the sides. The whole process of gene transfer, integration and expression is as follows:

(a) Cleavage of the plasmid molecule at a specific site with the help of enzyme restriction endonuclease,

(b) insertion of the foreign DNA segment carrying the desired gene under transfer (transgene) between the broken ends of the plasmid molecule (T$_i$ region),

(c) reformation of the circular plasmid molecule with the help of another enzyme DNA ligase,

(d) co-cultivation of **Agrobacterium** containing the transgene with suitable explant or protoplasts,

(e) incorporation of transgene into the host genome,

(f) regeneration of host plant,

(g) expression of transgene in the regenerated plants, and

(h) transmission and expression of the transgene in subsequent generations (in seed reproduced crop species through sexual reproduction and in vegetatively propagated crops through asexual reproduction).

This technique has been successfuly used in a large number of plant species (cotton, tomato, tobacco, sunflower, *Brassica,* etc.).

In addition to plasmid-mediated gene transfer, the transfer and integration of a transgene may be carried out using some plant viruses like **caulimo virus** (cauliflower mosaic virus) and **gemini virus** (wheat dwarf virus). Whereas caulimo virus infects members of Cruciferae family, gemini virus infects members of the Poaceae (Gramineae) family. The transfer of a desired gene through a virus can be described as under :

(a) Incorporation of DNA segment carrying the transgene into the viral DNA,

(b) the host plant is infected with virus containing the desired gene, and

(c) spread of virus in the host plant.

However, virus-mediated transfer of transgene has a serious drawback that it can cause only transient (unstable) transformation (usually limited to the same generation only) and thus has little practical utility in crop improvement.

Direct gene transfer or direct DNA uptake by cells or protoplasts : Since it is not possible to successfuly use the plasmid mediated gene transfer technique in many plant species, other methods like **incubation in PEG** (polyethylene glycol), **high voltage electric shock** (electroporation), and **insertion of DNA by particle gun** or by **microinjection**, are used to transfer the desired gene into the host genome. The first two techniques (PEG and electroporation) are more useful in transferring and integrating the desired gene into protoplasts than directly into cells. However, the regeneration of functional plants from protoplasts is mostly difficult than their regeneration from cells particularly in case of monocotylednous plant species. Further, in electroporation method, use of very high voltage current is required for a stable transformation of protoplasts. Low voltage current yields high rates of unstable transformation. The use of these two techniques has given good results in plant species like rice, tobacco, maize and wheat.

In plant species where regeneration of complete plants from protoplasts is not possible, either of the other two methods (DNA insertion by particle gun or by micro-injection) is used to transfer and incorporate the foreign DNA into the cell genome. In **particle gun technique**, the foreign DNA is coated onto small particles (**microcarriers**) of gold or tungsten and then these coated particles are projected into the cells or tissues under transformation with the help of a particle gun. These particles have a very high penetrating ability and thus can deliver the foreign DNA directly to the nuclei of the cells. Particle gun method has been successfully used in several plant species (soybean, maize, tobacco, etc.). Different gene transfer methods and their merits and limitations have been discussed in detail by Christou (1996).

The **microinjection technique** though very specific (since the foreign DNA is directly delivered into the nuclei of the cells using a microsyringe) and gives a very high rate of transformation in some plant species, is slow, tedious and needs a highly technical person to do the job of foreign DNA delivery into the nuclei of the cells. This technique has been successfully used in tobacco and lucerne.

Nevertheless, we have not to be very optimistic about the usefulness of genetic engineering to the plant breeder because of the following reasons :

(1) In many plant species, the regeneration of functional plants is still a very tedious job.

(2) Manytimes, the transgene either fails completely or has a very poor expression in the regenerated plants and/or in succeeding generations due to some genetic causes like faulty base sequence of foreign DNA, inadequate promoter/ enhancer sequences, and gene silencing.

(3) Till date, the transfer of foreign DNA into the host plant genome is confined to one gene at a time, whereas the plant traits of plant breeder's interest are mostly of quantitative type.

Molecular Markers

The third decade of the twentieth century was the beginning of the preparation of genetic maps of individual chromosmes on the basis of **morphological markers** (visible characters). The basis of these chromosome maps was the degree of recombination (or degree of linkage) between visible characters. The first organism used for this purpose was **Drosophila melanogaster**. Later on, the same procedure of chromosome mapping was used in many plant species (maize, peas, barley, wheat, etc.). The measurement of gene association or linkage has a great significance for the plant breeder as the linkage between two desirable genes greatly helps the plant breeder in the simultaneous selection of two characters (since the selection for one desirable character means automatic selection of another desirable character), whereas linkage between a desirable gene and an undesirable gene makes the job of plant breeder difficult (since, in this situation, the selection of one desirable character means automatic rejection of the other desirable character). This method of chromosome mapping, however, suffers from the following drawbacks :

(1) The procedure is based on a limited number of major genes only;

(2) the procedure is not applicable to quantitative characters; and

(3) the procedure is difficult and takes more time.

On the contrary, genetic mapping based on molecular markers has several advantages (discussed later) over the above mentioned traditional method of chromosome mapping. Let us discuss the three types of molecular markers (varietal differences at molecular level), namely, **isozymes** (complex proteins), **restriction fragment length polymorphisms** or RFLPs, and **random amplified polymorphic DNAs** or RAPDs, separately.

Isozymes : Several different forms of an enzyme are known as isozymes. They can be detected on the basis of their net electrical charges using electrophoresis. The maximum use of isozyme markers in plant genetics and breeding was perhaps done in 1970s and 1980s. Isozymes have several advantages :

(1) Their detection is simple.

(2) The procedure of their detection is less costly.

(3) Their detection is independent of environmental fluctuations.

(4) Isozymes have been successfully used in numerous genetic, taxonomic and evolutionary studies.

However, these genetic markers have some disadvantages too :

(1) Due to a small number of available marker loci, the isozymes provided a poor genomic coverage and thus, manytimes, failed to classify the breeding materials.

(2) Also, the limited number of available marker loci adversely affects the level of polymorphism and thus the procedure is not useful for broad-based breeding materials.

Nevertheless, the use of isozymes in genetic and plant breeding studies made the beginning of molecular markers.

RFLPs : Due to the drawbacks associated with isozymes, particularly the low level of polymorphism provided by these markers, the scientists shifted their attention to the use of better molecular markers like RFLPs and RAPDs. RFLPs provide detection of differences between individuals directly at the DNA level and, therefore, are an extremely powerful tool in the hands of the experimenter for the detailed evaluation of genetic diversity present in different plant species (both the cultivated and wild) and that present in plant pathogen populations.

Various steps of RFLP analysis in plant species are :

(1) Isolation of genomic DNAs from different varieties,

(2) digestion of these DNAs with restriction enzyme(s) for cleaving the plant's DNA at specific sites,

(3) separation of cleaved DNA fragments using gel electrophoresis (the rate of migration of a DNA fragment would depend upon its size, the smaller segements would move faster than the lengthy segements),

(4) denaturation of digested DNA to make it single-stranded, transfer of single-stranded DNA on to a nylon membrane from the agarose gel by southern blotting and hybridization of single-standard labelled probe DNA with this single stranded genomic DNA,

(5) washing-off the excess probe DNA,

(6) exposure of the membrane to X-ray film, and

(7) development of film for detecting hybridized DNA (i.e., probe DNA : cleaved DNA hybridization) through autoradiography.

The degree of polymorphism (differences) between plants can be determined on the basis of the degree of hybridization (between the probe DNA and the genomic DNA).

The **advantages** of RFLPs are :

(1) RFLPs provide a wide genomic coverage and a high level of polymorphism as a huge number of RFLP loci are available.

(2) RFLPs are independent of confounding environmental effects.

(3) RFLP mapping is largely independent of gene function.

(4) RFLP analysis is very rapid and convenient.

(5) Because of a spectacular advancement in the technology, RFLP analysis has now become a routine procedure.

(6) RFLPs are universal and can be detected in all living tissues.

(7) RFLPs are age-independent and thus can be detected at any stage of development.

(8) RFLPs facilitate evaluation and classification of germplasm.

(9) RFLPs may be used for strain identification for Plant Variety Right Protection.

(10) RFLPs have least pleiotropic effects.

(11) RFLPs facilitate mapping of quantitative trait loci, not feasible by conventional procedures.

(12) Phylogenetic relationships between different species can be known with the help of RFLPs.

(13) In addition to other markers (morphological, physiological and cytological) and biometrical analyses (D^2 analysis, principal component analysis, etc.), RFLPs can be used to assess genetic diversity in various types of plant populations.

(14) The linkage between an RFLP band and a particular gene can help in locating a desirable gene. Such gene can be transferred to a suitable genetic background through backcrossing. Therefore, RFLPs may play an important role in plant breeding as indirect selection criteria.

However, RFLP technique is quitely costly as a method of chromosome mapping.

RAPDs : RAPD is relatively a recent and simpler technique and facilitates rapid detection of polymorphisms among different plants. The technique is based on PR **(polymerase chain reaction)** and requires small amounts of DNA. No radioactivity is involved in the procedure. This technique may be used for tagging of major genes and quantitative trait loci, identification of somatic hybrids and for the conservation of genetic resources.

Grivet and Noyer (2003) discussed the usage of biochemical and molecular markers in detail.

Biotechnology in India

The realization that biotechnology is an important field for research and that this discipline is expected to make some spectacular contributions (not feasible by conventional methods) in future, took a practical shape in India in 1982 with the inception of **National Biotechnology Board** (NBTB) under the **Department of Science and Technology (DST)** on the recommendation of **Science Advisory Committee.** The objective of establishing NBTB was to coordinate and encourage research in the field of biotechnology. Another important development in this direction took place in 1986 with the creation of a separate **Department of Biotechnology (DBT)** in the Ministry of Science and Technology by the government of India. Also, New Delhi is one of the two centres (another centre at Triesta, Italy) of the **International Centre for Genetic Engineering and Biotechnology (ICGEB).** ICGEB was established in 1987 under the auspicies of United Nations for the developing countries of the world. Other biotechnology centres in India are at :

(1) I.A.R.I. (Indian Agricultural Research Institute), New Delhi,

(2) N.D.R.I. (National Dairy Research Institute), Karnal (Haryana), and

(3) I.V.R.I. (Indian Veterinary Research Institute), Izatnagar (Uttar Pradesh).

Department of Biotechnology, in collaboration with the University Grants Commission (UGC), has launched a massive programme in terms of physical facilities and trained manpower to give a boqst to biotechnology research in India. Nine distributed information centres (JNU, New Delhi; I.A.R.I., New Delhi; National Institute of Immunology, New Delhi; Institute of Microbial Technology, Chandigarh; Bose Institute, Calcutta; Poona University, Pune; Centre for Cellular and Moleclar Biology, Hyderabad; Indian Institute of Science, Bangalore; and Madurai Kamraj University, Madurai), 14 user centres and six centres for plant molecular biology (JNU, New Delhi; National Botanical Research Institute, Lucknow; Bose Institute, Calcutta; Osmania University, Hyderabad; Tamil Nadu Agricultural University, Coimbatore; and M.K.U., Madurai) were established by DBT. In addition, a large number of universities/institutes in India are now providing facilities for training and research in biotechnology to the students of M.Sc. and Ph.D. levels.

In view of the importance of biotechnology and the gains through biotechnology research, particularly in U.S.A., Japan and several European countries, public sector has also come forward and some private companies in India are doing useful work on biotechnology.

Expectations, Achievements and Constraints

Despite some seemingly impossible achievements through biotechnology research in the recent past, many scientists feel that crop biotechnology is still at the cross-roads. The opinions of the scientsits about the use of biotechnology, particularly about the uses and abuses of transgenic research, are certainly polarized in favour of and against the use of this discipline in crop improvement. The reasons for this polarization of opinions may be many, however, the assessment of risks involved in biotechnology research should be based on scientific grounds.

The expectations and claims of biotechnologists are endless. Of course, some of their expectations and claims have become realities in several countries of the world since a large number of doubled haploids (developed through tissue culture) and transgenics with novel traits of commercial value (developed through recombinant DNA technology) are being grown as commercial cultivars in these countries. The biotechnologists hope that biotechnology is a unique development with revolutionary chances in many different areas like resistance/toleracne to **biotic** (development of strains resistant to viral, bacterial and fungal diseases and to insect pests) and **abiotic** (development of strains tolerant to salinity, drought, etc.) **stresses, herbicide tolerance** (to confer selectivity and increase crop production), improvement in **protein quality** (transfer of 2S storage protein from *Amaranthus* to other target crops), improvement in **oil quality** (altering the fatty acid composition of oils according to human and industrial needs), **transgenic male sterility system** (Mariani *et al.*, 1990, 1992) (to facilitate hybrid seed production by pollination control and to check gene flow from transgenics to sexually compatible species), reduction in **post-harvest losses** (to slow down ripening by inactivating specific genes), **biopharming of vaccines** and a spectacular **increase in the production and productivity of crops**. However, the theoretical expectations and optimistic predictions of biotechnologists regarding spectacular increase in the production and productivity of crop plants has not yet come true.

Nevertheless, the first and the fundamental point that goes in favour of the biotechnologists is that they have erased the line of demarcation drawn by sexual reproduction between compatible hybridizations and non-compatible hybridizations (hybridization between genetically unrelated or distantly related organisms) and thus have been successful in circumventing all sexual barriers to distant hybridizations. Also, because of their untiring efforts, the identification, isolation, transfer, incorporation and expression of genes at will is now possible in some plant species. Many doubled haploids have performed excellently well (Chapter 23). Transgenics of cotton, rapeseed, soybean, maize, potato, tomato, etc. are being grown as commercial varieties in U.S.A. Similarly, scientists in Japan and several European countries have succeeded in developing many useful transgenics with novel genes of commercial value.

Perhaps the most powerful weapon discovered by the biotechnologists for the biological control of insect pests is the soil bacterium **Bacillus thuringiensis (Bt)**. There are several strains of this bacterium and depending upon the **cry** genes (genes coding for specific insecticidal crystal proteins) they carry, these strains are capable of synthesizing their own toxins. The alkaline environment found in the mid-guts of many lepidopteran insect pests (cotton boll worm, pod borers, stem borers, etc.) is a favourable environment for **Bt** toxin. **Bt** genes have been exploited in many crop plants for developing **Bt** transgenics. For example, **Bt** cotton is being grown at a commercial scale in U.S.A. The government of India has also approved the commercial cultivation of **Bt** cotton in the country.

Another important success achieved by the biotechnologists is the development of **herbicide-tolerant transgenics** in several crop species (cotton, rice, maize, sugarbeet, soybean, etc.). The need for developing herbicide tolerant transgenics was felt because though several good herbicides (non-toxic, quickly biodegradable and effective) are available to the farmers, but these herbicides have the drawback of non-selectivity (that is, inability of distinguishing weeds from crop plants). However, the development of herbicide tolerant transgenics is a debatable matter and suffers from two main criticisms :

(1) The main objective of producing herbicide-tolerant transgenics is that the herbicide used becomes selective (that is, it kills only the weeds and not the crop plants). But the transfer of herbicide-tolerant genes to some undesired plant (wild relative of the crop plant or a weed) through natural crossing, may create a very serious problem of the creation of **super weeds**.

(2) The control of weeds through the use of herbicides is not the best way of solving the weed problem since the herbicides are mostly costly and their use would ultimately pollute the environment. Therefore, use of some other means of weed control (biological control, mechanical control or some other type of control) should be preferred.

In addition to the development of **Bt** transgenics and herbicide-tolerant transgenics, another breakthrough in biotechnology research is expected in the form of **improved protein transgenics**. Efforts are being made in India at Jawaharlal Nehru University, New Delhi to transfer the gene coding for the **2S storage protein** (having higher levels of some essential amino acids) into potato. Efforts are also being made in Delhi to develop **transgenic tomatoes** by incorporating into them a gene from an edible fungus. The introduced gene makes tomatoes free from oxalic acid and increases the resistance of tomato plants against a disease-causing fungus. Oxalic acid is harmful if it is consumed in large amounts and causes kidney stones and hypocalcemia. Similar efforts are being made at Indian Agricultural Research Institute, New Delhi and some other biotechnology centres for improving protein quantity and quality of other crop species.

In additions to the risks involved in the development of transgenic crops discussed earlier (gene flow and environmental pollution), there may be some other drawbacks (like **erosion of biodiversity, expression of undesirable characters, gene silencing** and **terminator gene technology**). The erosion of biodiversity may lead to a potential future threat because of the dependence of scientists on a small amount of variability only. If there is expression of some undesirable phenotypic traits in a transgenic crop, the whole purpose of transgenic production is defeated.

Gene silencing (the **instability of gene expression** or **instability of transgenics**) is **the decline in the expression of transgene**. For example, in some transient transformations, the expression of the transgene is confined to the same generation only. In some other cases, the decline in the expression of transgene starts after a few or several plant generations. Such transgenics have no practical utility for the plant breeder.

Terminator gene technology is a type of **Genetic Use Restriction Technologies** (GURTs) for **restricting the use of seed of a transgenic cultivar to a single plant generation by introducing a genetic switch mechanism**. This genetic mechanism though does not affect the seed quality and its consumption suitability, it causes the death of the embryos and thus makes them non-viable so that they do not germinate in the second generation. Another type of GURTs is the **verminator gene technology** where, instead of killing embryos of seeds, **the plant growth is blocked in the second generation**. Therefore, the purpose of both these technologies is the same. Jefferson *et al.* (1999) presented an expert paper on GURTs and discussed their various aspects and potential implicatins in detail. However, both these technologies have been widely criticized all over the world, particularly by agriculture-based undeveloped and developing nations. In many countries like India, saving of seed for resowing (particularly in case of self-fertilizing crop species) and farmer-to-farmer seed exchange, are the easiest and most economical ways of obtaining seed for growing in the next crop season.

The terminator gene technology on one side is an effort to protect transgenes from piracy and from migrating to undesired plant species, but on the other side, it is an abuse of science in forcing the growers to buy seeds every season. However, the best way of nullifying the effect of terminator genes is to evolve indigenous promising transgenics. At this age and time, if the scientists of undeveloped and developing countries remain unrealistic, unimpressive and undemostrative in their approach, the things which are now seemingly unimportant and unconsequential, may prove to be very important and highly consequential in future.

In fact, it is difficult to forecast the path the recombinant DNA technology is expected to take. Therefore, it is necessary to make a realistic assessment of this novel technology. This technology should neither be treated as a panacea for all plant breeding problems nor it should be treated as an evil. Nevertheless, the biotechnologists must continue their efforts for a positive role of this discipline towards the improvement of plant species and strengthen their optimisms by orienting their expectations and claims at realities. Such an approach would certainly narrow down the gap between the overoptimistic biotechnologists and pessimistic plant breeders and would help in achieving their distant target.

CHAPTER 26

Intellectual Property Rights and Plant Breeding

Plant breeding began as a man's hobby. Then, it took the form of an art as the improvement in plant traits during that period was almost entirely dependent on the plant breeder's/farmer's skill (since the first breeders were farmers). But, as the knowledge of genetics and some other disciplines increased, this subject gradually became more of a science and less of an art. Now, it is also defined as the technology of producing improved crop plants. However, the job of the modern plant breeder is becoming : (1) more and more challenging because of several serious problems like population explosion, increasing complexity of the unholy triple alliance of phytopathogens, insect pests and weeds, and the loss of precious and irreplaceable genetic wealth day by day, to be faced by the plant breeder, (2) increasingly exciting because of biotechnological innovations, particularly the development of transgenic plants, and (3) more incentive oriented because of the enforecement of plant breeder's rights (PBR) by many countries of the world (Also see Chapter 1). Our discussion in this chapter will be restricted to the above mentioned third aspect of the plant breeder's job only.

However, the plant breeding subject could not remain as man's hobby for a long time because of the

(1) awakening and awareness of farmers all over the world about the profits from their plant products,

(2) increasing importance of scientific principles in the breeding of crops,

(3) increasing involvement of private companies in the production and distribution of improved seed, and

(4) tremendous increase in the cost of seed.

Therefore, plant breeding gradually became a costly profession, more so in private sector. As a result, a need was felt to encourage the plant breeders/inventors by creating incentives for them and granting them rights to protect their plant products. This need of the special economic return to the plant breeder for his innovation has become more relevant today because of the use of highly sophisticated techniques and equipments in crop improvement programmes.

Intelletual Property Rights (IPR) at International Level

Nevertheless, the idea of **Intellectual Property Right (IPR)** for plant products was not easily accepted in the beginning. The formal acceptance of this idea at international level took more than 60 years after the beginning of scientific plant breeding (that is, after the rediscovery of Mendel's laws in 1900) with the inception of the **International Union for the Protection of New Varieties of Plants** or **UPOV** (derived from its French name **Union Internationale pour la Protection des Obtentions Vegetales**) in the year 1961. UPOV, an intergovernmental organization with headquarters in Geneva, was established by the **International Convention for the Protection of New Varieties of Plants** which was signed in Paris in 1961. The convention came into force in 1968 and, according to the need of time, was revised at Geneva on 10th November, 1972, 23rd October, 1978 and on 19th March, 1991. By August, 1997, there were 34 member countries of UPOV.

Intellectual Property Right (IPR) is **a general term which includes patents, Plant Breeder's Right (PBR) or Plant Variety Protection (PVP), geographical indicators, copyrights, designs, trademarks and trade secrets. A patent is an exclusive right (a legal monopoly) for a defined period of time granted by the government to the inventor. Plant breeder's right (PBR) may be defined as an exclusive right granted to the plant breeder (inventor) to exploit his variety (that is, minimum right to produce his variety for the purpose of commercial marketing).** The minimum right to the inventor does not restrict a grower from producing seed of that variety for subsequent sowing on his own farm, nor it restricts any other breeder for the use of that variety as a breeding material in a plant breeding programme. Therefore, while granting right to the inventor, the interests of both the owner of the invention and that of the user (a farmer or some other plant breeder) are safeguarded.

Although PBR is its own kind of IPR, it has certain features in common with other forms of IPR. For example, both PBR and patents for industrial inventions grant their inventors a form of exclusive right so that they may pursue their innovative activity. Similarly, PBR enables the copying (reproduction) of the protected plant variety constrained by the inventor. Therefore, PBR has certain features in common with some other forms of IPR, but at the same time has some fundamental differences.

The need to protect plant products was felt because of the following reasons :

(1) The development of new varieties of plants has now become a costly affair as it requires lot of investment in terms of money, skill and material resoures and takes a long period of time (generally 10-15 years).

(2) Granting an exclusive right to the inventor for his innovation improves the economic condition of the inventor and encourages him to invest in plant breeding research and development of agriculture.

(3) The important characteristics of new varieties like substantial increase in the yield level, resistance to biotic stresses and tolerance to abiotic stresses, play an important role in increasing both the production and productivity of crop plants.

(4) Granting an exclusive right to the breeder to exploit his variety checks the piracy of his innovation by others and thus the inventor will not be deprived of the opportunity to get returns of his investment.

In view of the above points, creating incentives for innovations is fully justified.

The main functions of the UPOV convention are : (A) To establish standard international criteria for the grant of protection, and (B) to specify a minimum scope of protection.

Standard Criteria for the Grant of Protection

According to 1978 Act, a new variety must fulfil the following conditions for the grant of protection:

(1) The variety should be new (that is, **novelty** of the variety).

(2) The variety should be distinguishable from all other varieties (that is, **distinguishability or distinctness** of the variety).

(3) The variety should be homogeneous or uniform (that is, **homogeneity** or **uniformity** of the variety).

(4) The variety should be stable (that is, **stability** of the variety).

(5) The variety should have an appropriate generic designation (that is, **denomination** of the variety).

Novelty of Variety : The first condition for a variety to qualify for protection is that, based on the knowledge available in the member state and elsewhere, the variety should not have been commercialized before. However, depoending upon the country and the species, there are grace periods (ranging from one to several years) for commercialization. Novelty is generally determined by the superiority of the variety over other varieties in terms of one or more agronomic traits (yield, disease resistance, quality, etc.).

Distinctness of variety : The purpose of putting this condition for protection is to ensure the easy identification of the variety, that is, the variety should have its own characteristic identity amongst all other varieties. The identification of the variety (based on morphological, chemical, physiological or pathological characteristics) should be easy and clearcut.

Uniformity of variety : The variety to be protected should have a high degree of uniformity in terms of plant traits. However, this criterion can be easily satisfied by the genetically homogeneous varieties of the vegetatively propagated species, of highly self-fertilized species and by single cross hybrids, but synthetic and composite varieties and three-way and double cross hybrids cannot satisfy this condition so easily. For the genetically heterogeneous varieties (synthetics, composites, three-way crosses, double crosses and complex crosses) this condition has to be bit liberalized. Further, a high uniformity of the end product will ensure a high market value.

Stability of variety : According to this criterion of protection, the variety should breed true to type during repeated propagation, that is, the variety should remain identical to the description given at the time of its protection. Like uniforty of variety, this condition is easily satisfied by the genetically homozygous and homogeneous varieties of highly self-fertilizing plant species. However, genetically heterozygous varieties (hybrids, synthetics, composites, etc.) cannot satisfy this criterion unless they are vegetatively propagated.

Denomination of variety : The variety should have an appropriate name which should be different from names of other varieties of the species. Denomination of variety helps the breeder, users and traders in future dealings.

Minimum Scope of Protection

The second important function of UPOV is to prescribe minimum exclusive rights to be granted to the inventor by its member country. According to 1978 UPOV Act, the inventor's right is restricted to the commercial use of reproductive material of his variety, that is, the inventor's right includes only production for commercial marketing. However, production of propagating material which is not for commercial marketing falls outside the inventor's rights. Two kinds of liberties (exemptions or privileges) have been allowed in this Act (1978) :

(1) Farmers' rights, farmers' privilege or farmers' exemption : According to this privilege, farmers are permitted to produce seed of the protected variety for the purpose of subsequent sowing on their own farms. But, they are not allowed to produce seed of the protected variety for commercial marketing. However, limited farmer-to-farmer seed exchange is always there.

(2) Other breeders' rights, other breeders' privilege or other breeders' exemption : This privilege allows non-inventor breeders to use the protected variety as starting material in their breeding programmes, that is, free-of-charge use of protected variety by other breeders in their breeding programmes.

The minimum exclusive rights granted to inventor and the exemptions permitted to the farmers and non-inventor breeders, therefore, represent a careful balance between the interests of the inventor and those of the users (farmers and other breeders). But, unfortunately, the privileges allowed to both the farmers and other breeders have been withdrawn in the UPOV Act revised in 1991. According to this Act (1991), the inventor's right includes commercial use of all material of the protected variety. This withdrawal of rights of users would adversely affect the economic condition of small farmers.

International harmonization in the concept of variety protection and variety testing procedures became increasingly necessary because of the following conditions :

(1) Problems faced by plant breeders in the variety protection because of the lack of international harmonization and cooperation in this aspect as the criteria for the protection of new varieties and their testing procedures in different countries were different before the establishment of UPOV system,

(2) a high rate of development of plant varieties in different countries, and

(3) an abrupt increase in the activities of professional seed trade in the world.

The text of the UPOV Convention was prepared after a detailed discussion realizing that to achieve more success in the long run, the process of harmonization of variety protection should be gradual and step-by-step and the system should be based on recommendations rather than on binding rules. This well thought step towards harmony benefitted plant breeders on one hand and facilitated active cooperation between member nations both at the technical level and the administrative level, on the other. As a result, the cooperation shown by the member nations for the testing of varieties for the three most important criteria, namely, distinctness or distinguishability, uniformity or homogeneity and stability (DUS or DHS) provided a common basis for granting PBR as well as facilitated international harmonization and coopration for developing and evaluating plant varieties.

The formulation of **FAO International Undertaking on Plant Genetic Resources** in 1980s with a broad range of subjects like collection, conservation and utilization of plant genetic resources for food and agriculture, farmer's rights and benefit sharing, provided greater strength to the popular ideology of that time that plant genetic resources were the **common heritage of mankind** and that there should be a full and free exchange of these resources in the entire world. However, the **Convention on Biological Diversity (CBD)** which came into force on December 29, 1993, recognizes that nations have sovereign rights over their genetic resources. As a result, the ownership over biological resources became a point of debate between the believers of the two ideologies, namely, the ideology of **common heritage of mankind** reflected by the International Undertaking of Plant Genetic Resources and ideology of **national sovereignty** recognized by CBD. However, in the modern age of biotechnology, it is becoming difficult day by day to consider germplasm as common heritage of mankind. This is particularly true for private sector where most of the pharmaceutical and seed companies are not ready to share their profits with any one. Further, though several proposals have been put forward by different countries regarding improvement in the system, there are big differences in the contents of these proposals. For example, according to some countries, the rules for major crops should be different from those for the remaining crops, whereas some other countries say that rules for all the crops should be uniform. Similarly, there are differences in the proposals submitted for profit sharing. Again, there are problems in framing rules for the pre-CBD period and taking decision regarding some controversial patents taken during that period.

The CBD was signed by 173 countries including India. The convention, an international legal instrument, has following objectives :

(i) Conserving agrobiodiversity of both the cultivated and wild species,

(ii) utilizing biological diversity for sustainable development, and

(iii) sharing profits in the most appropriate way.

The **Global Plan of Action (GPA)** or **Leipzig plan** was adopted by 150 nations in the year 1996. The plan was endorsed by international bodies like FAO Commission on Genetic Resources for Food and Agriculture and the Conference of Parties to the CBD and is supported by **Consultative Group for International Agricultural Research (CGIAR)**. The main aim of the plan is the conservation (both *in situ* and *ex situ*) and utilization of plant genetic resources.

The enforcement of **Trade Related Intellectual Property Rights (TRIPs)** agreement on January 1, 1995 may be considered as one of the best efforts so far made towards harmonization of intellectual property rights rules. The agreement an outcome of the GATT (General Agreement on Tariffs and Trade) Uruguay round, is a comprehensive multilateral agreement for establishing minimum standards for intellectual property rights protection to commercial breeders and biotechnologists (and their companies) of all member countries of WTO (World Trade Organization). As per the broad framework of TRIPs, each member nation has to suitably ammend its intellectual property rights rules within a prescribed period of 5 years (10 years for least developed nations).

Biodiversity Act and Plant Variety Protection and Farmer's Right Act in India

Biodiversity is and will remain the main source of genes and gene combinations for plant improvement (both the conventional and the biotechnology-based) for long times to come. It is also the raw material for the fast-growing industries related to plant breeding and biotechnology. Further, since different types of genes and gene combinations are the building blocks of living matter, biodiversity has provided the basis for our health and well-being. It is, therefore, necessary to frame a suitable legislation for the protection of this product of millions of years of evolution. A very positive step in this direction was taken by the Ministry of Environment and Forests, Government of India by setting up a committee headed by Dr. M.S. Swaminathan for this purpose. The committee has suggested that various bodies/boards starting from the grass root level (village panchayat level) to the national level should be constituted so that there is participation of people at all levels (Chopra, 2000).

As per the conditions of TRIPs Agreement, India has adopted a *sui generis* system (a system of its own kind) of Plant Breeder's Rights. Efforts have been made in this direction and a legislation known as **'The Protection of Plant Varieties and Farmers' Rights Act'** has been prepared by the Ministry of Agriculture, Government of India. The cabinet has already approved the draft act.

The draft act has been prepared by considering the important features of UPOV 1978 Act. The main objectives of the act are :

(1) Protecting farmers' rights;

(2) establishing an effective system for protecting varieties of plants; and

(3) increasing facilities for seed industry.

Since farmers are involved in the conservation, improvement and the supply of plant genetic resources for evolving new improved varieties, more emphasis has been given on farmers' rights.

In addition to the farmers' rights, some other important features of the act are : compulsory license, compulsory certification, user's right, compulsory deposit in the national gene bank, period of protection (15 years for annual crops and 18 years for fruit trees), and establishment of a National Authority for protecting the rights of farmers, breeders and users (Rana, 1995). The variety to be protected should fulfil UPOV 1978 Act conditions for the grant of protection (novelty, distinctness, uniformity, stability and denomination). Sufficient amount of seed of the variety should be produced. Also, the variety should not contain gene(s) involving any kind of Genetic Use Restriction Technologies like terminator gene technology or verminator gene technology.

However, according to Sidhu (1994), the implementation of a PBR system may create following problems :

(1) PBR system will encourage monopolization of knowledge.

(2) It will adversely affect the exchange of germplasm.

(3) It will encourage marginalization of farmers of resource poor nations because of the

(a) expected increase in the cost of seed,

(b) delayed spread of new varieties among poor farmers due to restriction on seed exchange,

(c) gradual dilution in farmers' exemption, and

(d) production of less seed by the holder of PBR title.

(4) PBR system will put restriction on the choice of research problems.

References

Allard, R.W. (1960). *Principles of Plant Breeding*. John Wiley and Sons, Inc., New York, London.

Anderson, D.C. (1938). The relation between single and double cross yields of corn. *J. Amer. Soc. Agron.* **30**, 209-211.

Anderson, E. (1949). *Introgressive Hybridization*, John Wiley, New York.

Anderson, E. and de Winton, D. (1931). The genetic analysis of an unusual relationship between self-sterility and self-fertility in *Nicotiana. Ann. Mo. Bot. Garden* **18**, 97-116.

Anderson, E. and Hubricht, L. (1938). Hybridization in *Tradescantia*: The evidence for introgressive hybridization. *Amer. J. Bot.* **25**, 396-402.

Athwal, D.S. and Virmani, S.S. 1972. In *Rice Breeding. Int. Rice Res. Inst.,* Los Banos, Philippines, pp. 615-620.

Atkins, A.E. (1953). Effect of selection upon bulk barley populations. *Agron. J.* **45**, 311-314.

Auerbach, C. (1976). *Mutation Research (Problems, Results* and *Perspectives)*. London: Chapman and Hall.

Auerback, C. (1978). Forty years of mutation research: A pilgrim's progress. *Heredity* **40**, 177-187.

Auerbach, C. and Robson, J.M. (1947). The production of mutations by chemical substances. *Proceedings of the Royal Society,* Edinburgh, UK, Section B, **62**, pp. 271-283.

Avery, D.T., MacLeod, C.M. and McCarty, M. (1944). Studies on the chemical nature of the substance inducing transformation of pneumococcal types: Induction of transformation by a deoxyribonucleic acid fraction isolated from Pneumococcus type III. *J. Exp. Medicine* **79**, 137-158.

Baenziger, P.S., Kudirka, D.T., Schaeffer, G.W. and Lazar, M.D. (1984). The significance of double haploid variation. In *Gene Manipulation in Plant Improvement. Proc. 16th Stadler Genetic Symposium,* J.P. Gustafson (Ed.), pp. 385-414.

Bailey, L.H. 1895. *Plant Breeding*. Macmillan, New York.

Bajaj, Y.P.S. 1990. *In vitro* production of haploids and their use in cell genetics and plant breeding. In *Biotechnology in Agriculture and Forestry* Vol. 12. Haploids in Crop Improvement I, Y.P.S. Bajaj (Ed.). Springer-Verlag Berlin Heidelberg, pp.3-44.

Barabas, Z. (1992). A new era in the production of hybrid varieties ? *Hungarian Agricultural Research* **1**(1), 17-21.

Baur, E. (1921). *Die wissenschaftlichen Grundlagen der Pflanzenzuchtung,* Gebr. Borntraeger, Berlin.

Beal, W.J. (1976-1882). *Report Michigan State Board Agric.*

Boerma, H.R. and Cooper, R.L. (1975). Comparison of three selection procedures for yield in soybean. *Crop Sci.* **15**, 225-229.

Bollich, C.N. (1990). Report of Technical Committee on Seed Release and Increase: Texmont Rice. *Texas Agricultural Experiment Station Farm*, Beaumont, Texas, pp. 67-72.

Borlaug, N.E. (1959). The use of multilineal or composite varieties to cotnrol airborne epidemic diseases of self-pollinated crop plants, *Proc. 1st Int. Wheat Genet. Symp.* Winniped, Canada 1958 University of Manitoba, Winnipeg.

Bos, I. (1983a). The optimum number of replications when testing lines or families on a fixed number of plots. *Euphytica* **32**, 311-318.

Bos, I. (1983b). About the efficiency of grid selection. *Euphytica* **32**, 885-893.

Bourgin, J.P. (1978). Valine-resistant plants from *in vitro* selected tobacco cells. *Mol. Gen. Genet.* **161**: 225-230.

Bowman, D.T. and Weaver, J.B. (1979). Analysis of a dominant male sterile character in upland cotton, II Genetic studies. *Crop Sci.* **19**, 628-630.

Breakwell, E.J. and Hutton, E.M. (1939). Cereal breeding and variety trials at Roseworthy College 1937-38. *J. Agric. S.A.* **42**, 632-641.

Briggs, F.N. (1930). Breeding wheats resistant to bunt by the backcross method. *J. Amer. Soc. Agron.* **22**, 239-244.

Brim, C.A. (1966). A modified pedigree method of selection in soybean. *Crop. Sci.* **6**, 2206.

Brim, C.A. and Stuber, C.W. (1973). Application of genetic male sterility to recurrent selection schemes in soybeans. *Crop Sci.* **13**, 528-530.

Brooks, W.K. (1899). *The Foundations of Zoology*, Columbia University Press.

Browning, J.A. and Frey, K.J. (1969). Multiline cultivars as a means of disease control. *Ann. Rev Phytopathol.* **7**, 355-382.

Browning, J.A., Simons, M.D., Frey, K.J. and Murphy, H.C. (1969). Regional deployment for conservation of oat grown rust resistant genes. In *Disease Consequences of Intensive and Extensive Cultures of Field Crops,* J.A. Browning (Ed.). *Iowa State University agric. Home Eco. Spl. Rep.* **64**, 49-56.

Bruce, J.F. (1910). The Mendelian theory of heredity and the augmentation of vigour. *Science* **32**, 627-628.

Buringh, P., Heemst, H.D.J. van and Staring, G.J. (1975). Computation of the absolute maximum food production of the world, Publication Department of Tropical Soil Science, Agricultural Univesity, Wageningen 598 p. 59.

Camerarius, R.J. (1694). Academiae caesareo-Leopold. N.C. Hectoris. Rudolphi Jacobi Camerarii, Professoris Tubingensis, ad Thessalum, D. Mich. Bernardum Valentini, Professorem Giessensem, De sexu plantarum epistola, Tiibingen **8**, 110.

Camp, W.H. (1944). A preliminary consideration of the biosystematy of *Oxycoccus. Bull. Torrey Bot. Club* **11**, 426-437.

Camp, W.H. (1945). The North American blueberries with notes on other groups of *Vacciniaceae. Brittonia* **5**, 203-275.

Chaleff, R.S. (1983). Isolation of agronomically useful mutants from plant cell cultures. *Science* **219**, 676-682.

Chaleff, R.S. and Parsons, M.F. (1978). Direct selection *in vitro* for herbicide-resistent mutants of *Nicotiana tabacum. Proc. Acad. Sci. U.S.A.* **75**, 5104-5107.

Cherry, J.P., Katterman, F.R.H. and Endrizzi, J.E. (1970). Comparison of seed proteins of species of *Gossypium* by gel electrophoresis,. *Evolution* **24**, 431-447.

Choo, T.M. (1980). Double haploids for estimating additive epistatic genetic variances in self-pollinated crops. *Can. J. Genet. Cytol.* **22**, 125-127.

Choo, T.M. and Kannenberg, L.W. (1978). The efficiency of using double haploids in a recurrent selection program in a diploid, cross-fertilized species. *Can. J. Genet. Cytol.* **20**: 505-511.

Chopra, V.L. (2000). Intellectual property right issues and plant breeding. In *Plant Breeding: Theory and Practice*, 2nd edn., V.L. Chopra (Ed.), Oxford and IBH Publishing Co. Pvt. Ltd., New Delhi, Calcutta, India, pp. 461-472.

Chopra, V.L., Jain, S.K. and Swaminathan, M.S. (1960). Studies on the chemical induction of pollen sterility in some crop plants. *Indian J. Genet.* **20**, 188-199.

Christou, P. (1996). Transformation technology. *Trends in Plant Science* **1**, 423-431.

Coe, E.H., Jr. (1959). A line of maize with high haploid frequency. *Amer. Nat.* **93**, 381-382.

Collins, G.B. and Genovesi, A.D. (1982). Anther culture and its application to crop improvement. In *Applications of Plant Cell and Tissue Culture to Agriculture and Industry*, D.T. Tomes, B.E. Ellis, P.M. Harvey, K.J. Kasha and R.L. Peterson (Eds.). University of Guelph, Guelph, Ontario, pp.1-24.

Collins, G.N. (1921). Dominance and the vigour of first generation hybrids. *American Naturalist* **55**, 116-133.

Comstock, R.E., Robinson, H.F. and Harvey, P.H. (1949). A breeding procedure designed to make maximum use of both general and specific combining ability. *Agron. J.* **41**, 360-367.

Copp, L.G.L. (1957). Bulk and pedigree methods of wheat breeding. *Wheat Information Science*, Biological Laboratories, Kyoto University, Japan, **5**, 7.

Crow, J.F. (1948). Alternative hypotheses of hybrid vigour. *Genetics* **33**, 477-487.

Crow, J.F. (1952). Dominance and overdominance in heterosis. *Heterosis*, J.W.Gowen (Ed.), Iowa State College, Press, Ames, Iowa.

Culbertson, J.O. (1954). Seed-flax improvement. *Adv. in Agron.* **6**, 143-182.

Darwin, C. (1859). *The Origin of Species*, John Murray, London.

Darwin, C. (1868). *The Variation of Animals and Plants under Domestication*. John Murray, London.

Darwin, C. (1876). *The Effects of Cross and Self Fertilization in the Vegetable Kingdom*. John Murray, London.

Davenport, C.B. (1908). Degeneration, albinism and inbreeding, *Science* **28**, 454-455.

Davis, R.L. (1929). Report of the Plant Breeder. *Annual Report of Puerto Rico Experimental Station* for 1927. Puerto Rico Experiment Station, Rio Piedras, pp. 14-15.

de Buyser, J., Henry, Y., Lonnet, P., Hertzog, R. and Hespel, A. (1987). "Florin": A doubled haploid wheat variety developed by the anther culture method. *Plant Breed.* **98**, 53-56.

Demerec, M. (1929). Cross sterility in maize, *Zeitschrift fur inductive Abstammungs-und Vererbungslehre* **50**, 281-291.

Dhawan, N.L., Bhatt, B.K. and Jaggi, H.L. (1966). The role of cytoplasm in the manifestation of heterosis and other quantitative traits in maize, Proc. 3rd Intn. *Symp. on The Impact of Mendelism on Agriculture, Biology and Medicine* held at New Delhi in February 1965 *Indian J. Genet.* **26A**, 60.

Dobzhansky, T. (1940). Speciation as a stage in evolutionary divergence. *American Naturalist* **74**, 312-321.

Dobzhansky, T. (1947). Adaptive changes induced by natural selection in wild population of *Drosophila. Evolution* **1**, 1-16.

Dobzhansky, T. (1951). *Genetics and the Origin of Species* (3rd. edn.). Columbia University Press, New York, London.

Dobzhansky, T. (1952) Nature and origin of heterosis, In: *Heterosis*, J.W. Gowen, (Ed.). Iowa State College Press, Ames, Iowa, pp. 330-335.

Dobzhansky, T. (1964). How do the genetic loads affect the fitness of their carriers in *Drosophila* populations? *American Naturalist* **98**, 151-166.

Doggett, H., and Eberhart, S.A. (1968). Recurrent selection in sorghum. *Crop Sci.* **8**, 119-121.

Donald, C.M. (1968). The breeding of crop ideotypes, *Euphytica* **17**, 385-403.

Driscoll, C.J. (1972). XYZ system of producing hybrid wheat. *Crop Sci.* **12**, 516-517.

Driscoll, C.J. (1985). Modified XYZ system of producing hybrid wheat. *Crop Sci.* **25**, 1115-1116.

Driscoll, C.J. and Jensen, N.F. (1964). Characteristics of leaf-rust resistance transferred from wheat to rye. *Crop Sci.* **4**, 372-374.

Du Rietz, G.E. (1930). The fundamental units of botanical toxonomy. *Sevnsk. Bot. Tidskrift* **24**, 333-428.

Duvick, D.N. (1977). Genetic rates of gain in hybrid maize yields during the past 40 years. *Maydica* **22**, 187-196.

East, E.M. (1908). Inbreeding in corn, *Rept. Connecticut Agric. Exp. Sta. for 1907*, pp. 419-428.

East, E.M. (1916). Studies on size inheritance in *Nicotiana. Genetics* **1**, 164-176.

East, E.M. (1936a). Genetic aspects of certain problems of evolution. *American Naturalist* **70**, 143-158.

East, E.M. (1936b). Heterosis. *Genetics* **21**, 375-397.

East, E.M. (1940). The distribution of self-sterility in the flowering plants. *Proc. Amer. Phil. Soc.* **82**, 449-518.

East, E.M. and Jones, D.F. (1920). Genetic studies on the protein content of maize. *Genetics* **5**, 543-610.

East, E.M. and Mangelsdorf, A.J. (1925). A new interpretation of the hereditary behaviour of self-sterile plants. *Proc. Nat. Acad. Sci.* **11**, 116-183.

East, E.M. and Yarnell, S.H. (1929). Studies on self-fertility. *Genetics* **14**, 455-487.

Eckhardt, R.C. and Bryan, A.A. (1940). Effect of method of combining the four inbred lines of a double cross of maize upon the yield and variability of the resulting hybrid. *J. Amer. Soc. Agron.* **32**, 347-353.

Edwardson, J.R. (1970). Cytoplasmic male sterility. *Botan. Rev.* **36**, 341-420.

Elliot, F.C. (1957). X-ray induced translocation of *Agropyron* stem rust resistance to common wheat. *J. Hered.* **48**, 77-81.

Falconer, D.S. (1981). *Introduction to Quantitative Genetics* (2nd edn.), Longman, London.

Fasoulas, A. (1973). A new approach to breeding super yielding varieties, Publication 4, Department of Genetics and Plant Breeding, Aristotelian University of Thessaloniki, p.42.

Fisher, R.A. (1930). *The Genetical Theory of Natural Selection.* Clarendon Press, Oxford.

Fisher, R.A. (1932). The evolutionary modification of genetic phenomena. *Proc. 6th Intern. Congr. Genetics* **1**, 165-172.

Fisher, R.A. (1939). Selective forces in wild populations of *Paratetti texanus, Ann. Eugenics* **9**, 109-122.

Frankel, R. and Glun, E. (1977). *Pollination Mechanisms - Reproduction and Plant Breeding,* Springer-Verlag.

Frey, K.J., Browning, J.A. and Simons, M.D. (1975). Mulliline cultivars of autogamous crop plants. *SABRAO J.* **7**, 113-123.

Friebe, B., Raupp, W.J. and Gill, B.S. (2001). Alien genes in wheat improvement. In *Wheat in a Global Environment,* Z. Bedo and L.Lang (Eds.). Kluwer Academic Publishers, pp. 709-720.

Fujimaki, H. (1978). New techniques in backcross breeding for rice improvement, In symposium *Methods of Crop Breeding,* October 1977, Tokyo.

Fukasawa, H. (1953). Studies on restoration and substitution of nucleus in *Aegilotricum* I. Appearance of male sterile durum in substitution crosses. *Cytologia* **18**, 167-175.

Gabelman, W.H. (1956). Male sterility in vegetable breeding, In *Genetics in Plant Breeding.* symposium Biology Department Brookhaven National Laboratory Upton, New York.

Gale, M.D., Salter, A.M. and Law, C.N. (1986). *Annual Report* (Cytogenetics Department), Plant Breeding Institute Cambridge UK, pp. 53-55.

Gallais, A. (1989). Lines versus hybrids - The choice of the optimum type of variety. *Science for Plant Breeding.* Proc. XII Cong. EUCARPIA, 1989, pp. 69-80.

Gardner, C.O. (1961). An evaluation of effects of mass selection and seed irradiation with thermal neutrons on yield of corn. *Crop Sci.* **1**, 241-245.

Gardner, C.O. (1968). Mutations studies involving quantitative traits. In: *The Present State of Mutation Breeding.* Gamma Field Symposium No. 7, pp. 57-77.

Gardner, C.O.(1969). Genetic variation in irradiated and control populations of corn after ten cycles of mass selection for high grain yield. In *Induced Mutations in Plants.* Proceedings FAO/ IAEA Symposium, Pullman, USA), Vienna: IAEA, pp. 469-477.

Gardner, C.O. (1972). Development of superior populations of sorghum and their role in breeding programs. In *Sorghum in seventies,* N.G.P. Rao and L.R. House (Eds.). Oxford and IBH Publishing, New Delhi. pp. 180-196.

Gartner, C.F. (1849). Versuche und Beobachtungen Uber die Bastardzeugung im Pflanzenreich. Stuttgart.

Gastel, A.J.G. van and Nettancourt, D de (1975). The effect of different mutagens on self-incompatibility in *Nicotiana alata* Link and Otto II, Acute irradiations with X-rays and fast neutrons. *Heredity* **34**, 381-392.

Gerstel, D.U. (1953). Chromosomal translocations in interspecific hybrids of the genus *Gossypium. Evolution* **7**, 234-244.

Gerstel, D.U. and Sarvella, P. (1956). Chromosomal translocations in cotton hybrids. *Evolution,* **10**, 408-418.

Gerstel, D.V. (1950). Self incompatibility studies in *Guayule*, II. Inheritance. *Genetics* **35**, 482-486.

Giesbrecht, J. (1964). Gamete selection in two early maturing corn varieties. *Crop Sci.* **4**, 653-655.

Gill, K.S., Nanda, G.S. singh, G. and Anjla, S.S. (1979). Multilines in wheat - A review. *Indian J. Genet.* **39**, 30-37.

Gilmore, E.C. (1964). Suggested method of using reciprocal recurrent selection in some naturally self-pollinated species. *Crop Sci.* **4**, 323-325.

Goodchild, N.A. and Boyd, W.J.R. (1975). Regional and temporal variations in wheat yield in Western Australia and their implications in plant breeding. *Aust. J. Agric. Res.* **26**, 209-217.

Goulden, C.H. (1941). Problems in plant selection. *Proc. 7th Intn. Genet. Cong.* (1939).*Edinbugh,* pp. 132-133.

Gowen, J.W. (1952). *Heterosis.* Iowa State College Press, Ames, Iowa.

Grafius, J.E. (1965). Short-cuts in plant breeding. *Crop Sci.* **5**, 377.

Gregory, W.C. and Gregory, M.P. (1976). Groundnuts, *Arachis hypogea* - Leguminosae-Papilionatae. In *Evolution of Crop Plants*, N.W. Simmonds (Ed.), Longman, London, pp.151-154.

Grivet, L. and Noyer, J.L. (2003). Biochemical and molecular markers. In *Genetic Diversity of Cultivted Tropical Plants,* Hamon, P., Seguin, M., Perrier, X., and Glaszmann, J.C. (Eds.). Science Publishers, Inc. Enfield (NH). Plymouth, U.K., pp. 1-29.

Guha, S. and Maheshwari, S.C. (1964). *In vitro* production of embryos from anthers of *Datura. Nature,* **204**, 497.

Gustafsson, A. (1947). Mutations in agricultural plants. *Hereditas* **33**, 1-100.

Hagberg, G. and Hagberg, A. (1981). Haploid initiator gene barley. In *Barley Genetics IV*, Proc. of the 4th Int. Barley Genet. Symp. M.J.C. Asher (Ed.), Edinburgh, U.K., pp.686-689.

Hageman, R.H., Length, E.R. and Dudley, J.W. (1967). A biochemical approach to corn breeding. *Adv. Agron.* **19**, 45-86.

Haldane, J.B.S. (1933). The part played by recurrent mutation in evolution. *American Naturalist* **67**, 5-19.

Haldane, J.B.S. (1937). The effect of variation on fitness. *American Naturalist* **71**, 337-349.

Haldane, J.B.S. (1939). The theory of evolution of dominance. *J. Genet.* **37**, 365-374.

Hallauer, A.R. (1981). Selection and breeding methods. In *Plant Breeding*, K.J. Frey (Ed.). Symp. Iowa State University Press, Ames, Iowa, pp.3-55.

Hallauer, A.R. and Sears, J.H. (1972). Integrating exotic germplasm into Corn Belt maize breeding programs. *Crop Sci.* **12**, 203-206.

Hardy, G.H. (1908). Mendelian proportions in a mixed population. *Science* **28**, 49-50.

Harlan,H.V. and Martini, M.L. (1929). A composite hybrid mixture. *J. Amer. Soc. Agron.,* **21**, 487-490.

Harlan, H.V. and Martini, M.L. (1938). The effect of natural selection in a mixture of barley varieties. *J.Agric. Res.* **57**, 189-199.

Harlan, H.V., Martini, M.L. and Stevens, H. (1940). A study of methods in barley breeding. *US Dept.Agric. Tech. Bull.* **720**, 26.

Harlan, H.V. and Pope, M.N. (1922). The use and value of backcrosses in small grain breeding. *J. Hered.* **13**, 319-322.

Harlan, J.R. (1956). Distribution and utilization of natural variability in cultivated plants, In *Genetics in Plant Breeding*. Symposium Biology Department Brookhaven National Laboratory Upton, New York.

Harrington, J.B. (1937). The mass-pedigree method in the hybridization improvement of cereals. *J. Amer. Soc. Agron.* **29**, 379-384.

Hayes, H.K. and Garber, R.J. (1919). Synthetic production of high protein corn in relation to breeding. *J. Amer. Soc. Agron.* **11**, 309-318.

Hayes, H.K., Immer, F.R. and Smith, D.C. (1955). *Methods of Plant Breeding,*(2nd edn.). McGraw-Hill Book Comp. Inc. New York, London, Toronto.

Heiser, C.B. (1947). Hybridization between the sunflower species *Helianthus annuus* and *H. petiolaris*. *Evolution* **1**, 249-262.

Hermsen, J.G.T. (1974). Summation: Haploids in plant breeding. In *Haploids in Higher Plants: Advances and Potential*, K.J. Kasha (Ed.). Proc. 1st Intern. Symp., University of Guelph, Guelph (Canada), pp. 281-285.

Ho, K.M. and Jones, G.E. (1980). Mingo Barley. *Can. J. Plant Sci.* **60**, 279-280.

Ho, K.M. and Kasha, K.J. (1975). Genetic control of chromosome elimination during haploid formation in barley. *Genetics* **8**, 263-275.

Hoffman, W. Mudra, A. and Plarre, W. (Hrsg.) (1971). *Lehrbuch der Zuchtung landwirtschaftlicher Kulturpflanzen,* Bd. I. Paul Parey, Berlin and Hamburg.

Hopkins, C.G. (1899). Improvement in the chemical composition of corn kernel III. *Agri. Exp. Stn. Bull.* **55**, 205-240.

Hu, H. (1985). Use of haploids in crop improvement. In *Proc. International Seminar Biotechnology in International Agriculture*, IRRI, Los, Banos.

Hu, H. (1986). Wheat: Improvement through anther culture. In *Biotechnology in Agriculture and Forestry*, Vol. 2, Crops 1. Y.P.S. Bajaj (Ed). Springer-Berlin Heidelberg, New York, Tokyo, pp. 55-72.

Hughes, M.B. and Babcock, E.B. (1950). Self-incompatibility in *Crepis foetida* L., Subsp. rhoedifolia (Bieb.) Schinz et Keller. *Genetics* **35**, 570-588.

Hughes, W.G., Bennett, M.D., Bodden, J.J., Galanopoulon, S. (1974). Effects of time of application of Ethrel on male sterility and ear emergence in wheat. *Ann. Appl. Biol.* **76**, 243-252.

Hull, F.H. (1945). Recurrent selection for specific combining ability in corn. *J. Amer. Soc. Agron.* **37**, 134-145.

Hull, F.H. (1952). Recurrent selection and overdominance, In *Heterosis*, J.W. Gowen (Ed.), Iowa State College Press, pp. 451-473.

Ikehashi, H. and Fujimaki, H. (1980). Modified bulk population method for rice breeding. In *Innovative Approaches in Rice Breeding*, IRRI Manila, Philippines.

Jayasekara, N.E.M. and Jinks, J.L. (1976). Effect of gene dispersion on estimates of components of generation means and variances. *Heredity* **36**, 31-40.

Jefferson, R.A., Byth, D., Correa, C., Otero, G. and Qualset, C. (1999). Genetic Use Restriction Technologies. An expert paper presented before the Fourth Meeting of the UNEP/CBD Secretariat, Montreal, Canada. UNEP.CBD/SBSTTA/4/9/Rev.1.

Jenkins, M.T. (1934). Methods of estimating the performance of double crosses in corn. *J. Amer. Soc. Agron.* **26**, 199-204.

Jenkins, M.T. (1935). The effect of inbreeding and of selection within inbred lines of maize upon the hybrids made after successive generations of selfing. *Iowa State Coll. J. Sci.* **3**, 429-450.

Jenkins, M.T. (1940). The segregation of genes affecting yield of grain in maize. *J. Amer. Soc. Agron.* **32**, 55-63.

Jenkins, M.T. and Brunson, A.M. (1932). Methods of testing inbred lines of maize in crossbred combinations. *J. American Soc. Agron.* **24**, 523-530.

Jensen, C.J. (1974). Chromosome doubling techniques in haploids. In *Haploids in Higher Plants*. Proc. of the 1st Int. Symp. June 10-14, Univ. Guelph Press, Ontario, pp. 153-190.

Jensen, C.J. (1977). Monoploid production by chromosome elimination. In *Applied and Fundamental Aspects of Plant Cell, Tissue and Organ Culture*. J. Reinhart and Y.P.S. Bajaj (Eds.), Springer-Verlag, Belin, pp. 299-340.

Jensen, N.F. (1952). Intra-varietal diversification in oat breeding. *Agron. J.* **44**, 30-34.

Jensen, N.F. (1970). A diallel selective mating system for cereal breeding. *Crop Sci.* **10**, 629-635.

Jensen, N.F. (1978). Composite breeding methods and the DSM system in cereals. *Crop Sci.* **18**, 622-626.

Jensen, N.F. and Kent, G.C. (1963). New approach to an old problem in oat production. *Farm Res.* **29**, 4-5.

Jinks, J.L. (1983). Biometrical genetics of heterosis. In *Heterosis*, Reappraisal of theory and practice, Frankel, R. (Ed.). Springer Verlag, pp. 1-46.

Jinks, J.L. and Pooni, H.S. (1986). Description and illustration of the practical application of biometrical genetics to plant breeding. *Proc. 6th meeting EUCARPIA, Section Biometrics in Plant Breeding*. University of Birmingham, pp. 1-20.

Johannsen, W.L. (1903). Ueber Erblichkeit in Populationen und in reinen Leinen, Gustav Fischer, Jena.

Jones, D.F. (1917). Dominance of linked factors as a means of accounting for heterosis. *Genetics* **2**, 466-479.

Jones, D.F. (1918). The effect of inbreeding and crossbreeding on development. *Conn. Agric. Exp. Sta. Bull.* **207**, 1-100.

Jones, D.F. and Singleton, W.R. (1934). Crossed sweet corn. *Conn. Agric. Exp. Sta. Bull.* **361**, 489-536.

Jones, H.A. and Davis, G.N. (1944). Inbreeding and heterosis and their relation to the development of new varieties of onions. *US Deptt. Agric. Tech. Bull.* **874**, 1-28.

Karpechenko, G.D. (1928). Polyploid hybrids of *Raphanus sativus* L. × *Brassica oleracea* L. *Zeit. f. Indukt. Abst. u. Vereb Lehre* **48**, 1-85. (Radish × cabbage).

Kasha, K.J. and Kao, K.N. (1970). High frequency haploid production in barley (*Hordeum vulgare* L.). *Nature* **225**, 874-876.

Kasha, K.J. and Reinbergs, E. (1981). Recent developments in the production and utilization of haploids in barley. In *Barley Genetics IV, Proc. of the 4th Int. Barley Genet. Symp.*, M.J.C. Asher, (Ed.), Edinbergh, U.K., pp. 655-665.

Keeble, F. and Pellew, C. (1910). The mode of inheritance of stature and of time of flowering in peas (*Pisum sativum*). *J. Genet.* **1**, 47-56.

Kermicle, J.L. (1969). Androgenesis conditioned by mutation in maize. *Science* **166**: 1422-1424.

Kihara, H. (1951). Substitution of nucleus and its effects on genome manifestations. *Cytologia* **16**, 177-193.

Kihara, H. (1979). Artificially raised haploids and their uses in plant breeding. *Seiken Ziko* **27-28**, 14-29.

Kinman, M.L. and Sprague, G.F. (1945). Relation between number of parental lines and theoretical performance of synthetic varieties of corn. *J. Am. Soc. Agron.* **37**, 341-351.

Knight, T.A. (1823a). An account of some mule plants. *Transactions.* Horticultural Society of London, **5**, 292-296.

Knight, T.A. (1823b). Some remarks on the supposed influence of the pollen, in cross-breeding, upon the color of seed-coat of plants, and the qualities of their fruits. *Transactions,* Horticultural Society of London **5**, 377-380.

Knott, D.R. (1961). The inheritance of rust resistance, VI the transfer of stem rust resistance from *Agropyron elongatum* to common wheat. *Cand. J. Pl. Sci.* **41**, 104-123.

Knott, D.R. and Kumar, J. (1975). Comparison of early generation yield testing and a single seed descent procedure in wheat breeding. *Crop Sci.* **15**, 295-299.

Kolreuter, J.G. (1763). Vorlaufigen Nachricht von einigen das Geschlecht der Pflanzen Betreffenden Versuchen und Beobachtungen, nebst Fortsetzungen I, 2, und 3 (1761-66). W. Pfeffer, in Ostwald's Klassiker der exakten Wissenschaften, No. 41, Leipzig: Engelman.

Koelreuter, J.G. (1766). Vorlaufigen Nachricht von einigen das Geschlecht der Pflanzen Betreffenden Versuchen und Beobachtungen, Leipzig.

Kumar, J. 1973. Early generation selection and a comparison of two breeding methods in wheat (*T. aestivum* L.), Ph.D. thesis, University of Saskatchewan, Saskatoon.

Lacadena, J.R. (1974). Spontaneous and induced parthenogenesis and androgenesis. In *Haploids and Higher Plants*, Proc. of the 1st Int. Symp., K.J. Kasha, (Ed.) University of Guelph, Guelph, Ontario, pp. 13-32.

Law, J., Stoskopf, N.C. (1973). Further observations on ethaphon (ethrel) as a tool for developing hybrid cereals. *Can. J. Plant Sci.* **53**, 765-766.

Lawrence, P. (1974). Introgression of exotic germplasm into oat breeding populations, Ph. D. thesis Iowa State University, Ames, Iowa.

Lea, D.E. (1946). *Actions of Radiations on Living Cells.* Cambridge: Cambridge University press.

Leonard, K.J. (1969). Genetic equilibria in host-pathogen systems, *Phytopathol.* **59**, 1858-1863.

Lerner, I.M. (1953). The genotype in Mendelian populations, *Proc. Ninth Intern. Cong. Gen.* pp. 124-138.

Lerner, I.M. (1954). *Genetic Homeostasis*, Oliver and Boyd, London.

Lerner, I.M. (1958). *The Genetic Basis of Selection*, John Wiley and Sons Inc., London.

Lerner, I.M. (1950). *Population Genetics and Animal Improvement*, University Press, Cambridge.

Lewis, D. (1954). Comparative incompatibility in angiosperms and fungi. *Adv. in Gen.* **6**, 235-285.

Li, C.C. (1948). *An Introduction to population Genetics*, Peking University Press, Peiping.

Litzow, M. and Ascher, P.D. (1977). A potential method for producing F_1 hybrid seed with the sporophytic self-incompatibility system. *Incompatibility Newsletter* **8**, 83-86.

Livers, R.W. (1964). Fertility restoration and its inheritance in cytoplasmic male sterile wheat. *Science* **144**, 420.

Lonnquist, J.H. (1950). The effect of selection for combining ability within segregating lines of corn. *Agron. J.* **42**, 503-508.

Lonnquist, J.H. (1951). Recurrent selection as a means of modifying combining ability in corn. *Agron. J.* **43**, 311-315.

Lonnquist, J.H. (1967). Intra-population improvement: Combination S_1 and HS selection. *Maize 5, CIMMYT, Mexico.*

Lonnquist, J.H. and Mc Gill, D.P. (1954). Gametic sampling from selected zygotes in corn breeding. *Agron. J.* **46**, 147-150.

Loo, S. and Xu, Z.H. (1986). Rice: Anther culture for rice improvement in China. In *Biotechnology in Agriculture and Forestry*, Vol 2. Crops I. Y.P.S. Bajaj, (Ed.). Springer-Verlag Berlin Heidelberg, New York, Tokyo, pp. 139-156.

Lotsy, J.P. (1916). *Evolution by Means of Hybridization.* M Nijhoff, The Hague.

Love, H.H. (1927). A program for selecting and testing small grains in successive generations following hybridization. *Am. Soc. Agron. J.* **19**, 705-712.

Luedders, V.D., Duclos, L.A. and Matson, A.L. (1973). Bulk pedigree and early generation testing breeding methods compared in soybeans. *Crop Sci.* **13**, 363-364.

Lundquist, A. (1956). Self-incompatibility in rye. 1. Genetic control in the diploid. *Hereditas* **42**, 293-348.

Lupton, F.G.H. and Whitehouse, R.N.H. (1957). Studies on the breeding of self-pollinated cereals. *Euphytica* **6**, 169-184.

Mac Key, J. (1954). Breeding of oats, Handbuch fur Pflanzenzuchtung **2**, 512-517.

Mac Key, J. (1976). In *Heterosis in Plant Breeding* Proc. VII Congr. EUCARPIA, A. Janossy and FGH Lupton (Eds.). Budapest June 1974.

Malthus, T.R. (1798). *An Essay on the Principle of Population*, Oxford University Press, reprinted by Macmillan & Co. London.

Mangelsdorf, P.C. and Jones, D.F. (1926). The expression of Mendelian factors in the gametophyte of maize. *Genetics* **11**, 423-455.

Mariani, C., De Beuckeleer, M., Truettner, J.J.L. and Goldberg, R.B. (1990). Induction of male sterility in plants by a chimaeric ribonuclease gene. *Nature* **347**: 737-741.

Mariani, C., Gossele, V., De Beuckeleer, M., De Block, M., Goldberg, R.B., De Greef, W. and Leemans, J. (1992). A chimaeric ribonuclease-inhibitor gene restores fertility to male sterile plants. *Nature* **357**: 384-387.

Mashima, I. and Uchiyamada, H. (1955). Studies on the breeding of fertile tetraploid plant of rice. *Bull. Nat. Inst. Agric. Sci.* (Japan) Ser. D. 104-136.

Mather, K. (1943). Polygenic inheritance and natural selection. *Biol. Revs.* **18**, 32-64.

Mather, K. and Harrison, B.I. (1949). The manifold effects of selection. *Heredity* **3** (1-52), 131-162.

Mayr, E. (1940). Speciation phenomena in birds. *American Naturalist* **74**, 249-278.

Mayr, E. (1942). *Systematics and the Origin of Species.* Columbia University Press, New York.

Mayr, E. (1947). Ecological factors in speciation. *Evolution* **1**, 263-288.

Mayr, E. (1949). Speciation and systematics. In *Genetics, Paleontology and Evolution*, G.L. Jepsen *et al.* (Eds.), pp. 281-298.

Mayr, E. (1963). *Animal Species and Evolution.* Harvard University Press, Cambridge, USA.

Mc Daniel, R.G. and Sarkissian, I.V. (1966). Heterosis: Complementation by mitochondria. *Science* **152**, 1640-1642.

McGill, D.P. and Lonnquist, J.H. (1955). Effects of two cycles of recurrent selection for combining ability in an open-pollinated variety of corn. *Agron. J.* **47**, 319-323.

Mendel, G. (1866). Versuche uber pflanzen-hybriden. *Naturf. Ver. in Brunn, Verh.* **4**, 3-47.

Mertz, E.T., Bates, L.S. and Nelson, O.E. (1964). Mutant gene that changes protein composition and increases lysine content of maize endosperm. *Science* **145**, 279-280.

Micke, A., Donini, B. and Maluszynski, M. (1990). Induced mutations for crop improvement. *Mutation Breeding Review* **7**, 1-41.

Muller, H.J. (1927). Artificial transmutation of the gene. *Science* **66**, 84-87.

Murphy, R.P. (1942). Convergent improvement with four inbred lines of corn. *J. Amer. Soc. Agron.* **34**, 138-150.

Ndondi, R.V. (1987). Comparison of the single seed descent (SSD) method, the multiple seed descent (MSD) method and the bulk population method for yield and protein improvement in wheat. *Fifth Regional Wheat Workshop* October 5-10, 1987, Antsirabe, Madagascar.

Neal, N.P. (1935). The decrease in yielding capacity in advanced generations of hybrid corn. *J. Amer. Soc. Agron.* **27**: 666-670.

Nilsson-Ehle, H. (1909). *Kreuzunguntersuchungen an Hafer Und Weizen. Lund Univ. Arsskrift.* **5**, 1-122.

Ohta, Y. (1980). Mechanisms of male sterility in higher plants. *Gamma Field Symposia* **19**, 1-25.

Oka, H. (1955). Population genetics of rice and barley. *Ann. Rept. Nat. Inst. Genetics* (Japan) 1954, **5**, 40-50.

Osmanzai, M., Raja Ram, S. and Knapp, E.B. (1987). Breeding for moisture stressed areas. *Proc. Int. Workshop on Drought Tolerance in Winter Cereals* held at Capri, Italy from Oct. 27-31, 1985, pp. 151-161.

Owen, F.V. (1942). Male sterility in sugar beets produced by complementary effects of cytoplasmic and Mendelian inheritance. *Amer. J. Bot.* **29**, 692 (Abstr.).

Pandey, K.K. and Phung, M. (1982). 'Hertwig Effect' in plants: Induced parthenogenesis through the use of irradiated pollen. *Theor. Appl. Genet.* **62**, 295-300.

Parlevliet, J.E. (1979). Components of resistance that reduce the rate of epidemic development, *Annu. Rev. Phytopathol.* **17**, 203-222.

Pawar, I.S., Paroda, R.S. and Singh, S. (1986). A comparison of pedigree selection, single seed descent and bulk method in two wheat crosses. *Crop Improv.* **13**, 34-37.

Pawar, I.S., Paroda, R.S. and Singh, S. (1987). A comparison of range and phenotypic coefficient of variation in selected and unselected populations of two wheat crosses. *Seeds & Farms* **13**, 37-39.

Pawar, I.S., Paroda, R.S., Yunus, M. and Singh, S. (1985a). A comparison of three selection methods in two wheat crosses. *Indian J. Genet.* **45**, 345-353.

Pawar, I.S., Paroda, R.S., Singh, S. and Yunus, M. (1985b). Evaluation of early generation selection in two wheat crosses. *Seeds and Farms* **11**, 77-80.

Pawar, I.S., Redhu, A.S., Yunus, M. and Singh, I. 2000. Evaluation of selective intermated populations for yield and its component traits in wheat. *Haryana agric. Univ. J. Res.* **30**, 137-139.

Pawar, I.S., Singh, I. and Singh, S. (1990a). Traditional versus simplified pedigree selection in wheat. *Haryana agric. Univ. J. Res.* **20**, 134-138.

Pawar, I.S., Singh, S. and Yunus, M. (1990b). Studies on selective intermating in bread wheat. *Crop Improv.* **17**, 68-69.

Payne, K.T. and Hayes, H.K. (1949). A comparison of combining ability in F_2 and F_3 lines of corn. *Agron. J.* **41**, 383-388.

Peirce, L.C. (1977). Impact of single seed descent in selecting for fruit size, earliness and total yield in tomato. *J. Amer. Soc. Hort. Sci.* **102**, 520-522.

Penny, L.H., Russell, W.A., Sprague, G.F. and Hallauer, A.R. (1963). Recurrent selection, Symp. *Statistical Genetics and Plant Breeding* held at North Carolina State College Raleigh, March 20-29, 1961, W.D. Hanson and H.F. Robinson (Eds.). National Academy of Sciences - National Research Council Washington, pp. 352-367.

Phillips, L.L. (1963). The cytogenetics of *Gossypium* and the origin of new world cottons. *Evolution* **17**, 460-469.

Phillips, L.L. (1966). The cytology and phylogeny of diploid *Gossypium* species. *Am. J. Botany* **53**, 328-335.

Pinnell, E.L., Rinke, E.H. and Hayes, H.K. (1952). Gamete selection for specific combining ability. In *Hetersosis*. Iowa State College Press, pp. 378-388.

Poehlman, J.M. (1979). *Breeding Field Crops* (2nd ed.), AVI Publ. Co. Westport, Conn.

Poehlman, J.M. (1987). *Breeding Field Crops* (3rd ed.), Van Nostranid Reinhold.

Poehlman, J.M. and Sleper, D.A. (1995). *Breeding Field Crops* (4th edn.), Panima Publishing Corporation, New Delhi, Bangalore.

Pooni, H.S. and Jinks, J.L. (1981). The true nature of non-allelic interactions in *Nicotiana rustica* revealed by association crosses. *Heredity* **47**, 253-258.

Prabhakara, Rao, M.V. (1978). The transfer of alien genes for stem rust resistance to *durum* wheat, *Proc. 5th Int. Wheat Genet. Symp.* New Delhi, pp. 338-341.

Pring, D.R., Leving, C.S. and Conde, M.F. (1979). The organelle genomes of cytoplasmic male sterile maize and sorghum. In *Plant Genome*, D.R. Davies and D.A. Hapwood (Eds.). John Innes Charity, Norwich.

Putt, E.D. and Heiser, C.B. (1966). Male sterility and partial male sterility in sunflowers (*Helianthus annuus* L.). *Crop Sci.* **6**, 165-168.

Quinby, J.R. and Martin, J.H. (1954). Sorghum improvement. *Adv. in Agron.* **6**, 305-359.

Raeber, J.G. and Weber, C.R. (1953). Effectiveness of selection for yield in soybean crosses by bulk and pedigree systems of breeding. *Agronomy J.* **45**, 362-366.

Ramage, R.T. (1965). Balanced tertiary trisomics for use in hybrid seed production. *Crop Sci.* **5**, 177-178.

Ramage, R.T. (1975). Techniques for producing hybrid barley, 1974. *Barley Newsletter* **18**, 62-65.

Ramage, R.T. (1977). Varietal improvement of wheat through male sterile facilitated recurrent selection, *ASPAC Food and Fertilizer Technology.* Ctr. Tech. Bull. **36**, 6.

Ramage, R.T. (1983). Heterosis and hybrid seed production in barley. In *Heterosis: Reappraisal of Theory and Practice, Monogr. Theor. Appl. Genet.* No. 6, R. Frankel (Ed.), Springer-Verlag, Berlin, pp. 71-93.

Rana, R.S. (1995). Intellectual property right: protection of plant varieties. In: *Genetic Resources of Vegetable Crops: Management, Conservation and Utilization,* R.S. Rana, P.N. Gupta, M. Rai and S. Kochar (Eds.). N.B.P.G.R., New Delhi, pp. 421-427.

Rathjen, A.J. and Lamacraft, R.R. (1972). The use of computers for information management in plant breeding. *Euphytica* **21**, 502-506.

Rathjen, A.J. and Pugsley, A.T. (1978). Wheat breeding at the Waite Institute, 1925-1978. Biennial Report 1976-78 Waite Agricultural Research Institute, University of Adelaide, pp. 12-37.

Richey, F.D. (1927). The convergent improvement of selfed lines of corn. *Amer. Nat.* **61**, 430-449.

Richey, F.D. (1945). Isolating better foundation inbreds for use in corn hybrids. *Genetics* **30**, 455-471.

Richey, F.D. (1947). Corn breeding, gamete selection, the *Oenothera* method and isolated miscellany. *J. Amer. Soc. Agron.* **39**, 403-412.

Richey, F.D. and Sprague, G.F. (1931). Experiments on hybrid vigour and covergent improvement in corn. *U.S. Dept. Agric. Tech. Bull.* 267.

Rick, C.M. and Butler, L. (1956). Cytogenetics of tomato. *Adv. in Gen.* **8**, 267-382.

Riley, H.P. (1938). A character analysis of colonies of *Iris fulva, Iris hexagona* var. *gigantocaerulea* and natural hybrids. *Amer. J. Bot.* **25**, 727-738.

Riley, R. (1978). Application of recent advances in cytogenetics to breeding improved varieties in technology for increasing food production, Proc. 2nd FAO/SIDA seminar Field Food Crops in Africa and Near East, Lahore, Pakistan 18 Sept. to 5 Oct. 1977. Holmes, J.C., School of Agriculture, Edinburgh (Ed.). FAO Rome, 1978.

Riley, R. (1979). Methods for the future. In *Plant Breeding Perspectives.* J. Sneep and A.J.T. Hendriksen (Eds.). Centre for Agricultural Publishing and Documentation, Wageningen, Netherlands, pp. 321-369.

Rimpau, W. (1891). Kreuzungsprodukte landwirtschaftlicher Kulturpflanzen, *Landwirtschaftl., Jahrb* **20**, 335-371.

Roberts, H.F. 1965. *Plant Hybridization Before Mendel.* Hafner Publishing Company, New York & London.

Rosen, H.R. (1949). Oat parentage and procedure for combining resistance to crown rust including race 45 and *Helminthosporium* blight. *Phytopathol.* **39**, 20.

Rowell, P.L. and Miller, D.G. (1974). Effect of 2-chloroethylphosphonic acid (Ethephon) on female fertility of two wheat varieties. *Crop Sci.* **14**, 31-34.

Russell, W.A. (1974). Comparative performance of maize hybrids representing different eras of maize breeding. *29th Annu. Corn and Sorghum Res. Conf.* **29**, 81-101.

Ryder, E.J. (1963). An epistatically controlled pollen sterile in lettuce (*Lactuca sativa* L.). *Proc. Amer. Soc. Hort. Sci.* **83**, 585-589.

Sage, G.C.M. and Hobson, G.E. (1973). The possible use of mitochondrial complementation as an indicator of yield heterosis in breeding hybrid wheat. *Euphytica* **22**, 61-69.

Salaman, R.N. (1910). Male sterility in potatoes - A dominant Mendelian. *J. Linn. Soc.* (London), **39**, 301-312.

Sarkissian, I.V. and Srivastava, H.K. (1971). Mitochondrial polymorphism. III. Heterosis, complementation and spectral properties of purified cytochrome oxidase of wheat. *Biochem. Genet.* **5**, 57-63.

Sarkissian, I.V. and Srivastava, H.K. (1973). Some molecular aspects of mitochondrial complementation and heterosis. In *Genes, Enzymes and Population,* A.M. Srb (Ed.). Plenum Press, New York, pp. 53-60.

Sasakuma, T. and Mann, S.S. and William, N.D. (1978). EMS induced male sterile mutants in euplasmic and alloplasmic common wheat. *Crop Sci.* **18**, 850-853.

Saunders, J.H. (1965). Genetics of hairiness transferred from *Gossypium raimondii* to *Gossypium hirsutum.* *Euphytica* **14**, 276-282.

Schnell, F.W. (1961). On some aspects of reciprocal recurrent selection. *Euphytica* **10**, 24-30.

Schnell, F.W. (1969). Leistung und struktur heutiger Zuchtsorten. In *Vortrage des 5 Landw, Hochschultages* 56-75 Min. Landw Weinban Forsten Rheinland-Pfalz, Mainz.

Schnell, F.W. (1978). Progress and problems in utilizing quantitative variability in plant breeding. *Plant Res. Develop.* **7**, 32-43.

Schnell, F.W. (1981). Breeding methods: Synthesis, Proc. 4th meeting *Biometrics in Plant Breeding - Quantitative genetics and breeding methods*, A. Gallais (Ed.). Poitiers France September 2-4, 1981, pp. 9-30.

Schnell, F.W. (1982). A synoptic study of the methods and categories of plant breeding. *Planzenzuchtg* **89**, 1-18.

Scholl, R.L. (1974). Inheritance of isocitratase activity and its relationship to seedling vigour in a cross of upland cotton (*G. hirsutum* L.). *Crop Sci.* **14**, 296-300.

Schwartz, D. (1973). Single gene heterosis for alcohol dehydrogenase in maize: The nature of the sub unit interaction. *Theor. Appl. Genet.* **43**, 117-120.

Sears, E.R. (1956). The transfer of leaf rust resistance from *Aegilops umbellulata* to wheat. In *Genetics in Plant Breeding* symposium Biology Department Brookhaven National Laboratory Upton, New York.

Sharma, H.C. and Gill, B.S. (1983). Current status of wide hybridization in wheat. *Euphytica* **32**, 17-31.

Shull, G.H. (1909). A pure line method of corn breeding. *Rept. Amer. Breeders' Assoc.* **4**, 296-301.

Shull, G.H. (1914). Duplicate genes for capsule form in *Bursabursa pastoris Zeitschr. inducket. Abstamn-u Verenbungal* **12**, 97-149.

Sidhu, J.S. (1994). Plant breeders' rights in India. In: *Crop Breeding in India*, H.G. Singh, S.N. Mishra, T.B. Singh, H.H. Ram and D.P. Singh (Eds.). Int. Book Distribution Co., Lucknow, pp. 504-511.

Silvey, V. (1978). The contribution of new varieties to increasing cereal yield in England and Wales. *J. Nat. Inst. Agril. Bot.* **14**.

Simmonds, N.W. (1979). *Principles of Crop Improvement*, Longman, London, New York.

Singh, B.D. (1983). *Plant Breeding-Principles and Methods*, Kalyani Publishers, New Delhi, Ludhiana.

Singh, I., Chowdhury, R.K., Pawar, I.S. and Singh, S. (1987). A comparative study of three selection procedures in bread wheat. *Wheat Information Service* **65**, 16-18.

Singh, S. and Pawar, I.S. (1998). Recent advances in breeding of self-pollinated crops. In: *Genetics and Biotechnology in Crop Improvement*, P.K. Gupta *et al.* (Eds.), Meerut, India, pp. 255-275.

Singh, S. and Pawar, I.S. (2005). *Theory and Application of Biometrical Genetics*. CBS Publishers & Distributors, New Delhi.

Singleton, W.R. and Nelson, O.E. (1945). The improvement of naturally cross pollinated plants by selection in self-fertilized lines. IV. combining ability of successive generations of inbred sweet corn. *Conn. Agric. Exp. Sta. Bull.* **490**, 458-498.

Sinha, S.K. and Khanna, R. (1975). Physiological, bichemical and genetic basis of heterosis. *Adv. Agron.* **27**, 123-174.

Skovmand, B., Villareal, R., Van Ginkel, M., Rajaram, S. and Ortiz-Ferrara, G. (1997). Semi-dwarf bread wheats: Names, Pedigrees and Origins. CIMMYT, Mexico City.

Smith, D.C. (1966). Plant Breeding: Development and success. In: *Plant Breeding*, K.J. Frey (Ed.). Symp. Iowa State University Press, Ames. Iowa, pp. 3-54.

Smith, E.L. (1987). A review of plant breeding strategies for rainfed areas. *Proc. Int. Workshop on Drought toleracne in Winter Cereals* held at Capri, Italy from Oct. 27-31, 1985, pp. 79-87.

Smith, W.K. (1948). Transfer from *Melilotus dentata* to *M. alba* of the genes for reduction in coumarin content. *Genetics* **33**, 124-125.

Snape, J.W. and Riggs, T.J. (1975). Genetical consequences of single-seed descent in the breeding of self-pollinated crops. *Heredity* **35**, 211-219.

Sneep, J. (1977). Selection for yield in early generations of self-fertilizing crops. *Euphytica* **26**, 27-30.

Sneep, J., Murty, B.R. and Utz, H.F. (1979). Current breeding methods. In *Plant Breeding Perspectives*, J. Sneep and A.J.T. Hendriksen (Eds.). Centre for Agricultural Publishing and Documentation, Wageningen, Netherlands, pp. 104-233.

Sorrells, M.E. and Fritz, S.E. (1982). Application of a dominant male sterile allele to the improvement of self-pollinated crops. *Crop Sci.* **22**, 1033-1035.

Sprague, G.F. (1939). An estimation of the number of top-crossed plants required for adequate representation of a corn variety. *J. Amer. Soc. Agron.* **31**, 11-16.

Sprague, G.F. (1946). Early testing of inbred lines of corn. *J. Amer. Soc. Agron.* **38**, 108-117.

Sprague, G.F. (1952). Early testing and recurrent selection. In *Heterosis* Iowa State College Press, pp. 400-417.

Sprague, G.F. (1983). Heterosis in maize: Theory and practice. *Heterosis*. A reappraisal of theory and practice, R. Frankel (Ed.). *Springer Verlag*, pp. 47-70.

Sprague, G.F. and Brimhall, B. (1950). Relative effectiveness of two systems of selection for oil content of the corn kernel. *Agron. J.* **42**, 83-88.

Sprague, G.F. and Federer, W.T. (1951). A comparison of variance components in corn yield trials. II Error, year × variety, location × variety and variety components. *Agron. J.* **43**, 535-541.

Sprague, G.F. and Jenkins, M.T. (1943). A comparison of synthetic varieties, multiple crosses, and double crosses in corn. *J. Amer. Soc. Agron.* **35**, 137-147.

Sprague, G.F., Miller, P.A. and Brinhall, B. (1952). Additional studies on the effectiveness of two systems of selection for oil content of the corn kernel. *Agron. J.* **44**, 329-331.

Sprengel, C.K. (1793). Das entdeckte Geheimniss der Natur im Bau und in der Befruchtung der Blumen, In Ostwald's Klassiker der oxakten Wissenchaften, Paul Knuth (Ed.). 48 (4), Leipzig.

Srivastava, H.K. (1983). In *Heterosis: Reappraisal of Theory and Practices*, R. Frankel (Ed.). Springer Verlag, Berlin, New York.

Stadler, L.J. (1928). Mutations in barley induced by X-rays and radium. *Science* **68**, 186-187.

Stadler, L.J. (1942). Some observations on gene variability and spontaneous mutation, Spragg Memorial Lectures in Plant Breeding. Michigan State College.

Standler, L.J. (1944). Gamete selection in corn breeding. *J. Amer. Soc. Agron.* **36**, 988-989.

Stebbins, G.L. (1940). The significance of polyploidy in plant evolution. *American Naturalist*, **74**, 54-66.

Stebbins, G.L. (1945). The cytological analysis of species hybrids. *Bot. Reviews* **11**, 463-486.

Stebbins, G.L. (1949). Rates of evolution in plants, In *Genetics, Paleontology and Evolution*, G.L. Jepsen *et al.* (Eds.), pp. 229-242.

Stebbins, G.L. (1950). *Variation and Evolution in Plants*. Columbia Univerity Press, New York.

Stebbins, G.L. (1957). Genetics, evolution and plant breeding. Proc. Symp. *Genetics and Plant Breeding in South-East Asia. Indian J. Genet.* **17**, 129-141.

Stebbins, G.L. (1970). *The Process of Organic Evolution*, Prentice Hall, New Delhi.

Stebbins, G.L. (1971). *Chromosomal Evolution in Higher Plants*, Edward Arnold, London.

Stebbins, G.L.Jr. (1958). The inviability, weakness and sterility of interspecific hybrids. *Adv. in Gen.* **9**, 147-215.

Stephens, S.G. (1950). The genetics of "corky", II Further studies on its genetic basis in relation to general problem of interspecific isolating mechanisms. *J. Gen.* **50**, 9-20.

Suneson, C.A.(1945). The use of male sterility in barley improvement. *J. Am. Soc. Agron.* **37**, 72-73.

Suneson, C.A. (1949). Survival of four barley varieties in a mixture. *Agron. J.* **4**, 459-461.

Suneson, C.A. (1956). An evolutionary plant breeding method. *Agron. J.* **48**, 188-190.

Suneson, C.A. (1960). Genetic diversity a protection against plant diseases and insects. *Agron. J.* **52**, 319-321.

Suneson, C.A. (1968). Harland barley. *California Agr.* **9**.

Suneson, C.A. and Stevens, H. (1953). Studies with bulk hybrid populations of barley. *US Dept. Agric. Tech. Bull.* **1067**, 14.

Suneson, C.A. and Weibe, G.A. (1942). Survival of barley and wheat varieties in mixtures. *J. Amer. Soc. Agron.* **34**, 1052-1056.

Suneson, C.A., Bayles, B.B. and Fifield, C.C. (1948). Effects of awns on yield and market quality of wheat. *US Dept. Agric. Circ.* 783.

Swaminathan, M.S. (1955). Overcoming cross-incompatibility among some Mexican diploid species of *Solanum. Nature* **176**, 887-888.

Tee, T.S. (1971). Comparison of single seed descent and bulk population breeding methods and evaluation of single seed selection in wheat (*T. aestivum* L.), Thesis university of Califronia, Davis.

Tee, T.S. and Qualset, C.O. (1975). Bulk population in wheat breeding - Comparison of single seed descent and random bulk methods. *Euphytica* **24**, 393-405.

Thompson, K.F. (1964). Triple cross hybrid Kale. *Euphytica* **13**, 173-177.

Tollenaar, D. (1934). Untersuchungen uber Mutation bei Tabak. I. Entstehungsweise und Wesen Kunstlich erzeugter Gen-Mutation. *Genetica*, **16**, 111-152.

Tollenaar, D. (1938). Untersuchungen uber Mutation bei Tabak II. Einige Kunstlich erzeugte Chromosome-Mutation. *Genetica* **20**, 285-295.

Torrie, J.H. (1958). A comparison of the pedigree and bulk methods of breeding soybeans. *Agron. J.* **50**, 198-200.

Troyer, A.F. and Brown, W.L. (1976). Selection for early flowering in corn: Seven late synthetics. *Crop Sci.* **16**, 767-772.

Tschabold, E.E., Heim, D.R., Beck, J.R., Wright, F.L., Rainey, D.P., Terando, N.H. and Schwer, J.F. (1988). LY 195259, new chemical hybridizing agent for wheat. *Crop Sci.* **28**, 583-588.

U, N. (1935). Genome analysis in *Brassica* with special reference to the experimental formation of *B. napus* and peculiar mode of fertilization. *Jap. Jour. Bot.* **7**, 389-452.

Valentine, J. (1984). Accelerated pedigree selection: An alternative to individual plant selection in the normal pedigree breeding method in the self-pollinated cereals. *Euphytica* **33**, 943-951.

Van der Plank, J.E. (1968). *Disease resistance in plants*. Academic Press, New York.

Vavilov, N.I. (1926). Studies on the origin of cultivated plants. Leningrad, pp.129-238.

Vavilov, N.I. (1935). *Theoretical Bases of Plant Breeding*, Moscow.

Vavilov, N.I. (1951). The origin, variation, immunity and breeding of cultivated plants. *Chronica Bot.* **13**, 1-364.

Verhalen, L.M., Baker, J.L. and Mc New, R.W. (1975). Gardner's grid system and plant selection efficiency in cotton. *Crop Sci.* **15**, 588-591.

Vinall, H.N. and Cron. A.B. (1921). Improvement of sorghums by hybridization. *J. Heredity* **12**, 435-443.

Warner, R.L., Hageman, R.H., Dubley, J.W. and Lambert, R.J. (1969). *Proc. Nat. Acad. Sci. (USA)*, **62**, 785-792.

Weaver, J.B. (1968). Analysis of a genetic double recessive completely male sterile cotton. *Crop Sci.* **8**, 597-600.

Webster, G.T. (1955). Interspecific hybridization of *Melilotus alba* × *M. officinalis* using embryo culture. *Agron. J.* **47**, 138-142.

Weibe, G.A. and Briggs, F.N. (1937). The degree of bunt resistance necessary in commercial wheat. *Phytopathology* **27**, 313-314.

Weinberg, W. (1908). Uber den Nachweis der Vererbung bein Menschen, *Jh. Ver. vaterl Naturk. Wurttemb* **64**, 369-382.

Weiss, M.G. (1949). Soybeans. *Adv. in Agron.* **1**, 77-157.

Wellhausen, E.J. (1952). Heterosis in a new population. In: *Heterosis*.Iowa state, J.W. Gowen (Ed.), College Press, pp. 418-450.

Wellhausen, E.J. and Wortman, L.S. (1954). Combining ability of S_1 and derived S_3 lines of corn. *Agron. J.* **46**, 86-89.

Whaley, W.G. (1944). *Heterosis. Botan. Rev.* **10**, 461-498.

Whaley, W.G. (1952). Physiology of gene action in hybrids, In *Heterosis*, J.W. Gowen (Ed.), Iowa State State College Press, Ames, Iowa, pp.98-113.

Williams, W. (1959). Heterosis and the genetics of complex characters. *Nature*, **184**, 527-530.

Wilson, J.A. and Ross, W.M. (1962a). Male sterility interaction of *Triticum timophevi* cytoplasm. *Wheat Information Service* **14**, 29-31.

Wilson, J.A. and Ross, W.M. (1962b). Cross breeding in wheat, *Triticum aestivum* L.II. Hybrid seed set on a cytoplasmic male sterile wheat composite subjected to cross pollination. *Crop Sci.* **2**, 415-417.

Wright, G.M. and Thomas, G.A. (1976). An evaluation of the single seed descent method of breeding. *New Zealand Wheat Rev.* **13**, 46-49.

Wright, S. (1921). Systems of mating. *Genetics*, **6**, 111-178.

Wright, S. (1922). Coefficient of inbreeding and relationship. *Am. Nat.* **56**, 330-338.

Wright, S. (1930). The genetical theory of natural selection. *J. Heredity* **21**, 349-356.

Wright, S. (1931a). Evolution in Mendelian populations. *Genetics* **16**, 97-159.

Wright, S. (1931b). Statistical theory of evolution. *Amer. Stat. J. Supp.* pp. 201-208.

Wright, S. (1932). The roles of mutation, inbreeding, crossbreeding and selection in evolution, *Proc. 6th Intern Congr. Genetics* **1**, 356-366.

Wright, S. (1940a). The statistical consequence of Mendelian heredity in relation to speciation. In *New Systematics*, J.S. Huxley (Ed.), pp. 161-183.

Wright, S. (1940b). Breeding structure of populations in relation to speciation. *American Naturalist* **74**, 232-248.

Wright, S. (1948). On the roles of directed and random changes in gene frequency in the genetics of populations. *Evolution* **2**, 279-294.

Wright, S. (1949). Adaptation and selection. In *Genetics, Paleontology and Evolution*, G.L. Jepsen *et al.* (Eds.), pp. 365-389.

Wright, S. (1951). The genetical structure of populations. *Ann. Eug.* **15**, 323-354.

Yap, T.C., Poh, C.T. and Mak, C. (1977). Comparative studies on selection methods in longbean. 3rd Intn. Cong. SABRAO, Canberra, CSIRO, pp. 23-25.

Yonezawa, K. and Yamagata, H. (1981). *Japan J. Breed.* **31**, 35-42.

Yule, G.V. (1906). On the theory of inheritance of quantitative compound characters on the basis of Mendel's laws – A preliminary note. *Rept. third Intern. Conf. Gen.* pp. 140-142.

Zhang, A. and Huang, T.C. (1998). Progress of hybrid wheat breeding in China. In *Hybrid Wheat - A new crop going to farmers*. Proc. 1st International Workshop on Hybrid Wheat. Z. Aimin and T.C. Huang (Eds.). China Agricultural University Press, Beijing, pp. 9-14.

Zhou, C. and Yang, H.Y. (1990). *In vitro* production of haploids in rice through ovary culture. In *Biotechnology in Agriculture and Forestry*, Vol. 14, Rice, Y.P.S. Bajaj (Ed.) Springer-Verlag, Berlin, Heidelberg, New York, Tokyo.

Index

disharmonious because of large differences in the genes and their organization in the chromosomes of the two parental species. The criterion of taxonomic relationship between the species to be crossed also does not hold always. The crosses within between *Brassica* present the most striking example of the failure of this criterion. The species of *tobacco*, not crossable to its closely related species of same tribe *Nicotiana tabacum*, but is crossable to *Raphanus sativus* (radish), a species of different genus.

Very often several chromosomes are also, often accompanied by an equally failure. In the case of distant hybridization it is necessary to use polyploids for producing a hybrid possessing the chromosome complement of both the two parents. But the production of a hybrid does not guarantee the presence of ... and in the ... but ... the possibility of occurrence of hybrid sterility ... from ... either never be set.

Hybrid Disability and Reproductive Incapacity

The genetic differentiation between two species may occur at different stages of the separation of gene pools. If crosses are made between separated groups depending on the stage of genetic differentiation (not morphological differences), and ecological preference), two kinds of properties are to be observed in addition to normal segregation and recombination of genes. If the differentiation (genetic) is at an earlier stage, genetic disharmonies is found in the F and later generations. However the vigour of the new combination as compared to that of old ones would depend on the kind of genetic systems present in the parental species. If the genetic systems in the parental species are very well balanced, one should not expect new combinations superior to the old ones. But if the reverse is true, new combinations may be far superior to parental combinations.

In case the genetic differentiation is at an advanced stage, the hybrid may show reproductive incapacity or sterility in the F generation. However, as has been discussed earlier, the degree of hybrid sterility may vary from zero (hybrid is fully fertile) to 100 per cent.

The limitations of distant hybridization may be classified into two broad groups : (A) barriers to distant hybridization and (B) hybrid sterility.

Barriers to Distant Hybridization

There may be some general barriers (geographic isolation, failure of flowering due to differences in the day length, flowering of parental species at different times etc.) which may not allow the pollen to reach the stigma of the parent species used as female. Even though the pollen reaches the stigma there are a number of barriers which could adversely affect the success of distant hybridization. These barriers are as under.

(i) Barriers to zygote formation. In wide crosses, the zygote may fail to form either due to the failure of pollen germination on the stigma, slow pollen tube growth, bursting of pollen tube in the stylar region, short pollen tube, quick pollen tube or pollen tube reaches the embryosac but there no fertilization. Because of the cross incompatibility or some other genetic or physiological mechanism operating in some distantly related species, the pollen does not at all germinate on the stigma of the foreign species. For similar reasons, pollen tubes in some cases grow slowly in the styles of other species. One major reason of the failure of interspecific hybridization in Datura is the bursting of pollen tubes in the styles of other species. Hybridization between maize (as female) and gamagrass